Interdisciplinary Applied Mathematics

Volume 1

Martin C. Gutzwiller

Chaos in Classical and Quantum Mechanics

With 78 Illustrations

Springer

Martin C. Gutzwiller
IBM T.J. Watson Research Center
P.O. Box 218
Yorktown Heights, 10598
USA

ISBN 0-387-97173-4 Springer-Verlag New York Berlin Heidelberg
ISBN 3-540-97173-4 Springer-Verlag Berlin Heidelberg New York

Gutzwiller: Chaos in Classical & Quantum Mechanics

First Indian Reprint 2011

ISBN 978-81-8489-810-1

This edition is published by Springer (India) Private Limited, (a part of Springer Science+Business Media), Registered Office: 3rd Floor, Gandharva Mahavidyalaya, 212, Deen Dayal Upadhyaya Marg, New Delhi – 110 002, India

Printed in India by Rashtriya Printers, Delhi.

To Frances and Patricia

Faithful Companions on a Long Journey

Preface

Elementary mechanics, both classical and quantum, has become a growth industry in the last decade. A newcomer to this flourishing field must get acquainted with some unfamiliar concepts and get rid of some cherished assumptions. The change in orientation is necessary because physicists have finally realized that most dynamical systems do not follow simple, regular, and predictable patterns, but run along a seemingly random, yet well-defined, trajectory. The generally accepted name for this phenomenon is *chaos*, a term that accurately suggests that we have failed to come to grips with the problem.

This book offers a collection of ideas and examples rather than general concepts and mathematical theorems. However, an indifferent compilation of the most telling results can only discourage the novice. In order to focus on a central theme, I have singled out the questions that have a bearing on the connection between classical and quantum mechanics. In this manner we are led to ask whether there are chaotic features in quantum mechanics; the issue is still open, and all the preliminary answers suggest that quantum mechanics is more subtle than most of us had realized.

Reading this book requires a knowledge of both classical and quantum mechanics beyond a first introductory physics course. Advanced mastery of these subjects is not necessary, however, and probably not even desirable, since I am trying to appeal to the intuition rather than the analytical ability of the reader. Some of the more sophisticated concepts, such as the action function in classical mechanics and its analog in quantum mechanics, Green's function, are basic to the whole development as it is presented here. Their meaning and their use will be explained in the context in which they appear, and without the mathematical qualifications that would be necessary if I tried to offer general propositions rather than special cases.

In keeping with this informal style I have emphasized certain aspects of the whole story which are not usually found in scientific books. Whenever possible I base my arguments on elementary geometry rather than algebraic manipulations. In order to gain a better perspective on the more important results, references to the historical development are often helpful. In the same vein, related problems from different disciplines are mentioned in the same section, with particular attention to mathematics, astronomy, physics, and chemistry. Finally, I have taken the liberty to comment on the motivation behind certain efforts, to evaluate the validity and relevance of some results, and to consider future tasks, if not to speculate outright about possible developments in the field of chaos.

This book comes out of a course of the same title that I taught in the winter and spring of 1986 at the Laboratoire de Physique du Solide in Orsay, outside of Paris. I owe a debt of gratitude to my faithful audience, who helped me with a moderate amount of criticism; in particular, Francoise Axel, Oriol Bohigas, Alain Comtet, Marie Joya Giannoni, Bernard Jancovici, Maurice Kléman, Jean Marc Luck, Claude Itzykson, and André Voros provided many useful suggestions.

Most scientists have not participated in the recent development of ideas related to chaos in Hamiltonian systems; they are usually not aware of the many different viewpoints and interpretations, the new problems and methods for their solution, and the novel applications to important experiments. As far as this book is concerned, all of these ideas deal with relatively elementary questions in both classical and quantum mechanics; as soon as they are understood, some readers may be tempted to call them obvious because of their deceptive simplicity!

Since I have worked in this area for twenty years, I have benefitted from discussions with many colleagues who are interested in questions related to chaos. I want to thank them all, and apologize for not mentioning them by name. It is remarkable how many different personalities and individual tastes in scientific matters can be attracted to one central theme. I hope indeed that this book will appeal to all those who look for diversity in their pursuit of physics and its closest relatives, mathematics, astronomy, and chemistry. Thus we might eventually find harmony in chaos.

Contents

Introduction

Elementary mechanics is the model for the physical sciences. Its principles and methods are the ideal for most other disciplines that deal with nature. Substantial parts even of mathematics have been developed to deal with the problems of mechanics. Every scientist has a fairly well-defined picture of the way mechanics works and the kind of results it yields.

According to the prevailing views, a dynamical system runs along a predictable and regular course, and ends up in some periodic and steady state. If very many particles are involved and we are either unable or unwilling to follow each one individually, then we are satisfied with knowing the statistical properties of a dynamical system. The phenomena of *thermodynamics,* of *friction,* and of *diffusion* have a probabilistic character because we don't really need to know everything there is. The situation is subject to random processes because we choose to be ignorant. By contrast, the mechanical behavior of simple systems is assumed to be entirely comprehensible, easily described, perhaps even dull.

This erroneous impression is created by the special examples discussed in school, from elementary to graduate: two bodies attracting each other with an inverse-square-of-the-distance force, as in planetary motion and in the hydrogen atom; several oscillators coupled by linear springs; and a rotating symmetric rigid body in a uniform gravitational field, the gyroscope. The general methods for solving more difficult problems are presented as technical refinements best left to the experts -- astronomers working in celestial mechanics and theoreticians in atomic and molecular physics.

A century ago mathematicians discovered that some apparently simple mechanical systems can have very complicated motions. Not only is their behavior exceedingly sensitive to the precise starting conditions, but they never settle to any reasonable final state with a recognizable fixed pattern. Although their movements look smooth over short times, they seem to jump unpredictably and indefinitely when their positions and momenta are checked over large time intervals.

Astronomers became increasingly aware of this problem during the last 60 years, but physicists began to recognize it only some 20 years ago. The phenomenon, which now goes under the name *chaos,* has since become a very fashionable topic of investigation. Innocent onlookers might suspect one more passing fad. I do not think it will turn out that way, though. Chaos is not only here to stay, but will challenge many of our assumptions about the typical behavior of dynamical systems. Since mechanics underlies our view of nature, we will probably have to modify some of our ideas concerning the harmony and beauty of the universe.

As a first step, we will have to study entirely different basic examples in order to re-form our intuition. We must become familiar with certain novel specimens of simple mechanical systems based on chaotic rather than regular behavior. General abstract propositions do not serve that purpose, although they are desirable once we become knowledgeable about the issues involved.

This book is, therefore, committed to the discussion of specific examples, in particular the hydrogen atom in a magnetic field, the donor impurity in a semiconductor where the effective mass of the electron is different in different directions, and the motion of a particle on a surface of negative curvature. Other equally instructive systems, which are chaotic, yet simple enough to be understood thoroughly, will be mentioned without detailed discussion. Among them is the *hydrogen atom in a strong microwave field;* an adequate treatment would almost require a monograph by itself, if the recent experiments on this system are to be presented, and everything put into proper perspective.

These unfamiliar examples must be seen in full contrast with the familiar ones. A discussion of the regular behavior and of some borderline systems will therefore precede the main part of this book. In particular the *three-body problem of celestial mechanics* will be discussed in some detail with special attention to the *Moon-Earth-Sun* system. The important ideas of classical mechanics were first conceived and tested in this area, and their practical application can be observed in the sky without elaborate instruments. Chaos made its first appearance there.

Mathematicians have put a lot of effort into proving the formal equivalence between various abstract dynamical systems. Although one hopes that these endeavors will ultimately embrace all mechanical systems, the different kinds of chaotic behavior have not been characterized to the point where an exhaustive classification can be attempted. The study of further examples will eventually get us there. Meanwhile some simple practical distinctions are sufficient. I see no reason to split the phenomenon into more than two broad categories: *soft chaos,* which allows an approach starting from regular behavior by perturbation or breaking the symmetry as it were, and *hard chaos,* where each trajectory is isolated as the intersection of a stable and an unstable manifold. Such general features as bound states versus scattering, or conservative versus time-dependent forces remain important; but there will be *no discussion of friction, nor of any other dissipation of energy.*

Only systems with a *finite number of degrees of freedom* will be considered. This selection is dictated by a fundamental problem on which I have worked for two decades and which serves as the focus in this book: How can the classical mechanics of Newton, Euler, and Lagrange be understood as a limiting case within the quantum mechanics of Heisenberg, Schroedinger, and Dirac? Einstein in 1917 was the first, and for 40 years the only, scientist to point out the true dimensions of this problem when the classical dynamical system is chaotic. While we are still a long way from a satisfactory answer, I can think of no better issue to guide our thinking.

In this manner we are led straight into the main question of *quantum chaos:* Is there anything within quantum mechanics to compare with the chaotic behavior of classical dynamical systems? It seems unlikely, although there are cases of smooth chaos in quantum mechanics which border on the enigmatic, e.g., the scattering of waves on a two-dimensional box. Their discussion requires a certain degree of mathematical sophistication which is well worth the effort.

Mechanics, classical as well as quantal, with all the above restrictions in the choice of examples, seems almost simple enough to be within the grasp of purely algebraic and analytical methods. There are, however, striking examples where *numerical calculations* have given the investigator clues to the analytical solution of a problem. Furthermore, the ready availability of computers has led to many interesting numerical results, with as many intuitive interpretations, all in need of further sorting to find the relevant ideas. A somewhat arbitrary choice among these computational efforts is almost inevitable, particularly in view of the rapid accumulation of partial results. I have tried to concentrate on the work which tests the limit toward classical mechanics, and which is not the subject of some recent monograph, such as, for instance, the hydrogen atom in a microwave field, one of the few cases of quantum chaos with a wealth of experimental material.

Some readers may be disappointed because they do not find a satisfactory account of what interests them most. The number of articles in this general area has become overwhelming in the last decade. Since I think that new examples are of the essence, I regret that some particularly interesting ones are not dicussed here.

Among the most serious omissions I want to mention specifically almost everything connected with time-dependent Hamiltonian systems such as the *hydrogen-atom in a strong microwave field* (Bayfield and Koch 1974, for a review cf. Bayfield 1987) as well as the closely related *kicked rotator* (Casati, Chirikov, Izraelev, and Ford 1979; Fishman, Grempel, and Prange 1982, 1984). *Discrete maps and their quantum analog* get very little attention (Balazs and Voros 1989), and almost nothing is said about rotating bodies like gyroscopes and *coupled spins* (Magyari, Thomas, Weber, Kaufmann, and Müller 1987; Srivastava et al. 1988), nor about related work in *nuclear physics* (Meredith, Koonin, and Zirnbauer 1988; Swiatecki 1988). The above references are supposed to point to some of the seminal work in one of these areas.

Every author is entitled to use the lack of space as an excuse, although a lack of competence, interest, and/or hard work in certain subjects might occasionally provide a more adequate explanation for some shortcomings. Such obvious reasons, however, are based on the author's personal preferences and his perspective on the whole enterprise of theoretical physics. They are not easily condensed into simple declarations of intent, or statements explaining general views and methods, because they are always intimately mixed with the author's personal and professional experiences. The organization of this book and the choice of the several topics has to be seen as one possible, and to some extent coherent, approach to a novel and very active area in science.

CHAPTER 1

The Mechanics of Lagrange

The most general starting point for the discussion of any mechanical system is the *variational principle*. It was first proposed as a particularly concise formulation of Newton's laws of motion, and it turns out to be extremely useful for some simple manipulations such as the transition between different coordinate systems. Feynman (1948, 1965), with the inspiration of Dirac (1933, 1935), then found a complete analog for it in quantum mechanics. Indeed, the *path integral* provides the most direct link between the classical and the quantum regime (cf. Section 13.4).

The ideas concerning the variational principle of Lagrange are the backbone of this whole book. They will be explained in general terms assuming that the reader has met them before; they will also be illustrated explicitly using the example of space travel in the solar system.

The titles of this chapter and the next one are somewhat misleading. The historical development of mechanics is more complicated than the simple division into two kinds of mechanics, Lagrange's depending on time as the primary parameter, and Hamilton-Jacobi's depending on the energy. This distinction is important in quantum mechanics, however, and since our presentation is skewed in that direction, the relevant differences are brought out already at the classical level. The first two chapters are not meant to trace the origin of all the ideas back to their authors except where this is specifically mentioned.

1.1 Newton's Equations According to Lagrange

Let us think of the mechanical system to be studied as described at any fixed instant of time t by the collection of the position coordinates, $q = (q_1, q_2, ..., q_n)$, and the collection of the velocity components, $\dot{q} = (\dot{q}_1, \dot{q}_2, ..., \dot{q}_n)$, which are needed for a complete specification. We call q simply the position of the system, and the vector $\dot{q}$ its velocity. The number n of coordinates is called *the number of degrees of freedom* for the system. We will usually deal with $n = 2$ because that is the smallest for a system to be chaotic when its energy is conserved.

The variational principle is formulated with the help of a function L, called the *Lagrangian,* which depends on the position q, the velocity $\dot{q}$, and the time t, $L(q, \dot{q}, t)$. The most common form for L, and the only one to be used in this book, is deceptively simple: it is the difference $L = T - V$ of the kinetic energy T and the potential energy V.

The *kinetic energy* T is the product of half the mass, $m/2$, times the square of the velocity, which we will write simply as $\dot{q}^2$. Only when the particle is moving in a (Euclidean) plane and we are using Cartesian coordinates, can we write $\dot{q}^2 = \dot{q}_1{}^2 + \dot{q}_2{}^2$; when the motion takes place on a curved surface, the square of the velocity involves the Riemannian metric, as we will explain when the time comes.

The *potential energy* V is a function of the position coordinates only, q_1 and q_2, and possibly the time t. (In the learned language of modern mathematics the motion takes place on an n-dimensional manifold, and L is a function from the tangent bundle of this manifold into the reals.)

Newton's equations of motion are written in terms of the quantity p called the *momentum.* Its definition in terms of the Lagrangian is given quite generally by

$$p_j = \frac{\partial L}{\partial \dot{q}_j}. \tag{1.1}$$

The momentum specifies the state of motion of the system just as well as the velocity $\dot{q}$. In particular, if L is a quadratic function of $\dot{q}$, there is a simple linear transformation connecting p and $\dot{q}$. When $L = T - V$ and $T = m\dot{q}^2/2$, one has $p = m\dot{q}$ according to (1.1).

The classical form of *Newton's equations* is

$$\frac{dp_j}{dt} = -\frac{\partial V}{\partial q_j}, \tag{1.2}$$

where the rate of change with time of the momentum p is on the left, and the force as the derivative of the potential is on the right. *The momentum p has to be sharply distinguished from the velocity,* even though in many systems they are proportional to each other through the factor m, the mass. It is one of Newton's glories to have postulated the relation $\dot{p} = force$, rather than $m\,\ddot{q} = force$, which one finds in most elementary textbooks and many advanced presentations by mathematicians. Although Newton did not know at the time, his formulation correctly includes the motion of a charged particle in the presence of a magnetic field, the trajectory of a rocket whose mass decreases as it accelerates, and the motion of a particle close to the speed of light as in high-energy physics. (In modern mathematical parlance p is a cotangent on the manifold described by the coordinates q.)

The *Lagrangian equations of motion* are

$$\frac{d}{dt}\left(\frac{\partial L}{\partial \dot{q}_j}\right) - \frac{\partial L}{\partial q_j} = 0. \tag{1.3}$$

This formula combines the two preceding relations (1.1) and (1.2) in our special case $L = T - V$, but it has general validity even in the situations of which neither Newton nor Lagrange were aware. The author admits cheerfully that he does not have a good intuitive grasp of the Lagrange formalism as expressed in the Lagrangian $L = T - V$ and the equations of motion (1.3), although they are undoubtedly the foundation of physics. The next section may convey a better idea why the Lagrangian is the difference between the kinetic and potential energy; mechanics manages somehow to reduce this difference after averaging over a given time interval.

1.2 The Variational Principle of Lagrange

Before proceeding to the main topic of this section, we introduce three terms that will occur very frequently and have to be carefully distinguished. Although customary usage is not well defined for these terms, and the present assignment may seem arbitrary, it is helpful to pin down their meaning for the purpose of this book.

A *path* in the mechanical system is an arbitrary, continuous function $q(\tau)$ where the real variable τ varies from the initial value t' to the final value t'', i.e., $t' \leq \tau \leq t''$. We call the initial value $q(t') = q'$ and the final value $q(t'') = q''$. The functions $q_i(\tau)$, which describe the path of the system, are assumed sufficiently smooth so as to give us no trouble in the mathematical manipulations. A tremendous variety of such paths is usually available, even for the most intricate choice of the initial and final values, t', t'', q', and q''.

A *trajectory* of the mechanical system from the position q' at time t' to the position q'' at time t'' is a solution $q(t)$ of the equations of motion (1.1) with the time t in the interval (t', t'') and the stipulated initial and final values, $q(t') = q'$ and $q(t'') = q''$. Such a solution may not exist if we choose the values t', t'', q', and q'' awkwardly. On the other hand, there may be several or even infinitely many such solutions, although they are probably not easy to get explicitly.

A *periodic orbit,* occasionally abbreviated to *orbit,* is a trajectory whose final position and momentum coordinates q'' and p'' coincide with the initial position and momentum coordinates q' and p' so that $q'' = q'$ and $p'' = p'$. A periodic orbit is a trajectory that closes itself smoothly like a Kepler ellipse. Quite unexpectedly, even a mechanical system with the worst kind of chaos has a dense set of periodic orbits. They play a critical role in the transition from classical to quantum mechanics.

The trajectories are the essence of classical mechanics. Every dynamical system runs along a trajectory; the only choice in the matter concerns the initial and final values, getting from here q' at this time t' to there q'' at that time t''. In quantum mechanics, however, all possible paths contribute to the transfer of the system from q' to q'' in the time interval from t' to t''. The result of all these possibilities is a superposition of little wavelets, each associated with one of the paths. Needless to say, the relevant calculation is even more difficult than finding the classical trajectories.

The *variational principle* can now be formulated as follows: Given the initial values (q', t') and the final values (q'', t'') together with a trajectory $q_0(t)$ from here to there, look at all the neighboring paths whose position coordinates $q(\tau)$ are obtained by adding a function $\delta q(\tau)$, called a *displacement,* to the coordinates $q_0(\tau)$ so that $q(\tau) = q_0(\tau) + \delta q(\tau)$. Then calculate the integral of $L(\dot{q}, q, \tau)$ over τ from t' to t''. The value of this integral depends on the particular choice of the function $\delta q(\tau)$; it can be expanded in powers of $\delta q(\tau)$, always assuming that there is no analytical trouble with the chosen functions. The three first terms of this expansion are needed; they have special names, which will come up over and over again.

The lowest term depends only on the trajectory $q_0(t)$ because it is obtained by setting $\delta q(\tau) = 0$. It is called *Hamilton's principal function (HPF),* or somewhat indiscriminately, *the action integral* $R(q'', t''; q', t')$ from (q', t') to (q'', t'') along the trajectory $q_0(t)$. As a formula,

$$R(q'', t''; q', t') = \int_{t'}^{t''} d\tau \, L(\dot{q}_0, q_0, \tau). \tag{1.4}$$

The simplest example of R comes from a particle moving freely in (Euclidean) space. Its trajectories are straight lines, so in order to get from q' to q'' we have to draw the straight line connecting these two points. The velocity of the particle is the distance divided by the allowed time $t'' - t'$, and the kinetic energy T is half the mass times the velocity squared. Since T is the same along the entire trajectory, the integration over time simply multiplies with $t'' - t'$; therefore, we have in Cartesian coordinates for q, the formula

$$R(q'', t''; q', t') = \frac{m(q'' - q')^2}{2(t'' - t')} . \tag{1.5}$$

The next term in the expansion is called the *first variation* and is designated with the symbol $\delta \int d\tau\, L$. The variational principle demands that *the first variation vanish for any path which has the same initial and final values* as the trajectory $q_0(t)$, or in other words, has a displacement such that $\delta q(t') = \delta q(t'') = 0$. This principle is usually expressed by the formula

$$\delta \int_{t'}^{t''} d\tau\, L(\dot{q}, q, \tau) = 0. \tag{1.6}$$

This condition on the first variation of the integral over L can be reduced to the equation (1.3). The detailed argument for this conclusion can be found in every textbook on mechanics or on the calculus of variation (cf. Whittaker 1904 and 1944; Courant and Hilbert 1924 and 1953; Carathéodory 1935; Goldstein 1950; Landau and Lifschitz 1957; Arnold 1978). The history of the variational principle is more complicated than is generally realized, and shows that the principle is not nearly as easy to understand intuitively as its formal statement. Euler (1744) and Maupertuis (1744, 1746) gave the first valid proposition of this type in the history of mechanics (cf. Section 2.3); but it differed significantly from (1.6), and does not have a simple connection with Feynman's path integral (cf. the end of Chapter 13). The full impact on mechanics was pointed out to them some sixteen years later by the younger Lagrange (1760); but even he did not use what we call today the Lagrangian, nor did he go beyond his two predecessors to arrive at (1.6). That feat was left to Hamilton (1834, 1835) some 75 years later, who deserves the credit for using the function $L = T - V$, and deducing (1.3) from (1.6).

Jacobi (1842) was the first to offer definite conclusions as to the mathematical meaning of the condition (1.6). Even today, most scientists speak carelessly about the integral over L being a minimum, or they phrase their thoughts more cautiously by talking only about an extremum without any precise idea how to decide whether the trajectory is indeed minimal, maximal, or something in between. This issue is of central importance when we try to make the hazardous transition from classical to quantum mechanics. It requires the third term in the expansion, the so-called second variation, which will be discussed at the end of this chapter. Meanwhile we will try to put the action integral (HPF) itself into better perspective.

1.3 Conservation of Energy

Most of the dynamical systems we will study in this book 'conserve their energy', an expression that needs to be defined in this context. In terms of the Lagrangian L, *conservation of energy* comes about when L does not depend explicitly on the time variable t; the position and the velocity of the system determines the same value of L whatever the time happens to be. As a consequence, the quantity

$$E = \sum_j \dot{q}_j \frac{\partial L}{\partial \dot{q}_j} - L \tag{1.7}$$

stays constant along any trajectory, as is shown in any standard textbook with relatively few manipulations. A further, almost trivial calculation in the case where $L = T - V$, and $T = m\dot{q}^2/2$, shows that $E = T + V$, the sum of kinetic and potential energy, just what we would call the *total energy of the system.*

As another consequence, when we calculate the action integral (HPF), the precise timing of the initial and final state does not matter, as long as the time interval $t = t'' - t'$ is kept the same. Thus we will henceforth write simply $R(q''q't)$ instead of the previous $R(q'', t''; q', t')$. At the same time we simplify the writing by *leaving out the commas and the colons which serve as separators among the variables.* The order of appearance in the list of variables is enough to identify them. This convention will be used throughout this book as long as there is no possible confusion.

A trajectory is usually defined in terms of its initial position q', its final position q'', and the total time $t = t'' - t'$; the action integral R is written as if it were a function of exactly these quantities for a particular trajectory $q_0(\tau)$. What happens when q', q'', or t are allowed to vary? The first thing to make sure is that we still have a trajectory; moreover we want the original trajectory to go continuously into the new one. When these conditions are met, a number of tricky computations show that

$$\frac{\partial R}{\partial q''} = p'', \quad \frac{\partial R}{\partial q'} = -p', \quad \frac{\partial R}{\partial t} = -E, \tag{1.8}$$

where p' and p'' are the initial and final values of the momentum.

These relations are sometimes expressed in the form of a differential that reminds us of thermodynamics, namely

$$\delta R = p''\delta q'' - p'\delta q' - E\delta t . \tag{1.9}$$

The interpretation of this formula is straightforward: the use of $\delta q'$, $\delta q''$, and δt on the right-hand side indicates that q', q'', and t are the natural variables for R; moreover, p'' is the partial derivative of R with respect to q'', and so on. We will sometimes make use of this notation. The classical work of Lagrange and Hamilton is mostly phrased in formulas like (1.9), using *virtual displacements,* which seem to have lost their intuitive appeal in our time.

1.4 Example: Space Travel in a Given Time Interval; Lambert's Formula

Suppose that you leave the Earth in q' today at noon at t', and you want to arrive on Mars in q'' six months later at the time $t'' = t' + 6$ months. You will get a short boost from a rocket at departure, and you will then coast freely for six months. Most of the trip is made outside the gravitational field of either Earth or Mars, so that it suffices for this simple example to take into account only the gravitational attraction of the Sun. Our first task is to find the appropriate trajectory, and then we have to calculate the action integral R.

The known positions of the Earth in q' at the start and of Mars upon arrival in q'' determine a plane together with the central position of the Sun. Choosing polar coordinates (r, ϕ) in this plane with the Sun at the origin, we get q' as (r', ϕ') and q'' as (r'', ϕ''). The equation of the trajectory is given by the expression

$$r = \frac{a\,(1 - e^2)}{1 + e\cos(\phi - \phi_0)} , \tag{1.10}$$

where a is half the *major axis* of a *Kepler ellipse, e* its *eccentricity,* and ϕ_0 the angle in the direction of the *perihelion* (closest approach to the sun). The angle $\phi - \phi_0$ is called the *true anomaly* because it gives the true polar coordinate with respect to the perihelion; it does not increase uniformly with time, however, and is not a convenient parameter in our problem.

The *eccentric anomaly u* is the angle around the center of the trajectory in a special construction of the ellipse. In Cartesian coordinates with the Sun at the origin, the trajectory is given by

$$x = a\,(\cos u - e) , \; y = a\sqrt{1 - e^2}\,\sin u , \tag{1.11}$$

if the x-axis lies along the major axis. The values of x, y, and u are primed for the starting conditions, and double-primed for the arrival.

The time dependence of the eccentric anomaly is given by *Kepler's equation,*

$$u - e \sin u = n(t - t_0) \,, \tag{1.12}$$

where t_0 is the time of perihelion passage, and the right-hand side is called the *mean anomaly.* The *mean motion n* is the average angular speed of our spacecraft if it were allowed to make a complete journey around the Sun. It is given by *Kepler's third law* $n^2a^3 = GM$ in terms of the solar mass M and the gravitational constant G. (Kepler's third law expresses the average balance between the centrifugal force $m\,n^2a$ and the gravitational force GMm/a^2.) The relation between the time t and the eccentric anomaly u is transcendental, i.e., not algebraic, and that is the principal difficulty in treating planetary motions analytically.

The problem of finding the parameters of the elliptic trajectory from the known initial and final positions and times was solved by the eighteenth-century all-round genius Lambert, after the usual preliminary work of Euler. We shall simply record the main steps in this remarkable result because it is not easy to find in any textbook (cf. Battin 1964); the reader may check the algebraic manipulations, which are not difficult.

As a first step, angles α and β are defined by

$$\cos\alpha = e \cos(\frac{u' + u''}{2}) \,, \quad \beta = \frac{u'' - u'}{2} \,,$$

which are then combined into new angles γ and δ through $\gamma = \alpha + \beta$ and $\delta = \alpha - \beta$. If we write $r = ((x'' - x')^2 + (y'' - y')^2)^{1/2}$ for the distance between the points of departure and arrival, the following relations are easy to check with the help of elementary geometry and Kepler's equation (1.12):

$$\mu = r' + r'' + r = 4a \sin^2(\gamma/2) \,, \tag{1.13a}$$

$$\nu = r' + r'' - r = 4a \sin^2(\delta/2) \,, \tag{1.13b}$$

$$\sqrt{GM}\,(t'' - t') = a^{3/2}((\gamma - \sin\gamma) \mp (\delta - \sin\delta)) \,. \tag{1.13c}$$

These equations can be solved as long as the trajectory is indeed a Kepler ellipse, rather than a parabola or hyperbola, because the right-hand sides of (1.13a) and (1.13b) are then smaller than $4a$ by virtue of the triangle inequality. The double sign in (1.13c) comes from choosing the short ($-$) or the long ($+$) elliptic arc connecting the endpoints of our space trip.

The three equations (1.13) have three unknowns, γ, δ, and a ; but the first unknown can be eliminated by the simple expedient of writing

$\gamma - \sin\gamma = 2\arcsin\xi - 2(1-\xi^2)^{1/2}$ for $\xi = \sin(\gamma/2)$, and expanding both arcsin and $(1-\xi^2)^{1/2}$ in powers of $\xi = \sqrt{\mu/4a}$. A similar trick is used for $\eta = \sin(\delta/2) = \sqrt{\nu/4a}$. The necessary expansions are well known; but before writing the result, it is natural to use the energy E of the trajectory rather than its semimajor axis a; since $E = -GMm/2a$, we use $\varepsilon = 1/2a = -E/GMm$, where m is the mass of our space cabin. Finally, the result of Lambert is the series

$$4\sqrt{GM}\,(t'' - t') = \tag{1.14}$$

$$\sum_{j=1}^{\infty} \frac{(2j)!}{2^{2j} j! j!} \left(\frac{1}{2j-1} + \frac{1}{2j+1} \right) \left(\frac{\varepsilon}{2} \right)^{j-1} (\mu^{(2j+1)/2} \mp \nu^{(2j+1)/2})$$

which converges like a binomial expansion because both μ and ν are smaller than $4a$ as noted above.

This marvelous formula can be better understood if we insert concrete numbers; it is natural to use the semimajor axis of the Earth's orbit as the unit length, and the year as the unit of time; the mean motion of the Earth is then 2π/year, and Kepler's third law for the Earth then says that $\sqrt{GM} = 2\pi$. If the semimajor axis a is infinite, i.e., the trajectory is parabolic and its energy $E = 0$, then the formula (1.14) has only the first term $j = 1$, and gives directly the time of travel in years, $t = (\mu^{3/2} \mp \nu^{3/2})/12\pi$. This is the special case of (1.14), which Euler had already derived; it serves as the starting point for more realistic trajectories where $E < 0$ and $\varepsilon > 0$. Since the coefficients in the power series (1.14) are positive, the travel time increases monotonically with increasing ε, as one would expect as the trajectory has less energy E. Given $t'' - t'$, and the distances μ and ν, the necessary energy can be found by solving (1.14).

The whole calculation could have been carried out for hyperbolic trajectories where $E > 0$ and therefore $\varepsilon < 0$; the same formula (1.14) is still valid, but the series is now alternating, and the travel time shorter than for the parabolic orbit.

The action integral R is now calculated directly from its definition (1.4) with the Lagrangian $L = T - V$ as given in the first section, where $V = -GMm/\sqrt{(x^2+y^2)}$. With the trajectory (1.11) and Kepler's equation (1.12) the integral over time is reduced to an elementary integral over the eccentric anomaly u, and then to the parameters γ and δ. The result can be written as

$$R = \frac{m}{2}\sqrt{GMa}\,(3\gamma + \sin\gamma - 3\delta - \sin\delta)\,, \tag{1.15}$$

for the direct trajectory covering a polar angle less than π. The angles γ and δ can again be eliminated by the same trick as above; but the semimajor axis a, or equivalently the energy E, has first to be found from Lambert's equation (1.14).

1.5 The Second Variation

In trying to work out the *second variation,* all possible displacements have to be presented in some practical form, such as an expansion in a Fourier series over the time interval t,

$$\delta q(\tau) = \sum_{1}^{\infty} a_j \sin \frac{\pi j \tau}{t} . \quad (1.16)$$

The coefficients a_j are real, and we have chosen a sine series in order to get the boundary conditions $\delta q(0) = \delta q(t) = 0$. The second variation consists of the second-order terms in the expansion of $\int d\tau L$ in powers of $\delta q(\tau)$; it is a quadratic function of the numbers a_j. The nature of the extremum around the trajectory $q_0(\tau)$ is entirely determined by the character of this quadratic function. If it is positive definite, we have indeed a minimum; otherwise, we have a more complicated situation like a saddle-point.

This problem was first studied by Jacobi, mostly in the context of geodesic lines on a two-dimensional surface. Marston Morse (1934) in the 1920s and 1930s then came up with the definitive statements. The nature of the extremum can be determined without explicitly calculating the quadratic function of the preceding paragraph. The answer depends on the trajectories in the neighborhood of $q_0(\tau)$. Since they are given by solving the equation of motion (1.1) and their coordinates $q(\tau)$ do not differ much from the trajectory $q_0(\tau)$, it is natural to write again $q(\tau) = q_0(\tau) + \delta q(\tau)$, and find the equations to be solved by δq.

Let us take the simplest case where $p = m\dot{q}$, and use Newton's equation (1.2). The left-hand side is already linear in δq, but the right-hand side has to be expanded in powers of δq. Only the lowest terms are retained so that

$$m \frac{d^2 \delta q_i}{d\tau^2} = - \frac{\partial^2 V}{\partial q_i \partial q_j} \bigg|_{q_0(\tau)} \delta q_j , \quad (1.17)$$

where the so-called *Einstein convention* has been used: *indices occurring twice are to be summed unless specified otherwise.* The second derivatives of the potential at the time τ are evaluated by inserting for the coordinates q the values $q_0(\tau)$. In solving these ordinary linear differential equations we use the initial conditions $\delta q(0) = 0$, and

$m\delta\dot{q}(0) = \delta p'$, where $\delta p'$ are arbitrary numbers. The neighboring trajectories are thereby chosen to start in the same location q', but with momenta different from p'. These neighboring trajectories form a fan that spreads out as one moves away from the starting point.

This *fan of trajectories* is left to spread until the time t''. The endpoints differ from q'' by an amount $\delta q''$. The starting deviations $\delta p'$, and the final deviations $\delta q''$ are vectors that are linearly related through the formula $\delta q'' = N\,\delta p'$ where the matrix N is given by

$$N^{-1} = M = \frac{\partial(p'_1, p'_2, \ldots)}{\partial(q''_1, q''_2, \ldots)} = \left(- \frac{\partial^2 R}{\partial q''_i\, \partial q'_j} \right). \tag{1.18}$$

Instead of the matrix N we have calculated its inverse M, which can be expressed directly in terms of the second derivatives of the action integral $R(q''q'\,t''-t')$ along the trajectory $q_0(\tau)$ with respect to the coordinates q' and q'' because of (1.8). The matrix M and its determinant will play an important role later on.

The fan of trajectories starts out with a regular matrix N, or equivalently, with a regular matrix M. Any system moves like a bunch of free particles for the first few moments of its trajectory, provided that the forces acting on it are finite. The formula (1.5) for $R(q''q'\,t''-t')$ can, therefore, be used to calculate the matrix M, which becomes a multiple of the unit matrix.

As the swarm of trajectories moves away from its starting point, however, it may collapse occasionally; that happens when N becomes singular. These unfortunate occurrences are isolated as the swarm moves along, so that there is a sequence of times $0 < \tau_1 \leq \tau_2 \leq \tau_3 \leq \cdots$ when that happens. Usually, the rank of N gets reduced by 1, and the next time N becomes singular is strictly later. But sometimes the rank of N may be reduced by 2 at the same time; in that case we insert two values in the series of τ_i that are equal, and distinct from either their predecessor or their successor. The number of consecutive equal signs in the sequence of τ_i is limited to one less than the number of degrees of freedom. The times where the matrix N gets reduced in rank from its maximum are called the conjugate times, or, if we think of these events as occurring along the trajectory, the *conjugate points,* conjugate to the starting time or the starting point.

The second variation of the integral over the Lagrangian can now be characterized by the following proposition, due mainly to Marston Morse:

The second variation, considered as a quadratic form in the displacements $\delta q(\tau)$ of all the possible paths around a given trajectory *from q' to q'' in the time t, has as many negative eigenvalues as there are conjugate points along the trajectory.*

This simple theorem answers the question about the extremum in Lagrange's variational principle. For sufficiently short times the classical trajectory is indeed a minimum among all possible paths. But that nice feature gets lost as soon as the system has passed the first conjugate point. An obvious illustration of this situation is a *particle moving freely on the surface of a sphere.* Let the two endpoints q' and q'' not be antipodal so that there is a well-defined great circle connecting them. Moving from q' to q'' can be accomplished in two ways: the shorter route goes directly, whereas the longer one passes through the antipode of q'. All the great circles that start in q' go through its antipode. All particles starting out in q' with the same speed will meet in the antipode. The matrix N is singular there, because a non-vanishing value of $\delta p'$ yields a vanishing value for $\delta q''$; the antipode is, therefore, conjugate to q'. Thus, according to the above proposition, the route through the antipode is not a minimum, as we know already from elementary geometry, but can see now in the context of Lagrangian mechanics.

1.6 The Spreading Trajectories

The relation between classical and quantum mechanics depends on the way in which a swarm of classical trajectories spreads out. A picture for the typical situation in two degrees of freedom is shown in Figure 1. The fan of trajectories starting in q' first spreads, but then converges again so that the individual trajectories cut into one another. Actually, they form an envelope that looks as if the trajectories made a glancing reflection from a wall. The envelope is called a *caustic,* in contrast to the exceptional situation where all the trajectories starting in q' go exactly through the same point q_f which is then called a *focus.* Both terms are taken from the obvious analogy in optics. The antipode in the above example is obviously a focus for the trajectories on the sphere.

The spread of the trajectories which all start in q' can be watched very closely if one keeps track of the eigenvalues of N. The explicit expression (1.18) for M shows that it is symmetric and has, therefore, *real eigenvalues.* This symmetry is directly connected with the reversibility of the classical trajectories: Can the trajectory from q' to q'' in the given time t also serve to go from q'' to q' in the same allotted time t ? In general, this symmetry gets spoiled by the presence of a magnetic field, and the eigenvalues of N are then no longer real.

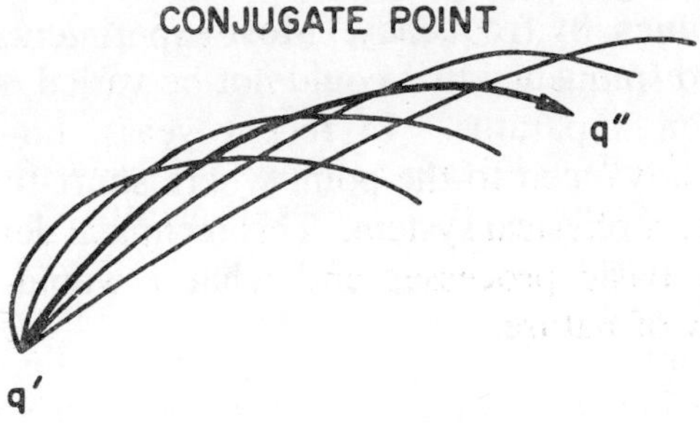

Figure 1 Fan of trajectories originating in the point q' and intersecting one another to form a caustic; the main (heavy) trajectory touches the caustic in the point *conjugate* to q'. There is a whole sequence of caustics and conjugate points along the main trajectory.

Conjugate points are the places along the trajectory where one of the eigenvalues vanishes. The rate at which the largest eigenvalue grows is called the *Lyapounoff exponent* of the trajectory. The growth of these eigenvalues is crucial for judging the long-term behavior of the dynamical system. In particular, we find instabilities when some of the eigenvalues grow exponentially with time.

A somewhat abbreviated characterization of the swarm of trajectories is obtained by looking at their *density* at any one given time. Consider a little volume in position space that is defined by as many initial values of $\delta p'$ as degrees of freedom. Each such $\delta p'$ leads to a $\delta q''$ after the time t. The density $C(q''q't)$ is defined as the ratio of the volume defined by $\delta p'$ over the volume defined by the $\delta q''$. The formula (1.18) now gives us the concise expression

$$C(q''\,q'\,t) \;=\; \left| \;-\; \frac{\partial^2 R}{\partial q''_i \partial q'_j} \right| . \tag{1.19}$$

We shall find that the *quantum mechanical amplitude* of a system that started out in q' is essentially given by the square root of C when the system is observed in q'' at the time t.

The mechanics of Lagrange, as presented very briefly in this chapter, comes closest to our most intuitive picture of the way in which things happen in nature. They start some place and then spread by small increments as time evolves. Situations which were close to one another may drift apart more and more, or they may come back together again, at least temporarily.

The next chapter will explain a different view which is not quite as appealing, but has a certain number of technical advantages both for experiments and for calculations. Time as the controlling parameter will be replaced by energy, or equivalently, as is well known from quantum mechanics, by frequency. Most experiments used to be conducted at a fixed frequency that could not be varied over a large interval in one given apparatus. In recent years, however, electronic technology has advanced to the point where short-time pulses can be used to stimulate a physical system. This technical development brings us back to the basic processes and what I would like to call the Lagrangian view of nature.

CHAPTER 2

The Mechanics of Hamilton and Jacobi

The transition from (what is here called) the mechanics of Lagrange, with time as the main parameter, to the mechanics of Hamilton and Jacobi, with energy as the controlling variable, is formally easy to carry out. Its importance becomes apparent when one tries to solve special problems. The test case is the motion of a body where the force decreases as the inverse square of the distance from the origin. It will be treated at the end of this chapter and will be given an appealing geometrical solution.

2.1 Phase Space and Its Hamiltonian

The state of a dynamical system at the time t is now specified by giving its momentum p and its position q, rather than its velocity $\dot{q}$ and its position q. If we start by describing the system with the help of its Lagrangian L, then we have to make the transition from the velocity $\dot{q}$ to the momentum p via the formula (1.1). This kind of transformation is well known from thermodynamics and is generally called a *Legendre transformation.* It implies a change in the function describing the system at the same time as using its derivatives as the new variables; in our case that means transforming the Lagrangian L, which is a function of $\dot{q}, q$, and t, into the *Hamiltonian* H, which is a function of p, q, and t, with the help of the formulas

$$p_j = \frac{\partial L}{\partial \dot{q}_j}, \quad H(p, q, t) = \sum \dot{q}_j \frac{\partial L}{\partial \dot{q}_j} - L. \tag{2.1}$$

Comparing with (1.7), the Hamiltonian can be interpreted as the energy of the dynamical system at the time t.

The space whose points are defined by the n momenta p and the n coordinates q is called the *phase space* of the dynamical system. For the familiar Lagrangian $L = T - V$ with $T = m\dot{q}^2/2$ and V depending only on q, one finds $H = T + V$ with $T = p^2/2m$. In mathematical terminology this is the cotangent bundle for the manifold in which the dynamical system moves.

The *Hamilton-Jacobi equations of motion* (1.2) become

$$\frac{dp_j}{dt} = -\frac{\partial H}{\partial q_j}, \quad \frac{dq_j}{dt} = \frac{\partial H}{\partial p_j}, \tag{2.2}$$

two sets of first-order equations, rather than one set of second-order equations in the Newtonian tradition. We can regard them as defined by a vector-field in phase space, which defines a *flow* in phase space. It is given by some kind of gradient of the Hamiltonian H. The opposite signs in the two sets of equations are crucial and cannot be eliminated by any simple device. The consequences of these differing signs are pursued in a special discipline, *symplectic geometry,* which is obviously important to the study of mechanics; but we will not discuss this field, except for a short excursion in Chapter 7.

Conservation of energy in Lagrangian mechanics follows from the fact that the Lagrangian L does not depend explicitly on the time t; in complete analogy, conservation of energy in Hamiltonian mechanics requires that $\partial H/\partial t = 0$. The value of $H(p, q)$ then remains constant along any trajectory. This constant value is usually designated by E as before, so that we will write $H(p, q) = E$. In a large measure, the value of E for a particular trajectory will replace the parameter t in many applications. The duality between E and t comes out quite naturally in Hamiltonian mechanics, but its full significance can only be appreciated in quantum mechanics.

2.2 The Action Function S

The replacement of t by E as the independent parameter requires that we use a new kind of action integral $S(q''q'E)$. It is again defined by a trajectory from q' to q''; but instead of the given time interval t, the energy E of the trajectory is now stipulated. In doing so, another Legendre transformation is made,

$$S(q''\, q'\, E) = R(q''\, q'\, t) + E\, t$$
$$= \int_0^t \sum p_j\, \dot{q}_j\, d\tau = \int_{q'}^{q''} \sum p_j\, dq_j\,. \tag{2.3}$$

A number of comments are necessary in order to clarify the meaning of the various expressions on the right. The first shows the Legendre character of the transition from R to S if we recall the relation (1.8), $E = -\,\partial R/\partial t$. The second expression is written as an integral over the trajectory with the time as the parameter of integration; that may be the practical recipe to adopt in many cases, but it is not in the spirit of Hamiltonian mechanics. Thus, we arrive at the third expression, which is most often quoted. The integral is calculated for a curve in phase space that coincides with the trajectory and whose parameter can be any monotonically increasing variable; the value of the integral does not depend on the particular choice of this parameter. The element of integration $p\, dq$, the scalar product of the vector p with the vector dq, is *the canonical 1-form* in the language of symplectic geometry.

This 1-form is important, among other reasons, because it guarantees the particularly simple form (2.2) of the equations of motion. More specifically, a second coordinate system $(\bar{p}, \bar{q})$ in phase space, and the transformation formulas $p = P(\bar{p}, \bar{q}, t)$; $q = Q(\bar{p}, \bar{q}, t)$ have to be such that $p\, dq - \bar{p}\, d\bar{q} = d\Phi$, with a total differential on the right-hand side. The interpretation of this relation is again the same as in thermodynamics: the variables in the function Φ have to be q and $\bar{q}$ so that $p = \partial\Phi/\partial q$ and $\bar{p} = -\,\partial\Phi/\partial\bar{q}$. The function Φ is called the *generating function,* because the explicit formulas for the change of coordinates can be written as derivatives of Φ.

The independent variables in Φ can be chosen in various ways; a useful choice is to write $p\, dq + \bar{q}\, d\bar{p} = d(\Phi + \bar{p}\, \bar{q}) = dW$, where $W(q, \bar{p}, t)$ is now considered to be a function of the mixed coordinates, the old position q and the new momentum $\bar{p}$; this form will be used in Chapter 5. The equations of motion in the new system have again the form (2.2) in terms of the new Hamiltonian $H'(\bar{p}, \bar{q}) = H(p, q, t) + \partial W/\partial t$. Coordinates in phase space for which the original 1-form $p\, dq$ is given by the same simple formula up to a total differential are called *canonical coordinates.* It should be emphasized that a particular system of canonical coordinates is not tied to any special Hamiltonian; but for some exceptional Hamiltonian, one might find canonical coordinates in which this Hamiltonian turns out to have a simple form. Whether the Hamiltonian can be simplified at all, by any possible choice of canonical coordinates, depends on the Hamiltonian;

if the simplification is possible, the system is called *integrable,* and will be discussed in the next chapter.

The formal relationships between the new action integral $S(q''q'E)$ and the momentum p' at the beginning as well as p'' at the end of the trajectory are contained in the formulas

$$p'' = \frac{\partial S}{\partial q''}, \; p' = -\frac{\partial S}{\partial q'}, \; t = \frac{\partial S}{\partial E}. \tag{2.4}$$

The partial derivatives with respect to q' and q'' are calculated for the fixed energy E, whereas in the corresponding formulas (1.8) these derivatives are taken at constant value of t. With this proviso the two sets of formulas are equivalent.

When going toward quantum mechanics it seems advisable to think of the integral over $p\,dq$ as a new and physically relevant 'length' of the trajectory. In the case of a particle moving freely in (Euclidean) space, the trajectory is a straight line from q' to q'', and its energy E is now $p^2/2m$; the momentum $p = m\dot{q}$ points in the direction of motion, as does the increment (differential) dq ; their scalar product $p\,dq$ is simply the length (absolute value) of p times the length (absolute value) of $dq, pdq = |p|\,|dq|$. Since $|p|^2 = 2mE$ is constant along the trajectory, we are left with the integral over $|dq|$ from q' to q'' which gives simply $|q'' - q'|$; therefore

$$S(q''\,q'\,E) = \sqrt{2mE\,(q'' - q')^2}\;. \tag{2.5}$$

This elementary calculation was carried out verbally rather than algebraically because the relationships between the quantities should be present in the mind of the reader at an almost instinctual level rather than as formal propositions to be written down when needed. Notice the big difference between this last expression and formula (1.5) for $R(q''q'\,t'' - t')$, which involves the square of the distance from q' to q'' rather than the distance itself as in (2.5).

2.3 The Variational Principle of Euler and Maupertuis

The variational principle of the last chapter is not easily transferred from Lagrangian to Hamiltonian mechanics. Nevertheless, we will mention at least one formulation because it is occasionally helpful. We assume again the most familiar situation where $H = T + V$ and the kinetic energy $T = p^2/2m$, while the potential energy V depends only on the position coordinates q. Now we can use the argument of the preceding paragraph: because the constant value E of H is given, we find $p^2 = 2m(E - V)$. Moreover, as before, the directions of p and dq

are parallel, and the integrand of (2.3) is given by the product of the absolute values of p and of dq. Thus,

$$S(q''\, q'\, E) = \int_{q'}^{q''} \sqrt{2m(E - V)}\ |dq|, \tag{2.6}$$

to be integrated along the trajectory that leads from q' to q''. The parameter along the trajectory is again irrelevant since the integrand is homogeneous of first degree in dq.

The argument that follows is quite similar to the one in Section 1.2; we start again from a trajectory $q_0(\tau)$, and we add a small deviation $\delta q(\tau)$ in order to get the path $q(\tau)$; then we expand the integral (2.6) in powers of $\delta q(\tau)$ so as to get the first and the second variation. The equations of motion for the trajectory $q_0(\tau)$ are again found to be equivalent with the proposition that the first variation $\delta S = 0$.

In this form Maupertuis and Euler gave the world the first variational principle of mechanics in 1744. This proposition can be given a purely geometric interpretation: If one defines a length for any curve in the space of position coordinates q by the integral (2.6), then the trajectories of a particle with energy E in the potential $V(q)$ have a vanishing first variation. They can be viewed as the locally shortest connection between the two endpoints when the distance is measured by the formula (2.6), exactly as geodesics in a manifold with a Riemannian metric.

2.4 The Density of Trajectories on the Energy Surface

As in Lagrangian mechanics one can ask what happens to a swarm of particles that starts from the initial position q' with the energy E, but taking off in directions $p' + \delta p'$. In contrast to the deviations in initial momentum of the preceding chapter, which are completely arbitrary, one has now lost one degree of freedom by specifying the energy to be the same for all trajectories. The trajectories are bound by the condition $H(p, q) = E$ to a $(2n - 1)$-dimensional surface on phase space, the *surface of constant energy*. This restriction leads to a slightly different definition for the density of trajectories in the neighborhood of the one that starts in q' and ends in q'' at the fixed energy E.

The original expression (1.19) for the density $C(q'q''t)$ in Lagrangian mechanics is used as starting point; but the derivatives of the action integral $R(q'q''t)$ are now written in terms of the action integral $S(q'q''E)$, which is defined by (2.3) together with $t = \partial S/\partial E$ from (2.4). The derivatives of R in (1.19) are taken at constant time

t, while the energy is allowed to vary. Thus, we have to allow for a variation $\partial E/\partial q$ while enforcing the condition

$$\frac{\partial t}{\partial q} = 0 = \frac{\partial^2 S}{\partial E\,\partial q} + \frac{\partial^2 S}{\partial E^2}\frac{\partial E}{\partial q},$$

which applies either to q' or to q''.

The entries in the determinant (1.19) become

$$-\frac{\partial^2 R}{\partial q'\partial q''} = -\frac{\partial^2 S}{\partial q'\partial q''} + \frac{\partial^2 S}{\partial E\partial q'}\frac{\partial^2 S}{\partial E\partial q''}\Big/\frac{\partial^2 S}{\partial E\partial E},$$

with the appropriate indices. The n by n determinant with these complicated entries can be written in terms of the $n+1$ by $n+1$ determinant

$$D(q''q'E) = (-1)^{f+1}\begin{vmatrix} \dfrac{\partial^2 S}{\partial q'_i\partial q''_j} & \dfrac{\partial^2 S}{\partial q'_i\partial E} \\ \dfrac{\partial^2 S}{\partial E\partial q''_j} & \dfrac{\partial^2 S}{\partial E^2}\end{vmatrix} = -\frac{\partial^2 S}{\partial E^2}C(q''q't)\,. \tag{2.7}$$

The n by n subdeterminant in D, which consists of the second derivatives of S with respect to q' and q'' alone, without the last row and the last column containing the derivatives with respect to E, vanishes. This can be seen if one writes the conservation of energy with the help of (2.4) in the form of a first-order partial differential equation for S as function of q'',

$$H(p'',q'') = H(\frac{\partial S}{\partial q''}, q'') = E\,. \tag{2.8}$$

If this equation is differentiated with respect to q', one finds

$$\frac{\partial}{\partial q'_i}(H(p'',q'')) = \frac{\partial H}{\partial p''_j}\frac{\partial^2 S}{\partial q''_j\partial' q_i} = 0\,. \tag{2.9}$$

The matrix of mixed second derivatives is singular, because the vector $\partial H/\partial p''_j$, the velocity vector according to (2.2), is mapped into 0.

The determinant (2.7) can be made more understandable if we use a local coordinate system in the neighborhood of the trajectory from q' to q'' ; further details will be worked out in Chapter 7. The coordinate axis for q_1 runs along this particular trajectory. The remaining coordinates, say q_2 and q_3, are transverse to the trajectory; for example, they are chosen at a right angle to the direction of motion along the trajectory. The velocity vector $\partial H/\partial p''_j$ has, therefore, the components $(\dot q_1, 0, 0)$. Equations (2.9) then tell us that the second derivatives $\partial^2 S/\partial q''_j \partial q'_i$ vanish whenever either $j = 1$ or $i = 1$. The first row and the first column in the determinant (2.7) vanish, except the last terms containing the partial derivatives with respect to E. Their values are obtained from differentiating (2.8) with respect to E, yielding $\dot{q}_1 \, \partial^2 S / \partial q_1 \partial E = 1$. In this way, we can write

$$D(q''q'E) = \frac{1}{|\dot{q}'| \, |\dot{q}''|} \left| \frac{-\partial^2 S}{\partial q'_i \partial q''_j} \right| , \qquad (2.10)$$

where the determinant excludes the indices $j = 1$ or $i = 1$. If the dynamical system has only one degree of freedom, only the first factor remains in (2.10), and D becomes simply the product of the inverse velocities at the beginning and at the end.

Although there is only the factor $\partial^2 S / \partial E^2$ between D and C in (2.7), the appropriate density of trajectories in the mechanics of Hamilton and Jacobi is undoubtedly D rather than C. The factor between them is the derivative $\partial t / \partial E$, which can be interpreted by going back to the original idea in (1.18). The n- dimensional volume of variations $\delta p'$ in the numerator is divided by the variation δE in energy, and the n-dimensional volume of variations $\delta q''$ in the denominator is divided by the variation δt in arrival time. If the energy E is changed, but the starting point q' and the endpoint q'' remain fixed, then the transit time t has to be changed, too. For this reason, one needs the peculiar division of the volume of endpoints by δt . The allowed swarm of trajectories has thereby been effectively constrained to the energy surface.

There is a more formal aspect to this argument, which betrays our ultimate goal of reaching the quantum-mechanical limit. The physical dimension of C is [volume in momentum space]/[volume in position space] as shown in (1.18); but (1.19) has the equivalent physical dimension of [volume in phase space]/[volume in position space]^2. If one takes the square root, the volume of position space appears in the denominator. Similarly, D has the physical dimension of [volume in phase space]/[volume in position space times energy]^2, which reduces after the square root to both [volume of position space] and [energy] in the denominator. As one goes to quantum mechanics, the [volume in phase space] in the numerator gets divided by the appropriate power of Planck's quantum h, leaving only the denominator. Thus, the square root of D is related to something like a density in energy as well as in position space that will be called Green's function in Chapter 12.

2.5 Example: Space Travel with a Given Energy

The same problem as in the preceding chapter will now be taken up, but the conditions are changed to fixing the available energy E rather than the available time interval $t'' - t'$ for the trip from Earth to Mars. We will first give an analytic expression for the action $S(q''q'E)$ using the same calculation as in Section 1.4. Then we will give a geometric construction of Jacobi for finding the trajectory from q' to q'' at the energy E. The main purpose of this exercise is to determine the number of conjugate points on a Kepler ellipse that will be important in quantizing the hydrogen atom.

The action $S(q''q'E)$ is obtained from the integral (2.3) using the eccentric anomaly u from (1.11) and (1.12) as the parameter of integration, exactly as in the calculation of $R(q''q'\,t'' - t')$ of (1.15). The result for the direct trajectory is

$$S(q''q'E) \;=\; m\sqrt{GMa}\;(\gamma + \sin\gamma - \delta - \sin\delta)\,, \qquad (2.11)$$

where the angle γ is given by (1.13a), and δ by (1.13b), while the semimajor axis $a = -\,GMm/2E$. Notice that this expression for the action is complete, i.e., it does not depend any longer on solving an equation like Lambert's (1.14). The right-hand side of (2.11) can be expanded exactly like (1.14) as a power series in the normalized energy ε and the given distances μ and ν, including the appropriate modifications for the indirect trajectory, and the positive energies.

The three expressions (1.14), (1.15), and (2.11) are not independent because of the definition (2.3) for S in terms of R, E, and t. More striking, however, is the last of the relations (2.4) which allows us to obtain Lambert's formula (1.14) directly from (2.11) by differentiating with respect to E.

Now to the *geometric construction* (Jacobi 1842, p. 48): Since the Sun at the origin O is one of the two foci for the Kepler ellipse of our space trip, the main problem is to find the other focus F from the information that the ellipse has to go through q' and q''. The construction of F, as shown in Figure 2, turns out to be quite simple, provided we have obtained the length of the semimajor axis $a = -\,GMm/2E$ from the given energy E.

The sum of the distance $r' = Oq'$ and the distance $d' = q'F$ equals $2a$, so that $d' = 2a - r'$; similarly, the distance $q''F = d'' = 2a - r''$. Therefore, we draw a circle of radius d' around q', a circle of radius d'' around q'', and find their intersections.

If the sum of the distances $Oq' + q'q'' + q''O$ is greater than $4a$, or equivalently, $q'q'' + q''O > 4a - r'$, the circles do not intersect, and there is no trajectory from q' to q'' at the given energy E. In other

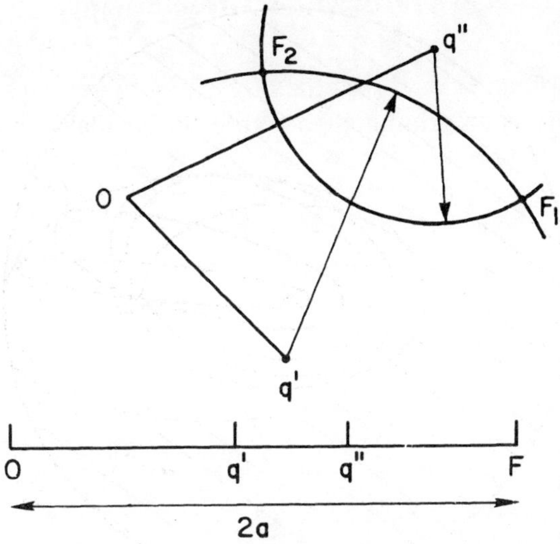

Figure 2 Jacobi's construction of the Kepler ellipse from q' to q'', if the energy E of the trajectory, i.e., its major axis $2a$, is given. Notice the completely different eccentricities, i.e., angular momenta, for the two solutions.

words, if the position of O and the initial position q' are given, the final position q'' has to be inside a *critical ellipse* with O and q' as foci and the major axis equal to $4a - r'$; otherwise, there is no trajectory from q' to q'' with the energy E. The action $S(q''q'E)$ is not defined outside this critical ellipse.

If q'' is inside the critical ellipse, the second focus F is obtained by intersecting the two circles, of radius d' around q', and of radius d'' around q''. For each of the two possible choices for F, the corresponding Kepler ellipse can be constructed, because not only the semimajor axis a, but also the eccentricty e as well as the perihelion is now known. The action $S(q''q'E) = \int p\, dq$ along each of the two ellipses can be calculated without explicit knowledge of the time-dependence. The integral contains no worse than the square root of a quadratic function, and can be worked out in terms of elementary functions to yield (2.11) without ever invoking Kepler's equation (1.12).

For q'' inside the critical ellipse there are always exactly two intersections, F_1 and F_2, for the two circles, and therefore, two Kepler ellipses from q' to q'' at the energy E. The two trajectories are well known to tennis players and artillery officers, the straight low shot and the indirect high shot, both with the same expenditure of energy or gun powder.

When q'' approaches the critical ellipse these two trajectories coalesce and give rise to a conjugate point. Equivalently, all the Kepler

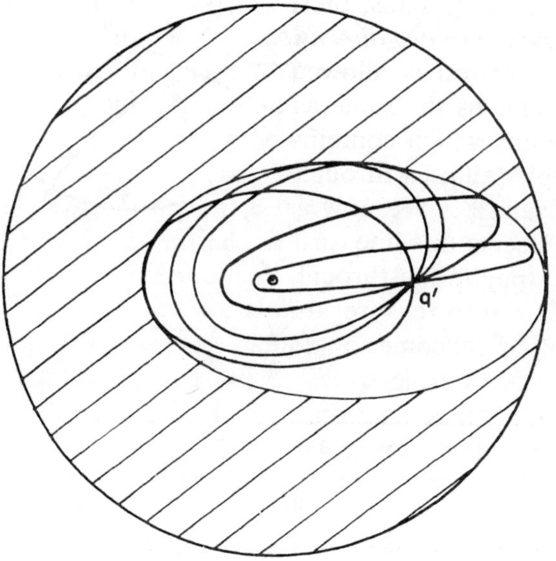

Figure 3 Critical ellipse for the point q': all the Kepler ellipses around the same center of attraction O, with the same energy (major axis), and going through the same point q' stay inside their common caustic, an ellipse with O and q' as foci.

ellipses through q' with the fixed semimajor axis a lie inside the critical ellipse and touch it. Indeed, consider the Kepler ellipse through q', with the focus F in addition to O, and with the semimajor axis a. We have both $Oq' + q'F = 2a$ and $Oq'' + q''F = 2a$, wherever q'' happens to lie. As we move q'' away from q' along this Kepler ellipse, we come to the point where q' and q'' lie on a straight line through F. This point lies on the critical ellipse because we have $Oq' + q'q'' + q''O = 4a$. Therefore, every trajectory through q' touches the critical ellipse in a conjugate point, and the critical ellipse itself is a caustic. Notice that it is not sufficient for the point q'' to satisfy the condition that $V(q'') < E$ to be reached from q' with the energy E. As is well known in space travel, and shown in Figure 3, a launch window is needed. The goal of the space journey has to be inside the critical ellipse at the time of arrival.

In the transition from classical to quantum mechanics, the number of conjugate points on a trajectory that closes itself smoothly like a Kepler ellipse is of great importance. The construction of the critical ellipse has shown the existence of one conjugate point on any given trajectory through q'. A second conjugate point is q' itself since all trajectories starting there return to it; q' is a focus rather than simply a point on a caustic. Thus, there are *two conjugate points along one closed Kepler ellipse in two dimensions.*

When the electron is allowed to move in three dimensions, the construction remains the same as above, but the plane of the Kepler ellipse can be any which contains both O and q'. Therefore, the collection of Kepler ellipses through O and q' can be rotated around the line joining O and q'. We now get a third conjugate point where the trajectory intersects this line, just as the first conjugate point was located on the straight line through q' and the focus F. Finally, all the Kepler ellipses return to q' regardless of their orientation in space, so that the point q' becomes a double conjugate point since a two-parameter family of trajectories all meet there again (cf. Gutzwiller 1967 and 1969). In conclusion, there are *four conjugate points on the Kepler ellipse in three dimensions.* This difference in the number of conjugate points accounts for the different energy spectra of the hydrogen atom, with half-integer quantum numbers in two dimensions and integer quantum numbers in three dimensions, as will be discussed in Chapter 14.

CHAPTER 3

Integrable Systems

Chaotic dynamical systems are the main topic of this book. Many readers, however, have grown up in the belief that most systems of interest are regular. The contrast between regular and chaotic behavior has to be well understood in order to appreciate the novel features in the chaotic systems. Therefore, this chapter is devoted to the discussion of the regular systems, and in particular to the display of some of the more challenging examples.

Since the discusssion of integrable systems has been the mainstay of all the advanced textbooks on classical mechanics, this topic is highly developed, and an adequate account would take up too much space. We shall try, therefore, to explain the salient features without proofs or elaborate calculations. Nevertheless, the ideas will be presented with greater care than might at first appear necessary, not so much for dealing with the examples in this chapter, as in view of the more difficult discussion of the three-body problem and the methods for its solution in Chapters 4 and 5.

3.1 Constants of Motion and Poisson Brackets

Integrable dynamical systems are characterized by the existence of constants of motion in addition to the energy. It may not always be easy to find an explicit expression for these, but their presence can immediately be recognized because they generate invariant tori in phase space. What constants of motion and invariant tori are, and what they do for the dynamical system, will be explained in the next few paragraphs. The adjectives 'integrable' and 'separable' designate systems that behave essentially in the same way. The small difference in meaning between these two words will be explained at the end of the chapter.

Although we started out with Lagrangian mechanics and declared it to be the more fundamental and natural approach to the secrets of the universe, most of the discussion concerning special examples will be done in the framework of Hamiltonian mechanics. First, the Hamiltonian $H(p, q)$ of the dynamical system will be written down, and then the equations of motion (2.2). Suppose that these two steps have been completed, and a function $F(p, q)$ is discovered, defined in the relevant part of phase space, with the following property: The value of F stays constant as a function of time t when its arguments, p and q, are replaced by a solution of the equations of motion (2.2). Such a function is called a *constant of motion* of the dynamical system.

The constant value along a trajectory in phase space requires that

$$\begin{aligned} 0 = \frac{d}{dt}F(p, q) &= \frac{\partial F}{\partial p}\frac{dp}{dt} + \frac{\partial F}{\partial q}\frac{dq}{dt} \\ &= \frac{\partial H}{\partial p}\frac{\partial F}{\partial q} - \frac{\partial H}{\partial q}\frac{\partial F}{\partial p} = [H, F]. \end{aligned} \tag{3.1}$$

The second line defines the *Poisson bracket* $[H, F]$, which can be computed for any two functions in phase space. The vanishing of the Poisson bracket between the Hamiltonian H and the function F throughout phase space makes F a constant of motion.

This last condition can be given a *geometric interpretation:* the vector field $(-\partial H/\partial q, \partial H/\partial p)$ in phase space is tangent to the surface $F(p, q) = constant$. Conversely, the vector field $(-\partial F/\partial q, \partial F/\partial p)$ is tangent to the energy surface $H(p, q) = E$. The trajectory lies in the intersection of these two surfaces in phase space.

A whole collection of constants of motion might be found for the dynamical system; call them $F_1, F_2, \dots$. Each constant has a vanishing Poisson bracket with the Hamiltonian in accordance with (3.1). Moreover, it is important that they are independent of one another; it should not be possible to express F_3 as a function of F_1 and F_2. The trajectory then lies in the intersection of all the surfaces $F_i(p, q) = constant$ with the appropriate values of these constants.

These restrictions on the constants of motion, however, are not sufficient as yet to assist in solving the equations of motion. The next paragraph will discuss additional conditions to be imposed on the constants of motion; if they are satisfied, a special system of coordinates can be constructed in phase space: the action-angle variables, which account for the special properties of *integrable* systems.

A dynamical system with n degrees of freedom has a phase space of $2n$ dimensions. Suppose that we have been able to find k independent constants of motion including the energy $H(p, q) = E$. The trajectory is then restricted to a $(2n - k)$- dimensional subspace of the whole phase space. This subspace contains k different vectorfields, one from each of the k constants of motion; *but these vectorfields are generally not compatible with one another.* One can choose a curve C_1 by following the first vectorfield, and take each point of C_1 as starting point for a set of curves C_2 along the second vectorfield; this set of curves defines a two-dimensional surface in phase space. It would be nice if this surface not only was filled with the curves C_2 along the second vectorfield, but also could be covered with curves of the type C_1 along the first vectorfield. This cannot be done, however, unless the two constants of motion F_1 and F_2 are *in involution,* which means that *their Poisson bracket vanishes.* In general, therefore, the existence of k integrals of motion does not leave the remaining $(2n - k)$-dimensional manifold with a simple internal structure, unless the constants of motion are in involution with one another.

3.2 Invariant Tori and Action-Angle Variables

The most favorable circumstances are achieved when there are at least *n constants of motion in involution,* as many as there are degrees of freedom. The trajectory is then confined to an n-dimensional manifold that is covered by n compatible vectorfields. Let us assume, moreover, that none of these vectorfields ever vanishes on the manifold of interest.. Then, by a remarkable theorem of topology, *this manifold has the shape of an n-dimensional torus;* i.e., by a smooth deformation the manifold can be transformed into an n-dimensional cube whose points on opposite sides are identified. Each trajectory of the dynamical lies inside such an *invariant torus.* The dynamical system is then called *integrable.*

The reader may be familiar with the fact that n constants in involution imply the integrability of the dynamical system, or equivalently, the possibility of finding explicit solutions for the equations of motion. But to demand the existence of these constants of motions is much more restrictive than is generally realized; when they do exist, however, the mechanical system can be described in very explicit detail. The next step in the treatment of such systems is the construction of *action-angle variables.*

Before getting into the technical details, the reader should try to imagine why, in a system with two degrees of freedom and two constants of motion in involution, the trajectories cannot lie on a sphere (or a surface with the same topology, like an egg or a pear); it is impossible to define a vectorfield on a sphere that does not vanish somewhere; just as it is impossible to comb down the hair on one's head without admitting a part or an eddy. On the surface of a torus that is covered with hair, however, there is no trouble in combing it down flat everywhere.

The deformation of the manifold into the n-dimensional cube can be used to define new coordinates $(w_1, w_2, \dots, w_n)$, called *angle variables,* designated as w collectively, each of which varies from 0 to 2π like an angle. If the n compatible vector-fields are taken as the basis for these coordinates, the vector-field $(-\partial H/\partial p, \partial H/\partial q)$ has constant components throughout the manifold, $(\omega_1, \omega_2, \dots, \omega_n)$, called the frequencies of the dynamical system, and the trajectories become straight lines. The set of frequencies ω depends on the values of the constants of motion for the particular invariant torus. Therefore, the ratios between the individual frequencies for a particular torus are, generally, irrational numbers; they become rational only for special values of the constants of motion. As a word of caution, the whole problem of solving the equations of motion is hidden in the few sentences of this paragraph and the next.

The coordinates w are used as the position coordinates in a new set of canonical coordinates for the whole of phase space (cf. Section 2.2); or as one half of a system of *canonically conjugate coordinates.* The other half of these coordinates will be called $(I_1, I_2, \dots, I_n)$, abbreviated by the single letter I. The action integral S in (2.3) as an integral over $p\,dq$ now becomes an integral over $I\,dw$. If the reader is willing to admit the existence of the variables I with the property that $p\,dq = I\,dw$ (its construction will be discussed quite generally in Chapter 7), then the explicit calculation of I proceeds as follows.

Suppose that only w_i varies, and that it runs from 0 to 2π; a closed loop C_i is thereby defined in the original coordinates (p, q) of phase space. Since the action integrals in the two coordinate systems have to agree, we find immediately the formulas

$$I_i = \frac{1}{2\pi} \int_{C_i} p\,dq\,. \tag{3.2}$$

The new variables I play the role of momentum with respect to the positions w ; they are called *actions,* while the variables w are called *angles.* Together they form the *action-angle variables.*

Most textbooks in classical mechanics present the transformation of an integrable system from the original coordinates (p, q) to the action-angle variables (I, w), using Jacobi's theory of first-order partial differential equations. Quite in contrast, the above discussion relies as a first step on finding the appropriate angular variables w ; they are obvious as soon as the invariant tori in phase space have been recognized, and vice versa. If the trajectories can be written in terms of as many angular variables as degrees of freedom, then the invariant tori have been obtained explicitly, as will be shown in the next section. Most of the recent work on chaos in classical mechanics starts from a search for the invariant tori, while most of the classical work in celestial mechanics starts out by writing the coordinates in terms of angular variables. Thus, our presentation stays close to these two important applications of Jacobi's general theory.

3.3 Multiperiodic Motion

The most important statement in the preceding section concerned the time-dependence of the angle variables: they could be chosen such that their derivatives with respect to time, the frequencies ω, were the same on the whole invariant torus. The equations of motion (2.2) in the action-angle variables now take the trivial form

$$\frac{dI_i}{dt} = -\frac{\partial H}{\partial w_i} = 0\,, \qquad \frac{dw_i}{dt} = \frac{\partial H}{\partial I_i} = \omega_i\,. \qquad (3.3)$$

The usual arguments have been reversed: since the actions I are constants of motion, the Hamiltonian can no longer depend on the angles w; also, the frequencies ω are simply the derivatives of the Hamiltonian with respect to the actions I. The Hamiltonian in the new coordinates does not depend on the angles, and is, therefore, written in the form $H(I_1, I_2, \ldots, I_n)$, not to be confused with the previous expression $H(p, q)$, which uses the same letter H to designate another function!

The action-angle variables give *the complete solution of the equations of motion.* The general solution of (3.3) is given by $w_i = \omega_i t + \phi_i$, where the frequencies ω are determined by the values chosen for the actions I, and the phase angles ϕ can be chosen arbitrarily. There are many useful angle-type variables in any special problem, like the true or the eccentric anomaly in the Kepler problem (cf. Section 1.4); but their variation with time is not linear in general; they are not angle variables as defined above, in contrast to the mean anomaly. The difficulty with integrating the equations of motion here consists in finding the angle-type variables whose time-dependence is indeed linear.

When transforming back from the action-angle variables to the original set (p, q), one sees that they are periodic with period 2π in each one of the angles w. It is, therefore, natural to make Fourier expansions in w,

$$(p, q) = \sum_{k_1, \ldots, k_n} (P, Q)_{k_1, \ldots, k_n} \exp[i(k_1 w_1 + \cdots + k_n w_n)], \quad (3.4)$$

where each integer k goes from $-\infty$ to $+\infty$. Formula (3.4) is shorthand for one such expansion for each component of p and q. The expansion coefficients P and Q are complex numbers and depend on the values of the actions I ; but they satisfy the simple relation $P^{+}_{\{k\}} = P_{\{-k\}}$, where + indicates the complex conjugate, to make the sum over the multi-index $\{k\} = (k_1, ..., k_n)$ real. If the motion of a Hamiltonian system can be written in the form (3.4), it is called *multiperiodic*.

How does a perturbation look in the action-angle coordinates? In most of the interesting cases the perturbation arises as an additional time-dependent potential energy $V(q, t)$ that is periodic with the frequency ω_0 with respect to t. In the action-angle variables of the dynamical system, $V(q, t)$ becomes again a Fourier series,

$$V(q, t) = \sum_{k_0, k_1, ..., k_n} V_{k_0, k_1, ..., k_n} \exp[i(k_0 w_0 + k_1 w_1 + \cdots + k_n w_n)], \quad (3.5)$$

where the complex coefficients $V_{k_0, k_1, ..., k_n}$ are functions of I. The same symmetry conditions among them apply as for P and Q in (3.4), to obtain a real value for the sum. The periodic time-dependence of the potential $V(q, t)$ has been included in the Fourier expansion through the angle $w_0 = \omega_0 t + \phi_0$; but in contrast to the other frequencies, ω_0 does not depend on the values of the actions I.

This whole theory of integrable systems will now be illustrated with three examples of a slightly more complicated kind than the ones ordinarily found in textbooks. The purpose of this presentation is not to offer a complete discussion of the solution, but to provide the reader with some material for practice and enjoyment on examples which can be treated without lengthy development. The frequencies ω depend on the actions I in a non-trivial manner, which means that the matrix of derivatives of ω with respect to I is not singular. This condition is not satisfied by the usual textbook examples, like the Kepler problem or a system of linearly coupled oscillators. Therefore, these two systems, although integrable, present special difficulties when perturbations are applied, or the transition to quantum mechanics is made.

The three examples, though typical of integrable systems, are special in other ways. All three of them are non-trivial in the sense that their integration came as a surprise to the specialists in the field, and the authors who first succeeded in solving the problem became famous among scientists for having done it. Moreover, each one of the three solves an important general question in physics or in geometry. They will be taken in their historic order.

3.4 The Hydrogen Molecule Ion

Two protons are fixed in the position A and B a distance 2c away from each other, in the locations $(0, 0, -c)$ and $(0, 0, c)$ of a Cartesian coordinate system. A single electron at (x, y, z) is subject to the Coulomb attraction of both protons. Let r_A be its distance from A, and r_B its distance from B; the Hamiltonian is then given by

$$H = \frac{p^2}{2m} - e^2\left(\frac{1}{r_a} + \frac{1}{r_B}\right) + \frac{e^2}{2c} , \qquad (3.6)$$

where the constant $e^2/2c$ has been added in order to account for the electrostatic repulsion of the two protons.

The solution of this problem can be rated, with only slight exaggeration, as the most important in quantum mechanics, because if an energy level with a negative value of E can be found, the chemical bond between two protons by a single electron has been explained. Classical mechanics yields all kinds of interesting trajectories for this problem, but none of them is safe from losing energy by some external perturbation and thereby leading to the collapse of the molecule. This situation remains even if one tries to replace the point charge of the electron by a charge cloud.

The classical problem was first solved by Euler in the context of two stationary heavy masses exerting a gravitational attraction on a third light mass. The solution is rather easy when one defines elliptical coordinates with the heavy masses at the two foci. The equations of motion can then be separated into three sets of second-order equations, with each set solved by elliptic functions.

Two of the constants of motion are obvious: the total energy E, and the angular momentum $M = xv - yu$ around the z-axis, where the Cartesian components of the momentum are called (u, v, w). The third constant of motion, however, is difficult to write down in a physically appealing manner; one expression is the following,

$$\Omega = \mathbf{L}_A \mathbf{L}_B + 2me^2c\,(\cos\theta_A - \cos\theta_B) , \qquad (3.7)$$

where $\mathbf{L}_A$ and $\mathbf{L}_B$ are the angular momentum vectors of the electron around A and around B, while θ_A and θ_B are the angular distances of the electron from the axis of the molecule, as seen from A and from B. The three variables H, M, and Ω are in involution.

The classical problem has three physical parameters: the distance $2c$ between the two protons, the mass m of the electron, and its electric charge e ; therefore, all dynamical variables can be normalized classically. The energy is measured in units of e^2/c, and Ω in units of

$2m\,e^2c$; the trajectories cannot be scaled with the energy E as in the Kepler problem.

Before discussing the results, a little history of the problem is well worth telling. After Bohr had given the first solution of a quantum-mechanical problem in 1913 by deriving the energy spectrum of the hydrogen atom, Sommerfeld generalized Bohr's idea, and then asked his precocious student, Wolfgang Pauli aged 19, to treat the hydrogen molecule ion for his Ph.D. thesis. At that time, in 1919, it was not known whether there exists such a molecule as the singly ionized H_2^+. Something like it had been seen, but under conditions which made it unlikely that the ion would be stable.

Pauli (1922) applied the rules which Bohr and Sommerfeld had proposed and concluded that the hydrogen molecule ion could only be metastable. He showed that its lowest energy level (3.6) was positive; but it would only decay because of collisions, and not by emitting light as most other unstable atoms or molecules. As soon as Schroedinger's equation had been established in 1926, the problem was solved correctly, and the energy of the lowest state was found to be negative (for the relevant references, cf. Strand and Reinhardt 1979).

Pauli worked exactly in the no-person's land between classical and quantum mechnics which is our main concern here. It would be very painful if his problem could not be done correctly at least to the point of obtaining a negative energy. It turns out that Pauli did not know about the role of conjugate points along the classical trajectories, and their importance for the approach to quantum mechanics. His quantization conditions were too restrictive; his wavelets made a hard reflection on the caustics instead of a soft one, changing their phases by π instead of $\pi/2$, as will be explained in Chapter 14.

The semiclassical quantization of the hydrogen molecule ion was carried out recently by Strand and Reinhardt (1979), using the improved quantization rules that will be discussed later. Their pictures of the classical trajectories are shown in Figures 4 and 5. The energy $\tilde{E}$ is defined such that $E = \tilde{E} + 1/2$, and $\Omega = -\tilde{E} - \tilde{\gamma}$. The three figures have the common values $E = 0$ and $M = 0$, whereas Ω has been given the values 0.4, 1.0, and 2.4.

Remarkably, the lowest quantum-mechanical state corresponds to the value 0.4 of Ω as in Figure 4. The trajectory for this value of Ω does not surround both protons as do the two other trajectories; it is confined to a region around one proton or the other. The electron takes advantage of an opportunity which is not available in classical mechanics, tunneling from one trajectory to another, from the neighborhood of one proton to the neighborhood of the other, back and forth.

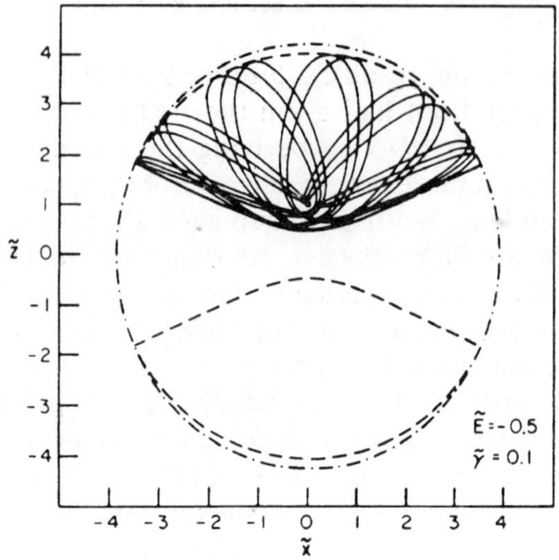

Figure 4 The classical trajectory of the electron for the relevant values of the constants of motion in the ground state of the hydrogen-molecule ion H_2^+; the quantum-mechanical ground-state requires the electron to tunnel between the two classically allowed invariant tori with the same constants of motion. [From Strand and Reinhardt (1979)]

The potential energy of the electron at the center of the molecule is quite negative when $r_A = r_B = c$; the need for the electron to tunnel is not due to its inadequate energy. The classical trajectory is limited by the dynamics of the problem, in much the same way the Kepler ellipses were confined to a critical ellipse. It is a big open problem to make classical mechanics in general cope with this quantum-mechanical phenomenon; some of the difficulties are discussed in the work of Strand and Reinhardt. Section 14.6 will briefly present one approach to the problem of tunneling in dynamical systems.

3.5 Geodesics on a Triaxial Ellipsoid

The surface of a two-dimensional ellipsoid can be imbedded in three-dimensional Euclidean space by the equation

$$\frac{x^2}{a^2} + \frac{y^2}{b^2} + \frac{z^2}{c^2} = 1 , \tag{3.8}$$

where the axes satisfy the inequality $a > b > c > 0$. When a particle moves freely on this surface, it is subject to a force that is always perpendicular to the tangent plane, and whose direction is, therefore, given

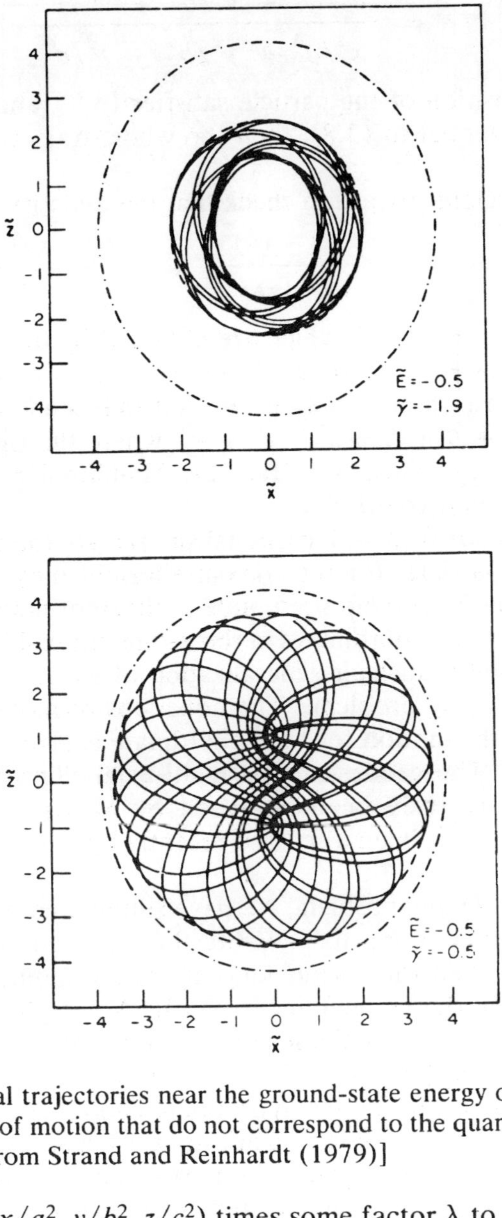

Figure 5 Classical trajectories near the ground-state energy of H_2^+ for values of the constants of motion that do not correspond to the quantum-mechanical ground state. [From Strand and Reinhardt (1979)]

by the vector $(x/a^2, y/b^2, z/c^2)$ times some factor λ to be determined. The equations of motion then become

$$\frac{du}{dt} = -\lambda \frac{x}{a^2}, \ldots, \quad \frac{dx}{dt} = \frac{u}{m}, \ldots, \tag{3.9}$$

where (u, v, w) is the momentum. By taking two derivatives with respect to time in (3.8) and replacing the second derivatives of (x, y, z) according to (3.9), one gets the condition,

$$m\lambda = \frac{u^2/a^2 + v^2/b^2 + w^2/c^2}{x^2/a^4 + y^2/b^4 + z^2/c^4} . \tag{3.10}$$

If the initial position of the particle satisfies (3.8), and the initial momentum is tangential to (3.8), then the whole trajectory stays on the ellipsoid.

It is now straightforward to check that the quantity

$$A = u^2 + \frac{(xv - yu)^2}{a^2 - b^2} + \frac{(xw - zu)^2}{a^2 - c^2} , \tag{3.11}$$

and similar ones, B and C, which are obtained by the cyclic permutation of the triples (x, y, z), (u, v, w), and (a, b, c), are indeed constants of motion. These three quantities are not independent since one has the relation $A + B + C = u^2 + v^2 + w^2$, where the right-hand side is the kinetic energy which is in fact the Hamiltonian of the system. A, B, and C are in involution.

The *geodesics* on a two-dimensional surface are the trajectories of a freely moving particle; for the triaxial ellipsoid they were found by Jacobi (1842, p. 212), who used quite a different method. He introduced triaxial elliptic coordinates in three-dimensional Euclidean space which constituted a difficult generalization of Euler's coordinate system in the preceding problem. The equations of motion can then be separated, and the motion for each coordinate is given again by an elliptic function. The geodesics on a multi-axial ellipsoid in higher dimensions have recently been discussed by Moser (1980); billiards inside an ellipse have been discussed recently by Chang and Friedberg (1988).

As in the preceding problem, we have simply quoted the constants of motion, without trying to integrate the equations of motion. As before, they turn out to be some kind of angular momenta, and are at most quadratic in the momentum (u, v, w). Although it took Euler in the first case, and Jacobi in the second, to find a method for integrating the equations of motion, the solutions they found are in a sense elementary. It suffices to find the appropriate coordinate system in position space, and the rest follows. Until some 20 years ago, that was the only successful method known. The next example shows that certain dynamical systems are integrable, but their constants of motion are much more involved; there are no special position coordinates where the equations of motion separate.

3.6 The Toda Lattice

The lattice or chain consists of n particles which move along a straight line and are coupled by non-linear springs. If the particle i has the momentum p_i and the position q_i, then the Hamiltonian becomes

$$H = \frac{1}{2m} \sum_1^n p_i^2 + V_0 \sum_1^{(n)} \exp\left(\frac{q_i - q_{i+1}}{a}\right), \tag{3.12}$$

where we have two possibilities which are indicated by the symbol (n) in the last summation: for the *open lattice* we set $(n) = n - 1$, while for the *periodic lattice* we set $(n) = n$ along with the convention $q_{n+1} = q_1$.

Several comments are necessary to appreciate the remarkable properties of this Hamiltonian. There are three physical parameters, the mass m of each particle, the strength V_0 of the potential, and the range a of the potential. It is natural to choose these parameters as the physical units, so that the Hamiltonian contains only dimensionless quantities. Once this has been accomplished, however, any new physical quantity cannot be reduced to 1 by the choice of the physical units. In particular, when going to quantum mechanics, Planck's quantum h now has some definite numerical value, small, medium, or large.

This situation is expected to prevail in any truly non-linear system; but no other integrable system known so far has this typical structure. All the others have at most two physical quantities, so that the transition to quantum mechanics does not allow the possibility of distinguishing the effects of a large or small Planck's quantum. The various domains of physics such as the motion of atoms in molecules and crystals, the motion of electrons in atoms and molecules, the motion of protons and neutrons in nuclei, and finally, the motion of elementary particles inside their little boxes, yield a Planck's quantum equal to 1, if one uses whatever units are natural to the system. The limit of a small Planck's quantum is a mathematical fiction when we come to the fields where quantum mechanics is important; but this fiction is apparently unavoidable if we want to form some intuitive picture of the events on the microscopic scale.

Toda (1967, 1970) invented his chain of particles and springs, and gave a special solution for it, a *soliton*. Up to that time, solitons had been found only in continuum mechanics, that is in non-linear systems where the masses are distributed continuously over the available space. Solitons are solutions of the equations of motion in which an isolated bump moves at constant speed. Such special solutions are not hard to find in continuous systems; but once found they must be shown to be stable against small disturbances. In discrete systems, however, it is very difficult to get such solutions in the first place, let alone to show their stability. Toda's solution requires an intimate knowledge of elliptic functions, and stability was not shown at first.

It occurred to Joseph Ford in 1973 to test numerically the integrability of the Hamiltonian (3.12), i.e., the presence or absence of invariant tori, using the standard method of surfaces of section to be explained in Chapter 7. He expected the system to be chaotic because it is close to the well-known Hénon-Heiles system, which will be the main topic of Chapter 8.

The numerical calculation was carried out by Ford, Stoddard, and Turner (1973), for the simplest case of interest, $n = 3$. The verdict was clear: the Toda lattice is integrable. Ford communicated this finding to Hénon (1974) and Flaschka (1974), who discovered independently, within less than a week and by two different methods, how to construct the missing constants of motion. We will discuss Flaschka's procedure because it is closely patterned after the general method that Lax (1968) invented for the discussion of the solitons in continuum mechanics.

The main idea is contained in the following statement whose proof is almost trivial. Assume that two matrices, L and M, have been constructed whose elements depend on the parameter t, and which satisfy the condition

$$\frac{dL}{dt} = LM - ML \,. \tag{3.13}$$

Then the eigenvalues of L, or equivalently, the coefficients in the characteristic polynomial, $\det(L - \lambda I)$, do not vary with t. The matrices L for different values of t are said to form an *isospectral series.* Methods for finding such a *Lax pair (L, M)* are discussed by Greene, Tabor, and Carnevale (1983). For the Toda lattice, the matrices L and M are

$$L = \begin{vmatrix} b_1 & a_1 & 0 & a_n \\ a_1 & b_2 & & 0 \\ 0 & & & \\ a_n & 0 & & b_n \end{vmatrix}, \quad M = \begin{vmatrix} 0 & a_1 & 0 & -a_n \\ -a_1 & 0 & a_2 & \\ 0 & -a_2 & 0 & \\ a_n & & & \end{vmatrix}, \tag{3.14}$$

where the diagonal elements b_i of L, and the off-diagonal elements a_i of L and M are given by

$$a_i = \exp\left(\frac{q_{i+1} - q_i}{2}\right), \quad b_i = p_i \,, \tag{3.15}$$

in terms of the momenta p_i and the positions q_i. The element a_n is zero for the open chain. The equations of motion for the Toda lattice are identical with the condition (3.13).

The constants of motion are the coefficients in the characteristic polynomial of L; or more explicitly, let F_1 be the trace, and let F_j be the sum of the j-th principal minors. In this manner one gets for the two first constants the expressions

$$F_1 = \sum b_i = \sum p_i = \text{total momentum } P, \qquad (3.16)$$

$$F_2 = \sum_{i<j} b_i b_j - \sum_i a_i^2 = \frac{1}{2}\sum_{i,j} b_i b_j - \frac{1}{2}\sum_i b_i^2 - \sum_i a_i^2 = \frac{P^2}{2} - H,$$

which are, of course, elementary, while the third constant is already rather complicated, and is left to the reader as an exercise. One notices immediately that the j-th constant is a polynomial of order j in the momenta p. This is an entirely novel situation compared with the previously known integrable systems with a finite number of degrees of freedom, where the additional constants of motion were at most quadratic in the momenta.

3.7 Integrable versus Separable

Both the motion of a small mass moving in the gravitational field or in the Coulomb field of two large masses, and the problem of the free motion on the surface of a triaxial ellipsoid could be solved after the appropriate position coordinates had been found. That procedure is called *separation of variables;* it results in constants of motion which are at most quadratic in the momenta; the equations of motion can be solved by 'quadrature', e.g., by elliptic integrals. Such a problem is called *separable.*

The Toda lattice cannot be treated in this way, and what is worse, the knowledge of the constants of motion does not automatically lead to a set of coordinates in which the equations of motion become very simple, such as in the action-angle variables. Such a system is called *integrable, but non-separable,* a somewhat misleading term, because it implies only the ability to integrate, but does not guarantee the explicit construction of the integrating variables.

The complete integration of the Toda lattice has been discussed by several authors, including Kac and van Moerbeke (1974) and Moser (1975a and b). The fascinating part of this work, in addition to the many beautiful mathematical propositions which appear there, is its implication for the corresponding quantum problem. It turns out that Schrödinger's equation always separates when the classical problem is separable. Nothing of the sort happens in the Toda lattice; there is no easy analog to the involved transformations that are necessary in the classical Toda lattice.

The main obstacle is the intimate mixture between position and momentum coordinates which finally yields the action-angle variables. The construction of the wavefunctions that are simultaneous eigenfunctions of the constants of motion is tricky. The author (Gutzwiller 1981b) has succeeded in carrying out the necessary transformation of Hilbert space for n up to and including 4; the results are easily generalized to $n \geq 5$. But nobody seems to have gone beyond $n = 4$, at least not to the point of proving an explicit algorithm for calculating the spectrum (cf. Sklyanin 1985).

It is general wisdom in modern physics that problems are solvable because they have symmetries which are responsible for their privileged situation with respect to all the other problems. But in none of the cases in this chapter are these symmetries particularly obvious, although they are known by now. It is not clear to what extent these known symmetries not only provide constants of motion, but are also helpful in constructing explicit solutions for the equations of motion or in finding simultaneous eigenfunctions.

In a more general context, the issue of symmetry is largely unresolved in view of the chaotic behavior of most dynamical systems. The traditional concept of symmetry is such that it leads to integrability wherever it applies. Its absence is then equivalent to chaos; but some chaotic mechanical systems exhibit simplifying features which have the same appeal as the usual idea of symmetry. One is tempted to set up a new goal in the study of dynamical systems, namely to find the new types of symmetry which allow the classical equations of motion and Schrödinger's equation to be solved in some effective manner, even when the system is not integrable and its behavior is chaotic by our present criteria.

CHAPTER 4

The Three-Body Problem: Moon - Earth - Sun

The problem of three interacting point-like masses continues to be at the center of physics. The latest version consists of three quarks making up a proton or a neutron; but there are also two hydrogens and one oxygen as the constituents of a water molecule, or an atom of helium built with one alpha particle and two electrons. Behind these modern examples are all the famous instances from celestial mechanics, particularly the most obvious as well as the most difficult of them all, the motion of the Moon in the combined gravitational field of the Earth and the Sun. Since chaos made its first though hardly recognized appearance in this context, and the case is far from closed, an abbreviated discussion of this special problem seems both instructive and topical.

4.1 Reduction to Four Degrees of Freedom

Newton was the first scientist to investigate the problem of three massive bodies attracting one another with a force which decreases as the inverse square of the distance. In the case of the Moon, deviations from the standard theory had already been correctly determined by Ptolemy of Alexandria in the second century a.C., and a number of further refinements had been introduced by Tycho Brahe at the end of the sixteenth century, all based on naked-eye observations. Newton succeeded in explaining why the Moon's orbital plane turns in a direction opposite to the motion of the planets, making one full turn every 18 years. But his calculation for the motion of the Moon's perigee in the same sense as the planets was too small by a factor of 2.

In 1878, almost two centuries later, George William Hill finally provided a natural explanation, and discussed in the process for the first time what has now become known as Hill's equation. His analysis is fundamental for the study of non-linear mechanics, integrable or chaotic. The three-body problem has provided the main inspiration for the development of mechanics, to the point where Poincaré eventually recognized the need to deal with chaotic systems, just about 100 years ago. The author has devoted several years to the study of the Moon's motion so that it seems entirely appropriate if the reader is offered some of this collective wisdom.

Astronomical observations from the surface of the Earth do not allow measurements to better than 1 second of arc, because the turbulence of the air causes the image of any object in the sky to blur. A telescope with an objective opening of 10 cm gives already this kind of resolution. The Moon moves over the background of fixed stars at a rate of about 1 second of arc in 2 seconds of time; her movements can be tracked quite adequately with an ordinary stop-watch to a precision of a fraction of a second. The perturbation of the planets can be neglected at this level of accuracy. The flattening of the Earth is accounted for by a small correction, as are several other minor effects. All that is left is the *combined action of the Moon, the Earth, and the Sun, considered as perfectly spherical structures with the masses L for the Moon (Luna), T for the Earth (Terra), and S for the Sun (Sol).*

There are nine degrees of freedom, which can best be described with the following set of coordinates:

(i) $\mathbf{R}_0 = (X_0, Y_0, Z_0)$ for the common center of mass of the three bodies;

(ii) $\mathbf{R} = (X, Y, Z)$ for the vector which points from S to the center of mass Γ of T and L;

(iii) $\mathbf{r} = (x, y, z)$ for the vector from T to L.

The phase space of this system has 18 dimensions, after we have added the momenta $\mathbf{P}_0 = (U_0, V_0, W_0)$, $\mathbf{P} = (U, V, W)$, and $\mathbf{p} = (u, v, w)$ to the positions. *Ten constants of motion* can be immediately recognized:

(i) Since the center of mass moves with a uniform motion which has no effect on the internal movements of the three bodies, we can eliminate both $\mathbf{R}_0$ and $\mathbf{P}_0$, thereby reducing the phase space by six dimensions.

(ii) The angular momentum of the whole system is constant; its direction determines an invariant plane through the center of mass. To a very good approximation, Γ moves around S in a Kepler ellipse whose (almost invariant) plane is called the *ecliptic* because the eclipses of the Sun and the Moon occur when all three bodies are in it. The absolute value of the angular momentum determines essentially the eccentricity of the orbit of Γ, the Earth-Moon center of mass, around the Sun.

(iii) The total energy of the system, assuming that $\mathbf{R}_0 = 0$, gives the Hamiltonian of the internal motion; its value determines basically the major axis for the motion of Γ around the Sun; the energy is the tenth and last constant of motion.

After reducing phase space to eight dimensions with the help of the ten constants of motion, the system is left with four degrees of freedom. The Hamiltonian is best written in terms of the reduced masses

$$\mu' = \frac{S(T+L)}{S+T+L}, \quad \mu = \frac{TL}{T+L}, \tag{4.1}$$

which yields the expression

$$H = \frac{\mathbf{P}^2}{2\mu'} + \frac{\mathbf{p}^2}{2\mu} - \tag{4.2}$$

$$- \frac{GST}{\left|\mathbf{R} - \dfrac{L}{T+L}\mathbf{r}\right|} - \frac{GSL}{\left|\mathbf{R} + \dfrac{T}{T+L}\mathbf{r}\right|} - \frac{GTL}{|\mathbf{r}|}.$$

The three components of angular momentum are still constants of motion at this stage, in addition to the Hamiltonian itself. Are there any others ?

This question is answered to some extent by a *theorem of Bruns and Poincaré:* There does not exist any constant of motion which is analytic in the variables $\mathbf{P}$, $\mathbf{p}$, $\mathbf{R}$, $\mathbf{r}$ as well as in the mass ratios, $(T+L)/S$ and L/T, except the angular momentum and the total energy. Notice that the constant of motion has to be analytic simultaneously in the coordinates and the mass ratios, to fit the assumptions of the theorem. A constant of motion could possibly be found which fails to be analytic in the mass ratios. (A detailed account of this theorem is found in the textbook by Whittaker (1904), Chapter XIV.) A modern discussion of this theorem was given by Benettin, Galgani, and Giorgilli (1985).

Such a case arises in the theory of the gyroscope in a uniform gravitational field, a rigid body of arbitrary shape and mass distribution that is suspended from one fixed point and is subject to a homogeneous gravitational field. *Sofia Kowalevskaya* found in 1890 an isolated case of integrability when the moments of inertia are $A = B = 2C$ with the suspension in the plane corresponding to the directions of the moments A and B. The only two other cases of integrability for the gyroscope are Euler's freely rotating rigid body and Lagrange's symmetric top with the suspension in the axis of symmetry. No such exceptional cases have been found for the three-body problem (cf. Leimanis 1965).

The absence of analytic constants of motion should not discourage us completely, because there could be some which depend on the momenta and positions like a real function with rather high-order bounded derivatives. All the observational, analytical, and computational evidence in the Moon-Earth-Sun motion encourages this view, although the mathematical proofs on the basis of the above Hamiltonian (4.2) are missing. We will, therefore, continue the discussion as if the problem were indeed integrable.

4.2 Applications in Atomic Physics and Chemistry

The three-body problem has a close analog in *atomic physics*. The helium-atom consists of an α-particle, a nucleus containing two protons and two neutrons, and two electrons; but there are other atoms with only two electrons, such as the negative ion of hydrogen, which was first found in the solar atmosphere, or the once ionized lithium, the twice ionized beryllium, and so on. The closest analog in celestial mechanics is the problem of the Jupiter-Saturn coupling in the solar system; these two most massive planets have their periods very close to a 2:5 ratio, and perturb each other's motion more effectively than a simple estimate of their interaction would indicate. A similar situation arises when two electrons in an atom get highly excited without being ejected; this effect has been studied recently in many different variations and is known as the *planetary atom* (cf. RS Berry 1986).

The *electrostatic forces* between the nucleus and the electrons, as well as between the electrons, are similar to the gravitational forces, since they vary with the inverse square of the distance; but these forces are not propotional to the masses of the particles on which they act, and they can be both attractive and repulsive, depending on the relative signs of the electric charges on the particles. Furthermore, the electrons obey Pauli's exclusion principle, which causes special restrictions in their freedom of motion.

The analogy between celestial mechanics and *chemistry* is more remote, although there is a great variety of interesting cases. The most immediate case consists of two protons and one electron, the hydrogen molecule ion of the preceding chapter, but this time without restriction on the motion of the protons. They are allowed to move, either by rotating around each other, or by oscillating like two masses connected by a spring. The forces are still of the inverse square type, as in the helium atom.

The most important situation in chemistry, however, involves the motion of three nuclei that interact with one another through the intermediary of electrons, such as in the water molecule H_2O. The forces between the three nuclei cannot be described mathematically without first solving a rather difficult problem in quantum mechanics. The procedure, called the *Born-Oppenheimer approximation,* can be described in the following simplified manner: each of the three nuclei is assigned a fixed position in space, such as $\mathbf{R}_1$, $\mathbf{R}_2$, $\mathbf{R}_3$, and the energy of the electrons $E(\mathbf{R}_1, \mathbf{R}_2, \mathbf{R}_3)$ is computed by solving the corresponding Schrödinger equation. This electronic energy, together with the electrostatic repulsion between the nuclei, yields the potential energy $V(\mathbf{R}_1, \mathbf{R}_2, \mathbf{R}_3)$ of the molecule, which then takes the place of the three last terms in (4.2).

This method of treating the motion of the nuclei inside a molecule assumes that the electrons are light enough to follow the motion of the nuclei almost instantaneously; it works well as long as there is no drastic reorganization of the molecule. Another way of stating this assumption is to say that the energy differences for an electronic transition are large compared with the energy quantum for a vibrational mode, which in turn is greater than the energy difference in a rotational transition.

In a typical chemical reaction, the atom A hits the molecule BC, with the result that the new molecule AB is formed while the atom C flies off. The new molecule AB might be left in an excited state; sometimes during the process, the electrons in the whole system found it advantageous to pass from their ground-state energy $E_0(\mathbf{R}_1, \mathbf{R}_2, \mathbf{R}_3)$ to the excited energy level $E_1(\mathbf{R}_1, \mathbf{R}_2, \mathbf{R}_3)$. The calculation of the classical trajectory starts with the potential energy V_0, and ends with the potential energy V_1; an additional calculation in quantum mechanics yields a probability P_{01} that depends again on the coordinates $\mathbf{R}_1, \mathbf{R}_2, \mathbf{R}_3$, and controls the transfer of the electronic system from the ground state to the excited state.

In all these cases, one treats a mechanical system consisting of three bodies, each moving in three-dimensional Euclidean space. The phase space has 18 dimensions, but there are always the same ten constants of motion, six from the uniform center-of-mass motion, three from the total angular momentum, and one from the total energy, kinetic plus potential; one is left with four degrees of freedom. The chemical problems require a major preliminary calculation to determine the potential energy of the electrons as a function of the nuclear positions.

As soon as this step has been completed, however, the analogy with celestial mechanics becomes quite striking. Some early calculations of classical trajectories for chemical reactions were carried out by Thiele and Wilson (1961) as well as Bunker (1962).

The spontaneous break-up of a three-star system, or its inverse, the capture of a third body, corresponds to the decay of a molecule, or the formation of a new one. The choice of coordinates in the lunar problem is entirely suitable if S represents an atom that scatters off the molecule TL. The result of such an interaction depends, of course, on the value of the total angular momentum and of the total energy, although their values remain the same throughout the whole event.

Celestial mechanics stays entirely within the confines of classical mechanics, whereas molecular mechanics requires that the system be viewed in the light of quantum mechanics. In addition, the motion of the nuclei in a molecule can only be fully treated by solving Schrödinger's equation. Since our understanding of quantum mechanics is based on the experience of classical mechanics, however, we gain a great deal of insight by studying some of the most instructive problems in celestial mechanics. They are also accessible to our daily observation if we are willing to watch the sky occasionally and notice some of its most obvious changes.

4.3 The Action-Angle Variables in the Lunar Observations

This section sketches the traditional view of the three-body problem under the assumption (as yet to be proved correct) that it is integrable. The Moon-Earth-Sun problem has been understood in this manner since time immemorial, and provides a highly non-trivial example where precise data have been available to be examined from different viewpoints.

What would be the consequences if there existed four more constants of motion to take care of the remaining four degrees of freedom? The remaining four degrees of freedom would behave essentially as four masses which are coupled by linear springs; there would be four independent harmonic motions. The movements in the three-body problem could be represented as a Fourier expansion exactly as in the formula (3.4) with the help of four frequencies. The numerical values of these frequencies depend on the initial conditions, but they remain constant in time. This state of affairs has been well understood from observation at least since 3000 years, going back to Babylonia.

The four relevant frequencies, or their periods, are conveniently measured in terms of the Earth's rotation, say in mean solar days, which are largely independent of the translatory motions of Moon, Earth, and Sun. Thus, we have the following periods (cf. Allen 1962):

(a) the *sidereal year* T_0 describing the Sun's yearly motion through the sky from a given fixed star back to the same star in 365.257 days, which is about 20 minutes longer than the tropical year, from equinox to equinox;

(b) the *sidereal month* T_1 of 27.32166 days taking the moon from a fixed star back to the same fixed star; by combining the sidereal year with the sidereal month one gets the better known synodic month of 29.53059 days, which takes our satellite from one full moon to the next;

c) the *anomalistic month* T_2 of 27.55455 days for the moon to complete its 'anomaly' (speeding up and slowing down) from one perigee, point of closest approach to the Earth, to the next;

d) the *draconitic or nodical month* T_3 of 27.21222 days, which takes the moon from one ascending (or descending) node, i.e., intersection with the ecliptic where it might get eaten by the dragon in an eclipse, to the next.

The values of these periods were known to the Greeks correctly with the accuracy quoted, corresponding to 1 second of time. These numbers constitute the first high precision results in the physical sciences.

The four periods are obtained by keeping careful records over long times; watching the Sun set on the horizon in early spring, and waiting till he sets exactly on the same spot over and over again, will eventually yield a good value for the tropical year. The reader should try to imagine what kind of observations will give the lunar periods.

Each period T_i (and the corresponding frequency $\omega_i = 2\pi/T_i$) gives rise to an angular variable w_i in the three-body problem, exactly as it was explained in Section 3.3. The coordinates of the Moon with respect to the Earth are, therefore, given by Fourier expansions like (3.4). The angle variables, however, do not appear as if by magic out of some general algorithm, but they arise out of a well-defined view of the problem at hand. In the lunar motion, the astronomical observations naturally suggest the angles that are associated with the primary periods. Similarly in chemistry, a particular model of the molecular motion in terms of bond angles and bond stretches leads usually to the good choice of variables.

The leading terms in the expansion (3.4) of the Moon's coordinates (x, y, z) with respect to the Earth can be interpreted geometrically. If (x, y) are chosen to lie in the ecliptic with the Earth at the origin, the most important term in (3.4) is a circular motion of radius a whose period is the sidereal month T_1. The next term increases and decreases the radius of this circular motion in the (x, y) plane sinusoidally at the rhythm of the anomalistic month T_2; its amplitude is a fraction $2e \simeq 1/9$ of the radius a ; this fraction e is essentially twice the eccentricity of the Moon's trajectory. The leading term in the motion along the z-axis (at right angle to the ecliptic) varies sinusoidally as the period of the draconitic month T_3, with an amplitude $k \simeq 1/11$ compared with the radius a of the basic circular motion, corresponding to the inclination $\simeq 5^0$ of the lunar trajectory.

The coefficients in these three basic terms determine the values of the actions that are associated with the angles. In a preliminary explanation, T_1 is associated with the 'binding energy' between Moon and Earth, T_2 with their relative angular momentum, and T_3 with its component at right angles to the ecliptic. The fourth angle w_0 gives the mean yearly motion of the Earth-Moon around the Sun, whose mean distance a' yields the fourth action.

Many other, and generally smaller, terms are added to these four basic terms in the Fourier expansions (3.4); their frequencies are combinations of the four basic ones. The problem is to find the exact values for the coefficients of these higher terms. Since the values of the four action variables are essentially determined by the four basic terms, all the coefficients of the higher terms are functions of these four basic ones. There is no room for fudging anything: Newton was the first to recognize clearly the existence of no more than four parameters besides the mass ratios and the values of the obvious constants of motion like the total angular momentum; he tried to obtain the coefficients of the higher terms in (3.4) from the knowledge of the lowest four. But he was not always beyond claiming that some coefficient had been calculated when, in fact, he had gotten it from fitting the data.

In the interest of simplifying the calculations, we will assume that Γ moves around the Sun in the ecliptic on a fixed Kepler ellipse with known semimajor axis a' and eccentricity e'. This is a very good approximation since the solar mass exceeds the combined mass of Earth and Moon by a factor of more than 300,000.

The polar angle of Γ as seen from the Sun is written as $\phi' = g' + f'$, where g' is the direction of the perihelion as referred to the spring equinox. The true anomaly f' differs from the mean anomaly ℓ' by the expansion (cf. Brouwer and Clemence 1961, Chapter II).

$$[2e' - \frac{e'^3}{4} + \cdots] \sin \ell' + [- \frac{5e'^2}{4} - \frac{11e'^4}{24} + \cdots] \sin 2\ell' + \cdots \tag{4.3}$$

The mean anomaly ℓ' increases linearly with time,

$$\ell' = n'(t - t_0) + \ell'_0 , \tag{4.4}$$

where t_0 is an arbitrary reference time called the *epoch,* and ℓ'_0 is the value of the mean anomaly at the epoch.

The normalized polar distance R/a' of Γ from the Sun is given by a similar expansion,

$$1 + \frac{e'^2}{2} + \cdots - [e' - \frac{3e'^3}{8} + \cdots] \cos \ell' - [\frac{e'^2}{2} - \cdots] \cos 2\ell' . \quad (4.5)$$

All the quantities referring to the Sun appear primed in accord with the traditions of celestial mechanics so that the analogous quantities for the Moon can remain unprimed. The mean angular speed $n' = \omega_0 = 2\pi/T_0$, also called the Sun's mean motion, is related to the semimajor axis a' by Kepler's third law,

$$n'^2 a'^3 = G(S + T + L) . \quad (4.6)$$

The Moon-Earth-Sun problem has now been reduced from four to three and one-half degrees of freedom, in the slightly facetious language of the practitioners in this trade. On the one hand, the movement of Γ around the Sun is no longer influenced by the internal complications between the Earth and the Moon; on the other, these internal motions are subject to the periodically changing distance of Γ from the Sun. The angle ℓ' is what is left over from the angle-variable w_0; it has been reduced to the role of a sinusoidally varying external perturbation of known period T_0 as in (3.5). We are left with the three action-angle variables corresponding to the periods T_1 , T_2 , T_3.

4.4 The Best Temporary Fit to a Kepler Ellipse

The location of the Moon as seen from the Earth at any instant t is determined in the following manner, which is basically the classic Greek description in modern form. The angular momentum of the Moon with respect to the Earth defines a direction in space, and an instantaneous orbital plane at right angles to it, which intersects the ecliptic in the nodal line and is inclined at an angle γ. The energy of the Moon-Earth system defines an instantaneous semimajor axis a, and the absolute value of the angular momentum determines an instantaneous eccentricity e. The angle from the reference direction in the ecliptic to the *ascending node* is called h; the angle from the ascending node to the instantaneous *perigee* is g; and the angle from the instantaneous perigee to the Moon is called f, corresponding to the *true anomaly.*

There is an advantage in using the mean anomaly ℓ instead of f with the help of the relation (4.3), but this time for the unprimed quantities. The instantaneous coordinates a, e, γ, h, g, ℓ are completely equivalent to the six Cartesian components of $\mathbf{p} = (u, v, w)$ and $\mathbf{r} = (x, y, z)$. They provide the best temporary fit of the real motion with the motion on a Kepler ellipse, and are called the *osculating elements.* A perspective drawing of this traditional picture for the Moon's trajectory is presented in Figure 6.

If it were not for the presence of the Sun the angles g and h would remain constant, while ℓ increases linearly in time exactly as in formula (4.4). The rate of increase n is 2π divided by the sidereal month T_1, and is related to the semimajor axis a again by Kepler's third law

$$n^2 a^3 = G(T + L)\,. \tag{4.7}$$

The six quantities ℓ, g, h, a, e, γ can be grouped in canonically conjugate pairs, called *Delaunay coordinates,* as follows: $\Lambda = \mu n a^2$ with ℓ, $M = \mu n a^2 (1 - e^2)^{1/2}$ with g, and $N = M \cos \gamma$ with h. Slightly clumsy expressions for Λ, M, N have been used to show that they have indeed the physical dimension of an angular momentum or action, ready to become multiples of Planck's quantum if we want to quantize the motion.

With the momentum $\mathbf{p} = (u, v, w)$, and the position $\mathbf{r} = (x, y, z)$, one has the mathematical identity $u dx + v dy + w dz = \Lambda d\ell + M dg + N dh$, which is rather troublesome to derive, although it is simply a consequence of the definition for Λ, M, etc. A careful stepwise proof was found by Whittaker, and is also given in the textbook of Brouwer and Clemence (1961, p. 279); the same result is derived in modern terminology by Abraham and Marsden (1978, p. 638).

If the Sun did not disturb the Earth and the Moon, Λ, M, N, as well as g and h would not change with time, and we would have a Kepler motion. If the effect of the Sun is small, the values of Λ, M, N vary slowly; also the angles ℓ, g, h vary almost exactly as linear functions of time, g and h rather slowly compared with ℓ.

Humanity has known for over 3000 years that h decreases by 2π in about 18 years, while the sum $h + g$ increases by 2π in about 9 years. The reader is encouraged to imagine how our ancestors were able to determine these facts without any instruments whatever, simply by keeping careful records of the events in the sky! The mean rates of change for h and $h + g$ follow directly from the various periods in Section 4.3. The main ingredients are the differences of the draconitic month and of the anomalistic month with the sidereal month; they are − 2 hours 37 minutes 29 seconds and + 5 hours 35 minutes 29 seconds, again known to this precision by the Greeks.

Until Newton no astronomer even suggested any reason for the slow change in h and $h + g$. It was accepted as an observational result, and since the moon's motion could not be explained without it, the planets were allowed to show similar slow migrations in their nodes and

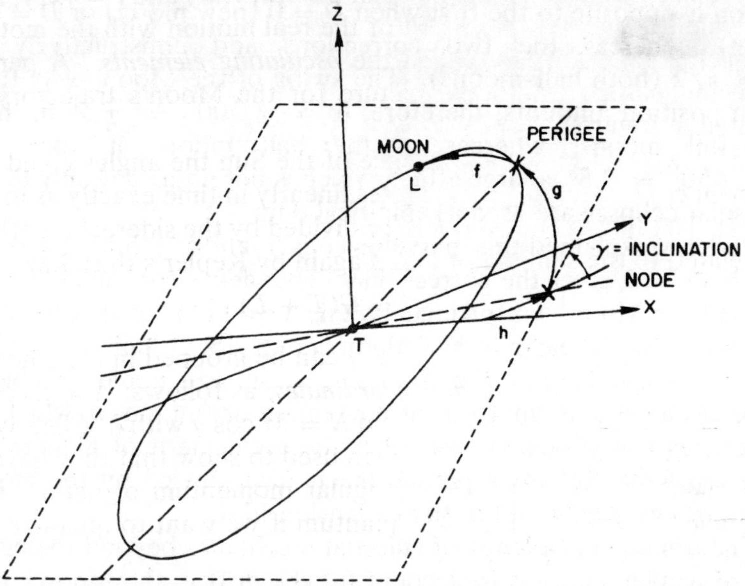

Figure 6 The traditional representation of the Moon's motion around the Earth: the x, y-plane is the ecliptic (plane of the Earth's orbit around the Sun); the orbital plane of the Moon (dashed lines) intersects the ecliptic in the nodal line at the angle of inclination γ; the Moon moves on a Kepler ellipse around the Earth whose point of closest approach (perigee) is at the angular distance g away from the node; the node moves backwards while the perigee moves forwards at roughly twice the speed, covering a full circle in about nine years.

perihelia, although they turned out to be extremely small. Kepler's Rudolphine tables are constructed exactly on this model, but using the formulas (4.3), (4.4), and (4.5) instead of the earlier epicycles and eccentric circles of Copernicus and Tycho Brahe.

Even this simple scheme based on the uniform change of the osculating elements with t does not do justice to the motion of the Moon. Ptolemy already had discovered a correction to the Moon's longitude $\phi = h + g + f$, called the *evection;* for the first time in science, it was found necessary to combine two different harmonic motions because of their non-linear coupling; a short explanation of this extraordinary discovery is quite in order.

The main difference between the true anomaly f and the mean anomaly ℓ is given by the first term in the expansion (4.3), namely $2e \sin \ell$, where in angular units $2e \simeq 375'$ for the Moon. The evection introduces a further correction to the mean anomaly which combines ℓ and the mean angular distance D from the Sun to the Moon, $\sin(2D - \ell)$, with an amplitude $\simeq 75'$. Notice that the second correction is opposite to the first when $D = 0$ (new moon) or $D = \pi$ (full moon), whereas the two corrections add constructively when $D = \pi/2$ (both half-moons). The swing of the Moon away from its mean position amounts, therefore, to $\simeq \pm 300' = \pm 5^o$ in the new and full moons, whereas in the half moons it goes up to $\simeq \pm 450' = 7.5^o$, a large effect, which is not noticed if only the lunar and solar eclipses are used as empirical data.

Ptolemy explained this marvelous observation with his crank model which, indeed, gives the correct time dependence for the polar coordinates of the Moon. Unfortunately, this model implies that the Moon's distance from the earth at half moon is only about half of what it is at new moon or full moon; the Moon's apparent size would then be about twice at half moon what it is at new moon or full moon, which is not correct, as everybody knows. It is the great merit of Copernicus to have replaced Ptolemy's pseudo-mechanical device by the more abstract, but very general Fourier expansion (3.4).

The principal challenge of celestial mechanics beyond the explanation of Kepler's laws is to account for the drift in the Moon's perigee and node. Newton solved the problem as well as he could without the use of extensive algebraic manipulations. He found that the nodical month is shorter than the sidereal month by the fraction $3n'^2/4n^2$, and that the anomalistic month is longer by the same fraction. The utter simplicity of this result is impressive, and is only marred by the fact that the second fraction is too small by a factor of 2. This discrepancy was finally explained after Newton's death by Clairaut and d'Alembert using a computational tour de force (for a full historical account cf. Waff 1975, 1976, 1977); but as mentioned earlier, it was left to Hill to find a simple convincing physical picture.

4.5 The Time-Dependent Hamiltonian

The Moon circles around the Earth as its main center of attraction with the Sun very far away. In practical terms, the distance R in the Hamiltonian (4.2) is larger than r by a factor of 400. Therefore, the first two terms in the potential energy are expanded in the ratio r/R. If we retain only the lowest order beyond the inverse first powers of R and r, we get the potential energy

$$GS\frac{(T+L)}{R} + \frac{GS\mu}{r^5}\left[\frac{3}{2}(\mathbf{R},\mathbf{r})^2 - \frac{1}{2}r^2R^2\right] + \cdots + \frac{GTL}{r}, \quad (4.8)$$

to replace the second line on the right of (4.2).

The first term in the kinetic energy of (4.2) and the first term in the potential energy of (4.8) yield the Hamiltonian for the Kepler ellipse of the Earth-Moon center of mass Γ around the Sun. This motion is described very closely by the formulas (4.3), (4.4), and (4.5) with Kepler's third law (4.6). As was already explained at the end of Section 4.2, one of the four degrees of freedom in the three-body problem is taken away by assuming that the position vector **R** depends on time exactly as given in (4.3) through (4.6). We are, therefore, left with the second term in the kinetic energy of (4.2), and the remainder of the potential energy (4.8) where **R** and R are to be replaced by their expressions (4.3) through (4.6).

The Moon's motion around the Earth in terms of the momentum **p** and the position **r** now results from the Hamiltonian

$$H = \frac{\mathbf{p}^2}{2\mu} - \frac{\mu n^2 a^3}{r} - \frac{\mu n'^2 a'^3}{R^5}\left[\frac{3}{2}(\mathbf{R},\mathbf{r})^2 - \frac{1}{2} r^2 R^2\right] - \cdots , \quad (4.9)$$

where we have used Kepler's third laws (4.6) and (4.7) to get rid of the masses and the gravitational constant G. Also a correction $(T + L)/S$ has been neglected compared with 1 at this stage of the development.

The influence of the Sun on the motion of the Moon around the Earth has been reduced to a simple *quadrupole* term whose size, however, depends on the distance R of the Sun, and the relative angle between Sun and Moon as seen from the Earth. Thus, we have a time-dependent Hamiltonian, and the energy is not conserved; the frequency of this time-dependence n' is small compared with the frequency n of the Moon's orbit; we have n'/n = sidereal month / year $\sim 1/13$. The quadrupolar term in (4.9) is smaller than the Earth's attraction by the square of this quantity, or about $1/180$.

Further terms in the expansion of the solar perturbation decrease with ever higher powers of the ratio r/R, or equivalently a/a' which is of the order $1/400$. As stated here, the time-dependent Hamiltonian (4.5) with its higher order terms in powers of a/a' constitutes the so-called *main problem of lunar theory,* whose solution accounts correctly for all the peculiarities of the Moon's motion. A systematic introduction to lunar theory was written by Brown (1896, 1960). It is surprising that such apparently small corrections to the Hamiltonian of a very well-behaved mechanical system lead to a complex motion.

CHAPTER 5

Three Methods of Solution

The search for a general method to solve problems in mechanics is a worthy endeavor; but the experience of two centuries shows that there is no such thing as a universal recipe, in particular now, when chaos has been recognized as an inevitable and even typical occurrence. The three methods to be discussed have each their advantages over the other two: Lagrange's variation of the constants provides a direct interpretation of the underlying physics; canonical transformations deal effectively with systems that are assumed to be integrable; Hill's method explores the neighborhood of an isolated periodic orbit. The emphasis is on the last because it is the least known and perhaps the best for understanding a limited portion of phase space. The example of the Moon's motion not only provides the first detailed application, but also gives a vivid picture of the ideas behind the mathematical formalism.

5.1 Variation of the Constants (Lagrange)

Lagrange was the first mathematical physicist to search consistently for underlying principles and to strive for the most economical and elegant solutions. He found a general method for solving the problems of celestial mechanics; it incorporates the description of the solar system that had been in use since antiquity. The basic pattern such as the Kepler ellipse is maintained, but it is allowed to modify and adjust itself under the influence of the perturbations from third, fourth, etc. bodies. These modifications and adjustments take place slowly compared with the motion in the basic pattern. They are defined in terms of the parameters that determine the basic pattern, namely the constants of motion such as the semimajor axis a, the eccentricity e, the inclination of the orbit γ, the direction of the ascending node h, the direction of the perigee $h + g$, and the mean anomaly ℓ_0 at the epoch.

Instead of solving the equations of motion for the coordinates of the body in some arbitrary reference system, one establishes equations for the change with time of the parameters a, e, γ, h, g, and ℓ_0. These equations take the general form

$$\frac{da}{dt} = K_a(a, e, \gamma, \ell_0, g, h) \, , \quad \frac{de}{dt} = K_e(a, e, \gamma, \ell_0, g, h) \, , \text{ etc.} \quad (5.1)$$

where the functions K_a , K_e , etc. contain only terms that come from the perturbations. While the idea is straightforward, it took the prodigious skill of Lagrange to give an efficient scheme for calculating these functions. In this context he invented the device of using what are now called the *Lagrange brackets,* the forerunners and actually the inverses of the Poisson brackets.

The reader is invited to spend an enjoyable few hours studying this method in the standard textbooks, provided the explanation of the method is followed by a non-trivial example. The textbook of Brouwer and Clemence (1961, Chapter XI) gives a self-contained presentation of Lagrange's formalism.

Just looking at the right-hand sides of the equations (5.1) conveys a lot of physics, because the causes for the change in the parameters are immediately visible, and the extent of these changes can be estimated. Nevertheless, Lagrange's method has severe limitations in the case of the Moon. Some parameters such as a may remain nearly constant, while others such as e vary by almost a factor of 2; the main culprit in this case is the evection which was discussed at the end of the last chapter. Figure 7 shows the variation of the Moon's eccentricity with time, and demonstrates that Lagrange's idea does not work well. Finally, h and g increase indefinitely with time, rather than varying around a constant mean value and become very tricky to handle.

5.2 Canonical Transformations (Delaunay)

Canonical coordinates in phase space were briefly discussed at the beginning of Section 2.2; now we shall consider more closely how they can be adapted to the solution of a particular problem. Canonical

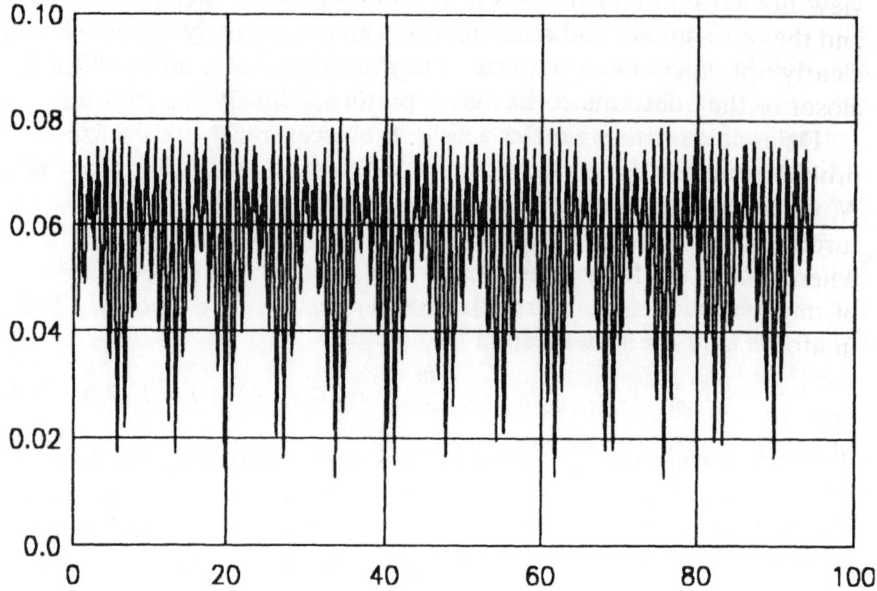

Figure 7 The effective eccentricity of the lunar trajectory as a function of time: the abscissa gives the time in synodic months starting with the year 1980 through 1986; the variations have a fairly simple spectrum, but show that the traditional picture of the Moon's orbit is not good.

transformations were invented by Hamilton in the first half of the nineteenth century, when he tried to find a new formalism for what we now call geometric optics. The basic ideas were then found to work for mechanics as well, and were elaborated by Jacobi. Eventually, they found their justification almost a century later in wave mechanics, which is very close to wave optics.

This explanation for the importance of canonical transformations in mechanics is well known, but it seems to imply that quantum mechanics is always lurking in the background, even when we deal with celestial mechanics where we can certainly do without Planck's quantum. Canonical transformations may just be one of several useful systematic procedures for solving problems in classical mechanics, but not the only one as it would appear from its presentation in many textbooks, to the exclusion of other methods.

The first large-scale application of canonical transformations was carried out by Charles Delaunay who set out in 1846 to solve the main problem of lunar theory as defined in the last chapter. His was a lonely and monumental endeavor, which ended in the publication of two large volumes in quarto of about 900 pages each as *Mémoires* of the (then) Imperial Academy of Sciences of France, 1860 and 1867. We will review his work briefly because it reveals the first indications of chaos, and the consequent limitations on this kind of effort. It also shows very clearly the appearance of the small denominators that again get us closer to the questions concerned with integrability.

Delaunay starts with the action-angle variables for the two-body problem which were explained in Section 4.4; the actions are Λ, M, and N, while the angles are ℓ, g, and h. The Hamiltonian without the perturbation has the somewhat unlikely form $H = -\mu(GTL)^2/2\Lambda^2$, where μ is the reduced mass $TL/(T+L)$. This expression takes care of the first two terms in the full Hamiltonian (4.9). The Bohr formula in atomic physics is obtained if one makes the following replacements: μ is the mass of the electron, GTL is the square of the electronic charge, and Λ is a multiple of Planck's constant divided by 2π. The remainder of (4.9) constitutes the perturbation where we now have to replace the position coordinates (x, y, z) by their expressions in terms of Λ, M, N, ℓ, g, and h.

That replacement is already a major enterprise which we will not try to carry out here. The new expression for the perturbation depends explicitly on time through the time dependence of the solar coordinates R which follows from the formulas (4.3) through (4.6). The Hamiltonian can be made formally time-independent if we use the fourth angular variable $k = \ell'$, and the corresponding action variable K, and if we add a term $n'K$ to the Hamiltonian, so that

$$H = -\frac{\mu(GTL)^2}{2\Lambda^2} + n'K + F(\Lambda, M, N; k, \ell, g, h) , \qquad (5.2)$$

where the function F represents the perturbation corresponding to the last term in (4.9) and does not contain the action variable K.

This little maneuver puts us back where we started, when we counted a total of four degrees of freedom in the three-body problem after eliminating the obvious constants of motion. The somewhat artificial occurrence of the fourth action-angle pair (K, k) in this Hamiltonian expresses *the reduction from 4 to 3 1/2 degrees of freedom,* which was mentioned at the end of Section 4.3.

The perturbation F has to be expanded, exactly as $V(q, t)$ in (3.6), into a multiple Fourier series in the angles k, ℓ, g, and h. The coefficients depend on Λ, M, and N through the small parameters $m = n'/n, e, \gamma, e'$, and a/a', as well as the mass ratios L/T, and $(T+L)/S$.

The values of these parameters are about 1/13, 1/10, 1/20, 1/60, and 1/400, as well as 1/80, and 1/330,000: small enough to suggest a power series expansion for each coefficient in the Fourier series of F. Delaunay carries out this expansion in algebraic form, keeping everywhere rational numbers! It is not as easy to carry out this tedious work on a modern computer as the reader might think. Barton (1966 and 1967) tried to duplicate Delaunay's work and barely succeeded in reproducing the function F, but he was not able to carry out the next and more difficult step.

Delaunay retains 461 terms in the Fourier expansion of F; the coefficients are polynomials in the various small parameters, some of which cover several pages. He proceeds systematically to get rid of one term after another, each time using a canonical transformation adapted to the particular term to be eliminated.

5.3 The Application of Canonical Transformations

The construction of such a transformation will now be demonstrated because it will show exactly in what form the solution of the three-body problem can be obtained in this standard procedure. After each transformation the old set of actions (I_1, I_2, I_3) and angles (w_1, w_2, w_3) is expressed in terms of a new set of actions (J_1, J_2, J_3) and angles (u_1, u_2, u_3). As explained already in Section 2.2, the transformation can be defined by a *generating function* $W(J_1, J_2, J_3, w_0, w_1, w_2, w_3)$ where $w_0 = u_0$ coincides always with the mean anomaly of the Sun ℓ', also called k in the preceding section. The transformation is given by the formulas

$$I_i = \frac{\partial W}{\partial w_i}, \quad u_i = \frac{\partial W}{\partial J_i} \quad \text{for } i = 1, 2, 3 . \tag{5.3}$$

These formulas are applied to the old Hamiltonian

$$\begin{aligned} H = {} & H_0(I_1, I_2, I_3) \\ & + \varepsilon V(I_1, I_2, I_3) \cos(m_0 w_0 + m_1 w_1 + m_2 w_2 + m_3 w_3 + \phi). \end{aligned} \tag{5.4}$$

All terms have been neglected except the ones that no longer depend on the angle variables, and the one special term of the perturbation that the transformation is designed to eliminate; it is characterized by the quadruple of integers (m_0, m_1, m_2, m_3) and the real coefficient $\varepsilon V(I_1, I_2, I_3)$, where the factor ε has been inserted to show its magnitude relative to the first term in (5.4). In general, the expansion (3.5) has complex coefficients; the argument of the cosine in (5.4) contains, therefore, a phase ϕ which depends on the multi-index (m_0, m_1, m_2, m_3). In the case of the lunar problem, however, the symmetry of the perturbation makes this phase vanish and yields the formula (5.4) with $\phi = 0$.

The *generating function* W is chosen so that the new Hamiltonian,

$$H' = H + \frac{\partial W}{\partial t}, \tag{5.5}$$

has no term in $\cos(m_0 w_0 + m_1 w_1 + m_2 w_2 + m_3 w_3)$ any longer; but in making the replacements of I and w by J and u, many other terms arise. If the consecutive elimination of terms is properly managed, however, the new terms are of higher order, and have more complicated angular arguments. The transformation has to be applied to all the terms in the complete Hamiltonian, not only to the particular term that appears in (5.4), before the next transformation can be carried out.

W is now set up in the form

$$W = J_1 w_1 + J_2 w_2 + J_3 w_3 + \varepsilon U(J_1, J_2, J_3, w_0, w_1, w_2, w_3), \tag{5.6}$$

which can be inserted into the formulas (5.3), and from there into (5.4) and (5.5) where one expands in powers of ε. The elimination of the term in $\cos(m_0 w_0 + m_1 w_1 + m_2 w_2 + m_3 w_3)$ leads to a first-order linear partial differential equation for the function U, which is easily solved in the form

$$U = -\overline{U}(J_1, J_2, J_3)\ \sin(m_0 w_0 + m_1 w_1 + m_2 w_2 + m_3 w_3), \tag{5.7}$$

with the function $\overline{U}$ given by the equation

$$\overline{U}(J_1, J_2, J_3) = \frac{V(J_1, J_2, J_3)}{4\,(m_0\omega_0 + m_1\omega_1 + m_2\omega_2 + m_3\omega_3)}. \tag{5.8}$$

The frequencies ω follow from the standard formula (3.3), $\omega_i = \partial H_0/\partial J_i$ in terms of the angle-independent part $H_0(J_1, J_2, J_3)$ in (5.4); ω_0 is always the mean solar motion n'.

5.4 Small Denominators and Other Difficulties

In the solution (5.8) for $\overline{U}$, the denominator in the generating function (5.7) is the cause of many difficulties. The division by a frequency corresponds to an integration over time which averages the perturbation and has to be carried out in one form or another in any solution; the problem of small denominators is unavoidable. In our case, ω_1 is close to n, corresponding to the sidereal month; ω_0 equals n', corresponding to the sidereal year, and therefore smaller than ω_1 by a factor $m = n'/n$ or about $1/13$; ω_3 is the motion of the node, which is close to $-3n'^2/4n$ according to Newton, and a factor $3m^2/4 \simeq 1/240$ below ω_1; finally, the combination $\omega_2 + \omega_3$ is the motion of the perigee, which is the same up to its sign as ω_3 according to Newton, although it is actually twice as much when all the higher-order terms are included.

If a term in the perturbation has $m_1 = 0$, it varies at a yearly rhythm; it has, therefore, a small denominator which boosts its importance by a factor of 13 compared with the monthly perturbations where $m_1 \neq 0$. If both $m_0 = 0$ and $m_1 = 0$, the boost is of the order 240. Finally, since ω_2 and ω_3 are empirically of the order of 3:1 with a change in sign, one has the exceptional case where $m_0 = m_1 = 0$ and $m_3 = 3m_2$, which gives a boost of 2000.

The effect of such small denominators is partially offset by the smallness of the perturbing term with that frequency. There is usually a direct relation, called the d'Alembert property, between the integers m_i and the powers of the expansion parameters m, e, γ, e', and a/a', which reduces the absolute value of the resonant perturbations. Poincaré (1908) gave a thorough discussion of this problem for the lunar theory, which has recently been taken up by Kovalewsky (1982); again Brouwer and Clemence (1961, p.317) provide a good first glance at the problem.

Two more comments are in order before we come to the discussion of Delaunay's results. The various frequencies in the lunar motion have empirical values, and we know nothing about their numerical character, rational, algebraic, or transcendental. Theorems about the convergence of expansions that are based on the number-theoretical nature of competing frequencies are, therefore, of little use. On the other hand, as the solution to the problem develops, one obtains a power series expansion with rational coefficients for the frequencies in terms of the small parameters in the problem; in our case, it is mainly the expansion in $m = n'/n$ which determines the resonances.

The occurrence of a resonance corresponds to the division by such a small parameter; the original ordering of the terms in the perturbation is obviated thereby. It is exactly this phenomenon which prevented Barton from carrying out Delaunay's solution on a computer of moderate size in the late 1960s, and inspired Deprit, Henrard, and Rom (1971) to try an entirely different approach yielding the same solution. Their method uses the so-called *Lie series* which had been known by some specialists but were not adapted to celestial mechanics until independently by Hori (1966) and Deprit (1969). Unfortunately, this calculation for the Moon has not been reported in its entirety because, with the coefficients in the expansion kept as rational numbers and the expansion carried to the order required by modern observations, the total output is huge.

Delaunay carried out a sequence of 505 canonical transformations, choosing terms in the perturbation to be eliminated according to his best guess about their importance. Thus, he went to the ninth order in m, but that yielded only the fourth decimal in the motion of the perigee while the Greeks already knew five decimals. The reason for this failure is the same which had defeated Newton; when the motion of the perigee $\dot{g} + \dot{h}$ and the motion of the node $\dot{h}$ are expanded to higher powers in m , the first two terms of this expansion are,

$$\frac{1}{n}\frac{d(g+h)}{dt} = +\frac{3}{4}m^2 + \frac{225}{32}m^3 + \dots , \qquad (5.9)$$

$$\frac{1}{n}\frac{dh}{dt} = -\frac{3}{4}m^2 + \frac{9}{32}m^3 - \dots , \qquad (5.10)$$

as first found by Clairaut and d'Alembert.

Newton was completely correct as far as the first term in each expansion is concerned, but he did not know that the second terms would be so vastly different. The coefficients in the first series keep on growing at about the above rate, while those in the second series decrease and alternate. Both series converge for the relevant value of $m \simeq 1/13$, but the first does so quite poorly. Schmidt (1979) has obtained over 30 terms for a precision of 10 decimals, but his coefficients are given as real numbers rather than rationals. Estimates of the higher order terms in expansions of this kind have been discussed among others by Bogomolnyi (1984a and b) as well as Llave and Rana (1989).

5.5 Hill's Periodic Orbit in the Three-Body Problem

The main idea of Hill's approach can be explained in relatively few words, and this very simplicity shows how radically different his method was from all its predecessors. Lagrange, Delaunay, and the other celestial mechanicians started from the Kepler ellipses, which are the most general solutions of the two-body problem, and modified them so as to accommodate the effect of the third body in the problem. Hill identifies a very special, but relatively simple solution of the full three-body problem, and then constructs the other solutions in the neighborhood of the particular one.

The special solution is a *periodic orbit* by which we mean that the trajectory closes itself smoothly, and the body keeps running around the same track indefinitely (cf. Section 1.2). Hill's original paper was first published separately in the USA (1877), and then reprinted in *Acta Mathematica 1886;* this work and the second series of articles (1878) are best available in the *Collected Works* (1905, p.243 and 284). They are quite readable. There are many other accounts in the textbooks and treatises on celestial mechanics, such as Brown (1896, 1960), and Poincaré (1907), as well as Brouwer and Clemence (1961).

The Kepler ellipses are periodic orbits; but the Moon's trajectory as it has been described since antiquity is not periodic, because both its perigee and its node rotate at rates that are not commensurate with its monthly circuit around the Earth. Periodic orbits in the full three-body problem are difficult to find, and Hill's choice is neither obvious nor easy to calculate. Its construction requires that one work in a special coordinate system in the ecliptic which had already been used in one of Euler's contributions to lunar theory.

The remainder of this chapter may be frustrating to the reader, because its pace seems slow. There are two separate endeavors both of which require special care: on the one hand, all equations are written in such a way that only the essential variables are left to deal with, and can be interpreted directly in terms of observations; on the other, the crucial first steps, in a long series of similar ones, are explained in sufficient detail to clearly perceive how the procedure is continued. Hill's method and its development by Brown (1897 to 1908) and Eckert (1966, 1967) are designed to construct directly the invariant tori without the usual perturbation approach starting from the Kepler ellipses.

The starting point is the Hamiltonian (4.9) with an additional simplification: the trajectory of the Earth-Moon around the Sun is assumed to be a circle, or in other words, a Kepler ellipse with the eccentricity $e' = 0$. The vector **R** in the ecliptic has a constant length a', and rotates at the constant rate n' around the Sun, exactly as in the formulas (4.3) and (4.4) with $e' = 0$.

The new coordinate system of the Moon (x, y, z), with the (x, y) in the ecliptic, is still centered in the Earth; but it rotates at the uniform speed n', so that the Sun is always seen in the x-direction. The Hamiltonian becomes

$$\begin{aligned} H = {} & \frac{1}{2\mu}\left[(u + \mu n' y)^2 + (v - \mu n' x)^2 + w^2\right] \\ & - \frac{\mu n'^2}{2}(x^2 + y^2) - \frac{\mu n^2 a^3}{r} - \frac{\mu n'^2}{2}(3x^2 - r^2)\,, \end{aligned} \tag{5.11}$$

where the momentum of the Moon has the Cartesian components (u, v, w) in the rotating frame of reference. The complete Hamiltonian in this coordinate system has many more terms, all of which, however, are proportional to powers of e' and a/a'. The distance of the Sun a' does not appear in (5.11) although the effect of the Sun is certainly represented in the last term, the quadrupole potential of the Sun in the neighborhood of the Earth.

The individual terms in this Hamitonian can be interpreted as if we were dealing with atomic physics: let μ be the mass of an electron, and the product $\mu n^2 a^3$ equal to the square of the electronic charge ε^2. Then the frequency n' becomes the Larmor frequency in the magnetic field B through the formula $n' = \varepsilon B/2\mu c$ with the speed of light c. Without the two terms in n'^2 we have the Hamiltonian of a hydrogen atom in a magnetic field, a problem to which we will devote Chapter 18, because it is an ideal case of a mildly chaotic system. The magnetic field is very strong since the Larmor frequency n' forms with the orbital frequency n the ratio $m = n'/n \simeq 1/13$; such a ratio corresponds to a Zeeman energy of 1 eV if the atom is in its ground state, or a magnetic field of 10,000 Tesla, far beyond anything technically available at present. If one deals with a *Rydberg atom,* however, where the outermost electron is in a very highly excited state, e.g., in a state with the principal quantum number 100, the magnetic field for $n'/n \simeq 1/13$ reduces to a manageable 1 Tesla = 10,000 Gauss.

The two terms in n'^2 add an 'electrostatic' potential with a parabolic shape, negative in the x-direction, neutral in the y-direction, and positive in the z-direction. The origin of this peculiar destabilizing force is easy to understand: the centrifugal force acts to drive the Moon away in the x-direction, but the solar attraction tends to pull the Moon toward the Earth along the z-axis. As a net result one finds the equations of motion in the rotating frame,

$$\begin{aligned} \ddot{x} - 2n'\dot{y} - n'^2 x &= -\frac{n^2 a^3}{r^3} x + 2n'^2 x , \\ \ddot{y} + 2n'\dot{x} - n'^2 y &= -\frac{n^2 a^3}{r^3} y - n'^2 y , \\ \ddot{z} &= -\frac{n^2 a^3}{r^3} z - n'^2 z , \end{aligned} \tag{5.12}$$

where the dot indicates differentiation with respect to time t. On the left we have the acceleration, the Coriolis force, and the centrifugal force, while on the right is the gravitational attraction of the Earth on the Moon, and the solar quadrupole force. No masses appear any longer, only the frequencies n and n', which are extremely well known from observation, and the dimensionless distance r/a of the Moon from the Earth.

The periodic orbit we are looking for is a solution of the equations (5.12) with the imposed period of 29.53059 days corresponding to the *synodic month,* from full moon to full moon, or equivalently, with a frequency $n - n'$. Therefore, we use the parameter $\tau = (n - n')t$, which is the mean angle from the Sun to the Moon as seen from the Earth, and we construct a solution of (5.12) with the period 2π in the variable τ . Moreover, we restrict the periodic orbit to the ecliptic, i.e., $z = 0$, and expand in a complex-valued Fourier series

$$x_0(\tau) + iy_0(\tau) = e^{i\tau} \sum_{\ell} a_\ell \, e^{i\ell\tau} . \tag{5.13}$$

As further requirement on the periodic orbit, we demand symmetry with respect to the x-axis in the form

$$x_0(-\tau) + iy_0(-\tau) = [x_0(\tau) + iy_0(\tau)]^+ , \tag{5.14}$$

where the upper index + indicates the complex conjugate. The coefficients a_ℓ are real because of (5.14), and conversely, the symmetry of the periodic orbit is assured when the coefficients a_ℓ are real.

Since the solar potential is approximated by a quadrupole field, there is also symmetry with respect to the origin, and we can require that $x_0(\tau + \pi) + iy_0(\tau + \pi) = -(x_0(\tau) + iy_0(\tau))$ which implies that all the odd coefficients $a_{2\ell+1} = 0$. The periodic orbit has the shape of an oval which is centered on the origin, and intersects the x-axis and the y-axis at right angles.

If the Sun is not pushed off to infinity with an infinite mass to make up for the distance, then the disturbing potential is not symmetric with respect to the y-axis, although it remains so with respect to the x-axis. The periodic orbit loses its symmetry correspondingly, and one needs both even and odd coefficients a_ℓ.

The equations (5.12) in terms of the variable τ contain only the parameter $m = n'/n$ and the normalized coordinates $(x_0 + iy_0)/a$. Thus one obtains a one-parameter family of ovals, which are sketched in Figure 8 for different values of m. We see that if the Moon had a synodic period of about 210 days, or seven calendar months, or a little less than eight sidereal months, her orbit would have a stationary point with respect to the Sun at half moon. According to a rough estimate using Kepler's third law for the Earth-Moon system (4.7), a reduction of the mean motion n by a factor of 8 would be compensated by an increase of the semimajor axis a by a factor of 4. The Moon would still be close to the Earth relative to the Sun since the ratio a/a' would be about 1/100.

As long as the Moon's orbit does not deviate too much from a circle, or equivalently, provided the term with $\ell = 0$ in (5.13) is dominant, Hill's periodic orbit can be roughly estimated. Let us, therefore, insert $x = a_0 \cos(\tau), y = a_0 \sin(\tau), z = 0$ into (5.12). The first and second equations become incompatible; by multiplying the first with $\cos(\tau)$, the second with $\sin(\tau)$, and adding, we get the radial aceleration on the left and the radial force on the right. Both sides are then integrated over τ from 0 to 2π to equate the average radial acceleration with the average radial force. Thus, we find that

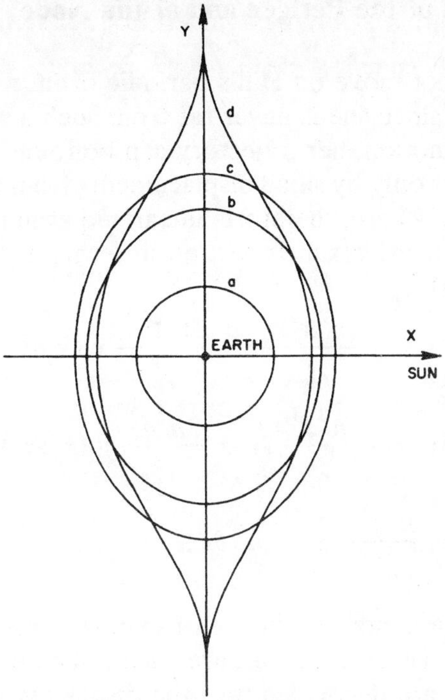

Figure 8 Hill's periodic orbits for the motion of the Moon: they lie in the ecliptic, and close smoothly in the rotating coordinate system where the (mean) Sun is on the x-axis; the closed orbits correspond to a: 12.37 periods, b: 4, c: 3, and d: 1.78 periods (lunations) per year; the sizes of the orbits are scaled properly.

$$a_0 = \frac{a}{(1 + m^2/2)^{1/3}} ; \tag{5.15}$$

the average distance of the Moon from the Earth has actually decreased compared with its Keplerian value. The Moon has to come closer to the Earth in order to maintain her frequency in spite of the Sun, which tries to tear the Moon away from the Earth.

The complete solution for the periodic orbit, i.e., the coefficients a_ℓ in (5.13), have to be obtained before one can go on to find the general solution of the equations (5.12) in its neighborhood. The ideal method is to write the coefficients as power series in the parameter $m = n/n'$; we shall assume that this feat has been completed with the help of some of the ingenious algorithms which we owe to Hill (1878), Poincaré (1907), and others.

5.6 The Motion of the Perigee and of the Node

The Moon does not move on Hill's periodic orbit, not even in the rotating frame; but since she is never far from such a special solution of the equations of motion, her trajectory can be found by looking at the motions that differ only by small displacements from the periodic orbit. The equations (5.12) are, therefore, linearized around the periodic orbit $(x_0(t), y_0(t), 0)$; the result is written in terms of the displacements $(\delta x(t), \delta y(t), \delta z(t))$,

$$\delta\ddot{x} - 2n'\delta\dot{y} - n'^2\delta x = -\frac{n^2a^3}{r_0^3}\delta x + \frac{3n^2a^3}{r_0^5}x_0(x_0\delta x + y_0\delta y) + 2n'^2\delta x\,,$$

$$\delta\ddot{y} + 2n'\delta\dot{x} - n'^2\delta y = -\frac{n^2a^3}{r_0^3}\delta y + \frac{3n^2a^3}{r_0^5}y_0(x_0\delta x + y_0\delta y) - n'^2\delta y\,,$$

$$\delta\ddot{z} = -\frac{n^2a^3}{r_0^3}\delta z - n'^2\delta z\,, \tag{5.16}$$

where the dots again indicate differentiation with respect to time t and $r_0 = (x_0^2 + y_0^2)^{1/2}$. These equations are a special case of (1.17).

The equations are decoupled into the displacements in the ecliptic $(\delta x, \delta y, 0)$, and the displacement normal to the ecliptic $(0, 0, \delta z)$. Since the periodic orbit lies in the ecliptic so that $z_0 = 0$, the value of the Hamiltonian (5.11) does not depend on δz to first order; a solution of the last equation (5.16) by itself is acceptable. This condition for δz is already in the form of a typical Hill's equation, since we can write

$$0 = \delta\ddot{z} + \left(\frac{n^2a^3}{r_0^3} + n'^2\right)\delta z \simeq \delta\ddot{z} + \left(n^2 + \frac{3}{2}n'^2\right)\delta z\,. \tag{5.17}$$

The factor of δz in the second member is a function of period 2π in the independent variable τ because of r_0 which is given by (5.13).

The third member in (5.17) uses the approximation (5.15) for the periodic orbit as a circle with the radius $r_0 = a_0$. It has the solution $\delta z = \kappa\cos(\omega t + \chi)$ with the frequency $\omega = n(1 + 3n'^2/4n^2)$, which is Newton's result for the motion of the node: The vertical motion of the Moon in the neighborhood of the periodic orbit has a higher frequency, so that the Moon returns to the ecliptic before she faces again the same fixed star; her node is *retrograde*. The first part of (5.17) was obtained and its consequences were discussed independently by J.C. Adams (1878), the codiscoverer (with Leverrier) of the planet Neptune, at the same time as Hill published his general method.

The equations for the deviations from the periodic orbit in the ecliptic are more difficult to treat. They reflect accurately the problems one encounters every time the neighborhood of a periodic orbit is investigated. There are two degrees of freedom since there are two position coordinates δx and δy ; but one degree of freedom is uninteresting. A shift in the time variable by some constant value δt makes no difference to the trajectory of the Moon, and a shift by δE of the Moon's energy, i.e., the constant value of her Hamiltonian (5.11), would change the frequency $(n - n')$ of the periodic orbit.

The relevant degree of freedom is a shift δs that is perpendicular in the (x, y) plane to the direction of motion of the periodic orbit. Therefore, we write

$$(\delta x, \delta y) = \delta s \frac{(\dot{y}_0, -\dot{x}_0)}{\sqrt{\dot{x}_0^2 + \dot{y}_0^2}}, \tag{5.18}$$

and then find the original *Hill's equation* exactly as first written down by its dicoverer,

$$\delta\ddot{s} + \Theta(t)\delta s = 0, \tag{5.19}$$

where the function $\Theta(t)$ has the period $2\pi/(n - n')$ of the periodic orbit.

The calculation of Θ in terms of the coodinates $x_0(t)$ and $y_0(t)$ can be made by inserting the expression (5.18) into the equations (5.16); but the necessary manipulations and the final result are surprisingly tricky. In the limit of a circular periodic orbit with the radius $r_0 = a_0$ given by (5.15), the time-independent part of Θ becomes $n^2 - 3n'^2/2$. Just as in the case of δz, we find now that $\delta s = \varepsilon \cos(\Omega t + \psi)$ where $\Omega = n(1 - 3n'^2/4n^2)$, again Newton's result for the motion of the Moon's anomaly, which is her motion toward and away from the Earth.

The coefficients of the Fourier series (5.13) depend only on the ratio $m = n'/n$ apart from some obvious scaling factors, and the same holds for the periodic functions which appear in Hill's equations (5.17) and (5.19). Various systematic procedures have been devised in order to find directly the expansions (5.9) and (5.10) for the motions of the node and of the perigee as power series in m. The most ingenious was first proposed by Hill, and involves the use of *infinite determinants,* a very daring idea at the time. These difficult series can thus be obtained, without worrying about the effect of the other parameters such as the e, γ, e', and a/a'.

5.7 Displacements from the Periodic Orbit and Hill's Equation

Hill's work has another aspect which is important for the sequel. The solutions of (5.16), (5.17), and (5.19) obey Floquet's theorem (1883) which was, of course, well understood by Hill five years earlier. They consist of a product of two functions: the first factor is the simple trigonometric function that we quoted earlier, and whose frequency gives the motions of the node and of the perigee, called ω and Ω in the preceding section. The second factor is periodic with the periodicity of the underlying periodic orbit; the second factor provides only a modulation of the amplitude for the first factor.

The symmetry of the periodic orbit suggests the following expression for the solutions of the equation (5.17):

$$iz_1(\tau) \;=\; e^{ig_0\tau}G(\tau) - e^{-ig_0\tau}G^+(\tau)\,, \tag{5.20}$$

and a slightly more complicated form for the equation (5.19):

$$x_1(\tau) + iy_1(\tau) \;=\; e^{ic_0\tau}F_1(\tau) + e^{-ic_0\tau}F_2(\tau)\,. \tag{5.21}$$

in terms of the mean synodic angle $\tau = (n - n')t$ from the Sun to the Moon. The functions $G(\tau)$, $F_1(\tau)$, and $F_2(\tau)$ are periodic with the same period 2π as the orbit (5.13), and can be expanded in the same manner with real coefficients,

$$F_1(\tau) = \sum_\ell \varepsilon_\ell e^{i\ell\tau}\,,\; F_2(\tau) = \sum_\ell \varepsilon'_\ell e^{i\ell\tau}\,,\; G(\tau) = \sum_\ell \kappa_\ell e^{i\ell\tau}\,. \tag{5.22}$$

The motion of the perigee (5.9) is given by $1 - (1 - m)c_0$, and the motion of the node (5.10) is $1 - (1 - m)g_0$, because the coordinate system for the periodic orbit is rotating with the angular speed $n(1 - m) = n - n'$, and the time parameter τ has been normalized to increase by 2π in one period.

The trajectory of the Moon can now be represented to first order by adding the displacements $(\delta x, \delta y, \delta z)$ to the coordinates $(x_0, y_0, 0)$ of the periodic orbit. These displacements are the solutions of the second-order linear equations (5.17) and (5.19), each of which has two constants of integration. The most convenient choice for them are the coefficients $k = \kappa_0$ for δz, and the difference $e = \varepsilon_0 - \varepsilon'_0$ for $\delta x + i\delta y$, as well as a phase angle in each of the exponentials of (5.20) and (5.21). The constant k defines the effective *inclination* of the lunar trajectory with respect to the ecliptic, and the constant e determines its effective *eccentricity*. The phase angles fix the position of the node and of the perigee at the epoch.

By taking the time derivatives we get for the momentum (u, v, w) as well as for the position (x, y, z) of the Moon formulas of the type (3.4) which describe the motion on an invariant torus. There are altogether three angle variables: the basic angle τ in the periodic orbit and in the functions F_1, F_2 of (5.21) and G in (5.20), which describes the synodic motion (with respect to the Sun); the angle $c_0\tau$ for the anomaly toward and away from the Earth; the angle $g_0\tau$ for the up and down motion with respect to the ecliptic. A particular torus is characterized by the effective eccentricity e and the inclination k.

Before concluding this chapter a few additional remarks about Hill's method may help to gain perspective on his work. I find it surprising that the trajectories in the neighborhood of a periodic orbit have frequencies that are completely different from the underlying period; after all, one could have expected that the frequency differences vanish as the trajectories get closer to their periodic progenitor, but the difference between the sidereal and the anomalistic month does not vanish even when the eccentricity e goes to 0, and similarly with the nodical month and the inclination k of the lunar trajectory.

The expansion around the periodic orbit can be pushed to higher order in a fairly straightforward manner. Instead of expanding the Hamiltonian or the equations of motion around the periodic orbit $(x_0, y_0, 0)$, one expands around the first-order trajectory $(x_0 + x_1, y_0 + y_1, z_1)$, which is determined by the values of e and k. Second-order corrections, $\delta x = x_2(\tau)$, $\delta y = y_2(\tau)$, $\delta z = z_2(\tau)$, now appear, which are proportional to the small parameters e' and a/a' , as well as to the higher powers of e and k.

These corrections and all the subsequent ones satisfy linear differential equations whose only difference from (5.16) is the presence of inhomogeneous driving terms on the right-hand side. They arise when $x = x_0(\tau) + x_1(\tau)$, etc. are inserted into the Hamiltonian (5.11), and its ancestors, (4.9) and (4.2). Each combination of small parameters e, k, e', a/a' leads to one such new set of inhomogeneous linear equations. The whole expansion (3.4) for the lunar trajectory is found by solving over and over again the same set of linear differential equations (5.16).

This program was carried out around the turn of the century by Brown (1897-1908), again by hand, with important technical improvements by Eckert (1954), and formed the basis for all lunar calculations before the landing of human beings on the Moon. By that time modern computers had become available, which allowed all the former computations to be pushed significantly beyond their original goals. Dieter Schmidt of the University of Cincinnati and the author (Gutzwiller and Schmidt 1986) completed in this manner the work that had been started by Hill, and carried on by Brown and Eckert.

The resulting series for the Moon's invariant torus is useful as far as practical calculations are concerned, but it has not been shown to converge for the relevant choice of parameters. Moreover, the incipient chaos of the Moon's trajectory shows up even in this traditional result and can perhaps be appreciated more nowadays than in the heroic precomputer times.

The first symptom is the great proliferation of small terms of higher order. If the lunar coordinates are required to a relative precision of 10^{-8}, it is not sufficient to know all the coefficients $\geq 10^{-8}$ in the Fourier expansion (3.4). The author (Gutzwiller 1979) has calculated the root-mean-square of the coefficients in the interval $(10^{-8}, 10^{-9})$; i.e., the squares of the coefficients in this interval were added up and the root of the sum was taken; it is almost 10^{-7}. Thus, the small terms generate a background noise which has to be controlled by pushing the expansion far beyond its original goal.

The second symptom is the very small denominator that was mentioned at the beginning of Section 5.4. The original intention of Schmidt and the author was to guarantee a relative precision of 10^{-10} by obtaining all coefficients down to 10^{-12} and calculating each to an accuracy of 10^{-14}, i.e., reals defined by 64 bits. The boost of 2000 by the small denominator, however, forced us to push down to 10^{-17}, and reals defined by 128 bits, an accuracy far beyond any measurement, since it corresponds to 4.10^{-7} cm, or 40 atomic radii, at the distance of the Moon.

CHAPTER 6

Periodic Orbits

The idea of everything returning eventually to its point of departure has a strong hold on humanity, with many historical, philosophical, and religious implications. Kepler's discovery of the elliptical orbits for the planets in the solar system seemed to give a scientific basis to this predilection for things running along a closed track and repeating their history over and over again. The motion of the Moon does not fit this picture in its most narrow interpretation; but we have seen in the past chapters how the invariant tori of an integrable system generalize the simple-minded view. Instead of one period after which all momenta and positions return to their initial values, one deals with as many different periods as degrees of freedom.

The various periods in the lunar problem are vastly different: a month for the main motion around the Earth, a year for the direct periodic influence of the Sun, as well as 9 and 18 years for the Sun's secular effect. It hardly matters whether the exact ratios of these periods are rational numbers or not, except when one has to set up a calendar.

All the great civilizations have established simple rational approximations for the ratios of these periods. The most useful of them is the Metonic cycle of 19 tropical years, which contains 235 synodic months, with a difference of only 1/4 of a day; on its basis the Jewish calendar has a well-defined sequence of 12 short years containing 12 months and 7 long years containing 13 months. Solar and lunar eclipses can be conveniently organized into Saros cycles of 223 synodic months ≅ 239 anomalistic months ≅ 241 sidereal months ≅ 242 draconitic months ≅ 18 years 11+1/3 days. The 1/3 day is remarkable, because it means that almost exactly the same type of eclipse is seen in the same place on Earth every 54 years and 34 days, i.e., with a delay of only one month.

The search for exactly periodic orbits may have been inspired by these amazing near coincidences; but their importance for modern physics has a quite different origin and was first recognized by Poincaré (1892, Chapter III). He found that periodic orbits, i.e., solutions of the equations of motion that return to their initial conditions, are densely distributed among all possible classical trajectories; and he suggested that the study of periodic orbits would provide the clue to the overall behavior of any mechanical system. In his words,

> "*what makes these periodic solutions so valuable, is that they offer, in a manner of speaking, the only opening through which we might try to penetrate into the fortress which has the reputation of being impregnable.*"

Periodic orbits form continuous families in phase space that can be investigated by varying either the energy of the system or some external parameter like the relative masses of the bodies involved. There are fascinating details, in particular the sudden bifurcation of a particular isolated periodic orbit, or its unexpected birth without further warning. The reader will find interesting examples in the work of Hénon (1966-70) on the Restricted Three-Body Problem, i.e., a light body moving in the same plane as two heavy bodies that are in a circular orbit around their center of mass; Hill's theory of the lunar perigee (cf. Section 5.5) is a special case. Contopoulos (1970) investigated the motion in the galactic gravitational field; Baranger and Davies (1987) with de Aguiar and Malta (1987) finally carried the good word to the quantum-chaos-minded physicists. A lot of work has been done about actually finding periodic orbits; cf. Helleman (1978), as well as Kook and Meiss (1989).

Poincaré's suggestion seemed, at first, to be valuable only as a general approach for the better understanding of some difficult problems in classical mechanics. Since the advent of quantum mechanics, however, the periodic orbits have turned out to be of special significance in the transition from the classical to the quantum regime; this idea is the essence of the trace formula which will be explained in Chapter 17. The present chapter will discuss how periodic orbits arise in an integrable system, how many of them there are, and how phase space looks in their neighborhood whether the system is integrable or not.

6.1 Potentials with Circular Symmetry

Periodic orbits in an integrable system are best handled with the help of the action-angle formalism. The main ideas come out clearly when studying a particle moving in a plane under the influence of a potential $V(r)$ that depends only on the distance r from the origin. Einstein (1917) gave this example in his paper on quantization conditions, which has inspired many researchers ever since it was "discovered" by Keller (1958). Although the circular symmetry of the problem makes the discussion almost trivial, the reader should pay close attention to some of the finer points such as the exact definition of the angle variables, and the calculation of the Hamiltonian in terms of the actions.

In polar coordinates (r, ϕ), the Hamiltonian becomes

$$H(p, M, r, \phi) = \frac{p^2}{2m} + \frac{M^2}{2mr^2} + V(r) = E\,, \qquad (6.1)$$

where we have used the conjugate pairs (p, r) and (M, ϕ) with the radial momentum p and the angular momentum M. The radial momentum p is not a good action variable for this problem because its conjugate variable r still occurs explicitly in H; but since we have succeeded in separating the variables, we can find the relevant actions with the help of (3.2). M is the action for the azimuthal motion, while the action N for the radial motion is given by the integral

$$N = \frac{1}{\pi}\int_{r_1}^{r_2} p\,dr = \frac{1}{\pi}\int_{r_1}^{r_2} dr\sqrt{2m(E - V(r)) - M^2/r^2}\;, \qquad (6.2)$$

where the limits of integration, r_1 and r_2, are the roots of the equation $E = V(r) + M^2/2mr^2$.

The invariant tori in the three-dimensional surface of constant energy E can now be represented, exactly as Einstein saw them for the first time, by using a cylindrical coordinate system with the polar coordinates (r, ϕ) in the horizontal plane and p along the vertical. The trajectories of the Hamiltonian (6.1) for a given value of M form a closed curve in the vertical (r, p) plane, which has to be rotated around the p –axis to generate a torus. These tori for different values of M are nested inside one another as shown in Figure 9.

The angular momentum M reaches its largest value M_0 at the core of these nested tori, when the two solutions, r_1 and r_2, coincide, and their common value is r_0; the corresponding value of $N = 0$. As M decreases, N increases, and reaches a maximum N_0 when $M = 0$. The values of M can be both positive and negative; but N is always positive, and depends only on $|M|$. For each value of E, there is a well-defined curve of N versus M.

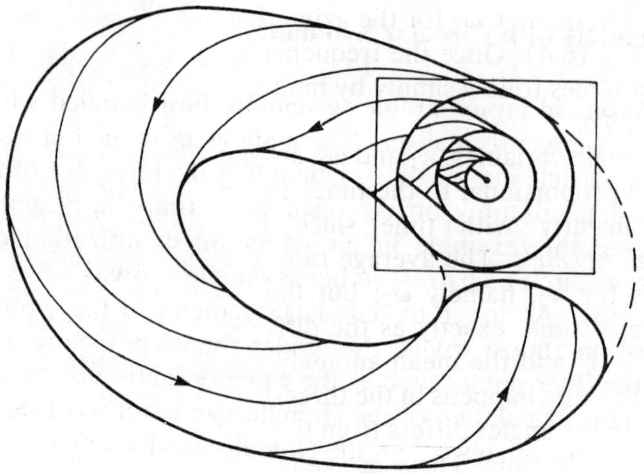

Figure 9 Invariant tori in the three-dimensional surface of constant energy.

In order to transform the system to the action-angle variables that are associated with N and M, one has to solve the equation (6.2) with respect to E. The resulting function $E(N, M)$ is the required Hamiltonian $H_0(N, M)$ whose derivatives give the frequencies according to (3.3). The equation (6.2) cannot be solved explicitly, except for some special potentials $V(r)$.

If $V(r)$ is the Coulomb potential $-e^2/r$, one finds

$$H_0 = \frac{me^4}{2(N + |M|)^2} ; \tag{6.3}$$

the two-dimensional harmonic oscillator of frequency ω has the Hamiltonian $H_0 = (N + |M|)\omega$. In both cases the matrix of second derivatives with respect to N and M is singular; the further investigations do not apply to these special cases. A similar fate awaits the case of two independent harmonic oscillators; it is not a good example for studying the effect of perturbations on typical integrable systems.

The frequency ω_1 which is associated with N according to (3.3), is obtained by taking the derivative of (6.2) with respect to N at constant M, while considering E to be a function of both N and M. The resulting expression is more easily written in terms of the period

$$T_1 = \frac{2\pi}{\omega_1} = 2\int_{r_1}^{r_2} \frac{m\,dr}{\sqrt{2m(E - V) - M^2/r^2}} ; \tag{6.4}$$

this formula gives the elementary expression for the time to cover the radial motion from r_1 and r_2 and back, since the equations of motion give $dt = m\,dr/p$ and p is obtained from (6.1). A similar argument gives the frequency ω_2 for the azimuthal motion as the ratio of two integrals like (6.4). Once the frequencies are known, the corresponding angle variables follow simply by multiplying with t which is the integral over $m\, dr/p$.

The angle variables, w_1 and w_2, are not very interesting because they are both proportional to the time. The angle ϕ, however, does not increase linearly with time since the equations of motion give $d\phi = dt\, M/mr^2$. The average rate of increase with time for ϕ is the same as for w_2, namely ω_2; but the difference $\phi - w_2$ is a periodic function of time, exactly as the difference between the true anomaly $f' = \phi' - g'$ and the mean anomaly ℓ' in the Kepler motion of (4.3). The same thing happens in the three-body problem: the intuitive choice of the various angles differs from the final angle variables by terms that are periodic in time. These are called the *inequalities* in the traditional astronomical literature because they give the real motion of the celestial bodies through the sky an irregular appearance.

Periodic orbits arise when the ratio of the two frequencies, ω_1/ω_2, is a rational number ν/μ where ν and μ are assumed to be relatively prime. One can then write $\omega_1 = \nu\omega_0$ and $\omega_2 = \mu\omega_0$ in terms of the overall period $T_0 = 2\pi/\omega_0$. In the case of *orbital precession,* the polar angle ϕ has increased beyond 2π by a rational fraction of 2π, in the time it takes the radial motion to complete its period; Figure 10 shows a case of orbital regression.

Let us now go back to the surface of constant energy, as shown in Figure 11d, and look at two consecutive intersections of a trajectory with the vertical plane $\phi = 0$. The intersection of an invariant torus with $\phi = 0$ forms a simply closed loop. An initial point on the r-axis, with $r > r_0$, does not complete the closed loop as it returns to $\phi = 0$; it may advance by some angle less than 2π, in a rough manner of speaking. The greater the loop in Figure 11d, the less the advance of the representative point.

The periodic orbits occur on those tori, where upon μ consecutive intersections with $\phi = 0$, the representative point completes exactly ν turns around its loop. For a fixed value of the energy E, each loop is characterized by its value for M, or equivalently by N, or again by the ratio of the two frequencies, all of which vary continuously in some interval. The rational values of this ratio form a dense set in this interval, just as Poincaré pointed out.

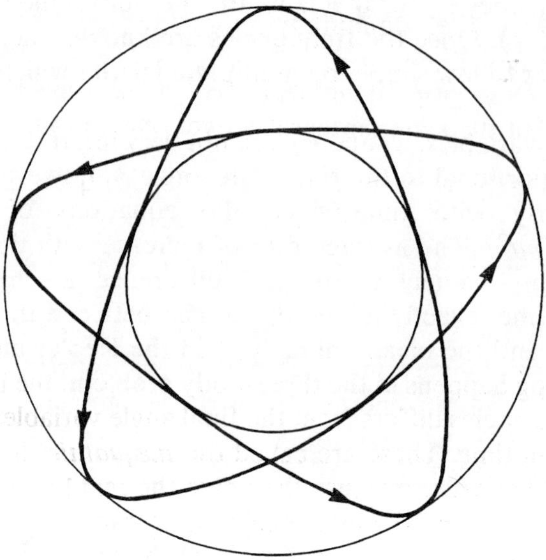

Figure 10 Precessing trajectory: the angular motion covers $4\pi/5$ while the radial motion goes through one cycle, from its minimum r_1 to its maximum r_2 and back.

6.2 The Number of Periodic Orbits in an Integrable System

The arguments of the last section are now generalized to an integrable system with n degrees of freedom. We assume that the transformation to action-angle variables has been carried out, and the new Hamiltonian is a function of the actions $I_1, I_2, \ldots, I_n$. The frequencies are then given by the formulas (3.3), $\omega_j = \partial H/\partial I_j$ in terms of the actions. But it is now crucial that these formulas can be inverted: one needs the map $\omega \to I$ rather than the map $I \to \omega$, which the Hamiltonian yields. This inversion is possible only if the matrix of the second derivatives $\partial\omega_i/\partial I_j = \partial^2 H/\partial I_i \partial I_j$ is regular. This last condition fails for the Kepler problem and the harmonic oscillator, as shown in (6.3).

Let us now proceed to the frequency space $(\omega_1, \omega_2, \ldots, \omega_n)$, and consider in it a straight line through the origin in the direction $(\kappa_1, \kappa_2, \ldots, \kappa_n)$. This line cuts through the energy surface $H(I_1, \ldots, I_n) = E$ in the point $(\omega_{01}, \ldots, \omega_{0n})$. In particular, if we have chosen integer values for the κ's, i.e., $\kappa_j = k_j$, we find that $\omega_{0j} = k_j \omega_0$; all the frequencies are integer multiples of some basic frequency ω_0. The corresponding trajectory has the period $T_0 = 2\pi/\omega_0$. Inside the allowed region of frequency space, the ratios of the integers k can be given arbitrarily; the more complicated they are, the smaller is the common frequency ω_0 and the longer is the overall period T_0.

The use of classical periodic orbits in quantum mechanics requires the calculation of the action integral S over one period,

$$S = \int_0^{T_0} (I_1 dw_1 + \cdots + I_n dw_n) = \int_0^{T_0} (I_1\omega_1 + \cdots + I_n\omega_n) dt$$
$$= (I_1\omega_1 + \cdots + I_n\omega_n)\frac{2\pi}{\omega_0} = 2\pi\,(k_1 I_1 + \cdots + k_n I_n)\,. \tag{6.5}$$

The last expression is easily misunderstood, because the integers k do not only appear as the explicit factors of the actions I, but the actions depend implicitly on the chosen values of the integers. Let us, therefore, find out how the value of S changes when the integers k are changed by some small amounts δk.

In order to carry out this computation, we go back to the original real numbers $(\kappa_1, \ldots, \kappa_n)$, which are to be changed by $(\delta\kappa_1, \ldots, \delta\kappa_n)$. The energy $E = H(I_1, \ldots, I_n)$ has to remain the same, and this leads to the condition

$$\delta H = \sum_{i,j} \frac{\partial H}{\partial I_j}\frac{\partial I_j}{\partial \omega_i}(\delta\kappa_i\,\omega_0 + \kappa_i\,\delta\omega_0) = 0\,,$$

where $\omega_j = \partial H/\partial I_j = \kappa_j\,\omega_0$ has to be inserted. Now we can work out

$$\delta S = 2\pi\,\delta \sum_j \kappa_j I_j = 2\pi \sum_j I_j\,\delta\kappa_j + 2\pi \sum_{ij} \kappa_j \frac{\partial I_j}{\partial \omega_i}(\delta\kappa_i\omega_0 + \kappa_i\delta\omega_0)\,;$$

but the second term on the right vanishes because of $\delta H = 0$.

The remaining first term can be interpreted in the light of Bohr's correspondence principle. The κ's are replaced by the large integers k, as in a highly excited state, and the real increments $\delta\kappa$ become the small integer increments δk, as in the transition to a neighboring state. The above calculation is still valid since the δk's are small compared with the k's. Thus, one finds that

$$\delta S = 2\pi\,(\,I_1\delta k_1 + \ldots + I_n\delta k_n\,)\,. \tag{6.6}$$

The action is a first-order homogeneous function of the κ's, and also of the k's as long as they are large.

For the purpose of counting how many periodic orbits there are whose action S is smaller than some positive number σ, we can treat S as if it were a linear function of the positive integers k_i with positive coefficients I_i. Therefore, *the number of periodic orbits for a fixed energy E, and for which $S < \sigma$, grows as the n-th power of σ.*

This polynomial growth in the number of periodic orbits for an integrable system is in sharp contrast to the exponential growth in a mechanical system with hard chaos, a condition that will be discussed at length in the later chapters. The polynomial growth can be directly traced to the one-to-one relation between n-tuples of integers $(k_1, ..., k_n)$ and periodic orbits. Such a simple characterization does not work any longer in a chaotic system.

6.3 The Neighborhood of a Periodic Orbit

Hill's theory of the motions of the Moon (cf. Sections 5.5 and 5.6) was based on the idea of investigating the neighborhood of a particular simple periodic orbit of the full Hamiltonian. The neighboring trajectories were calculated as Fourier series in the basic angle variable τ of the periodic orbit; the computations can be carried to high accuracy. The study of chaotic systems requires a less detailed, but more general understanding of periodic orbits, which can eventually be related directly to the variational principle and Feynman's path integral. The present section is intended to lay the foundation for this complementary approach.

Without restriction as to the integrability or lack thereof in the mechanical system, let us latch onto one particular periodic orbit, which starts with the momentum $\bar{p}$ in the position $\bar{q}$ and returns to this place in phase space after the time $\bar{T}$. We shall assume that all the trajectories that start at (p', q') in some sufficiently small neighborhood of $(\bar{p}, \bar{q})$ will return to that neighborhood after some time t close to $\bar{T}$. None of these trajectories ever strays far from the periodic orbit during the time t.

We will examine the neighboring trajectories in function of their initial conditions (p', q') and their endpoints (p'', q''), all in the neighborhood of $(\bar{p}, \bar{q})$. Although (p', q') and (p'', q'') are in the same neighborhood of $(\bar{p}, \bar{q})$, they are joined by a long trajectory; when either (p', q') or (p'', q'') are allowed to vary, the whole trajectory connecting them will change in order to join correctly the initial and the final points while running close to the periodic orbit.

The whole calculation will be carried out for a system with three degrees of freedom, because that represents already the most general case, whereas such is not true for two degrees of freedom. We will also avail ourselves of a special set of coordinates in phase space whose existence and benefits will be established in the next chapter. The momentum coordinate p_1 is simply the energy E, quite generally; the position coordinate q_1 is the time variable along the periodic orbit for the points on it. For the points off the periodic orbit, however, the coordinate q_1 is a continuously differentiable function of the original coordinates. The points with $q_1 = constant$ and $p_1 = E$ form a four-dimensional submanifold transverse to the periodic orbit. The remaining coordinates (p_2, p_3, q_2, q_3) vary in such a submanifold, but are otherwise chosen arbitrarily with the only restriction that (p_1, q_1), (p_2, q_2), (p_3, q_3) are canonically conjugate, or equivalently that the action integral S is given by $\int (p_1 dq_1 + p_2 dq_2 + p_3 dq_3)$ from q' to q''.

The action integral $S(q'' q' E)$ is taken along the trajectory that starts at q' in the neighborhood of $\bar{q}$, runs close to the periodic orbit for one, two, or more turns, whichever is of interest, and ends up at q'' again in the neighborhood of $\bar{q}$, while moving at the energy E. We could have left out the explicit mention of E since in our special coordinate system $p''_1 = p'_1 = E$, but the energy E is listed explicitly to make the connection with the formulas in Chapter 3. In particular we need (2.4) in order to calculate the momenta p' and p''.

This formula is now used to get the displacements $\delta p = p - \bar{p}$ and $\delta q = q - \bar{q}$, while p_1 and q_1 are kept fixed,

$$\begin{aligned} \delta p'_i &= -\sum_{j=2}^{3} \frac{\partial^2 S}{\partial q'_i \partial q'_j} \delta q'_j - \sum_{j=2}^{3} \frac{\partial^2 S}{\partial q'_i \partial q''_j} \delta q''_j \\ \delta p''_i &= \sum_{j=2}^{3} \frac{\partial^2 S}{\partial q''_i \partial q'_j} \delta q'_j + \sum_{j=2}^{3} \frac{\partial^2 S}{\partial q''_i \partial q''_j} \delta q''_j , \end{aligned} \tag{6.7}$$

where the index i takes only the values 2 and 3, while the second derivatives are evaluated at $q'' = q' = \bar{q}$. These formulas can be written in the abbreviated form

$$\delta p' = -a\,\delta q' - b\,\delta q'' , \quad \delta p'' = b^+\,\delta q' + c\,\delta q'' , \tag{6.8}$$

where a, b, and c stand for the 2 by 2 matrices of second derivatives of S with respect to q' and q'', and b^+ is the transpose of b.

In order to understand the geometry of the trajectories in the neighborhood of the periodic orbit, one would like to have formulas for $\delta p''$ and $\delta q''$ in terms of $\delta p'$ and $\delta q'$, such as

$$\delta q'' = A\,\delta q' + B \delta p' , \quad \delta p'' = C\,\delta q' + D \delta p' , \tag{6.9}$$

where $A = -b^{-1}a$, $B = -b^{-1}$, $C = b^+ - cb^{-1}a$, $D = -cb^{-1}$. These 2 by 2 matrices are well defined provided the matrix b of mixed second derivatives of S with respect to q' and q'' is regular. A similar matrix was discussed in Chapter 2 when trying to find the formula (2.7) for the density of trajectories; but the relevant expression contained the second derivatives with respect to all position coordinates and the energy, rather than only the position coordinates of index larger than 1. The relation between these two matrices and their determinants will become clear when the trace formula is discussed in Chapter 17.

6.4 Elliptic, Parabolic, and Hyperbolic Periodic Orbits

The behavior of the trajectories near the periodic orbit is described by the linear transformation $(\delta p', \delta q') \to (\delta p'', \delta q'')$ of (6.9). Naturally, one looks for the eigenvalues λ of this linear transformation, and calculates, therefore, the characteristic polynomial $F(\lambda)$, which is given by the determinant

$$F(\lambda) = \begin{vmatrix} A - \lambda I & B \\ C & D - \lambda I \end{vmatrix} = \begin{vmatrix} - b^{-1}a - \lambda I & - b^{-1} \\ b^{+} + \lambda I & - \lambda I \end{vmatrix} , \quad (6.10)$$

where I is the 2 by 2 unit matrix. The second determinant was obtained from the first by replacing A, B, etc. by their expression in terms a, b, and c as given in (6.9). Moreover, the first two lines in the first determinant have been multiplied with c, and then subtracted from the last two lines.

This kind of manipulation is continued by first multiplying the first lines in the second determinant of (6.10) by b, and compensating this change by dividing the whole determinant with $\det | b |$, which yields the expression

$$\frac{1}{|b|} \begin{vmatrix} - a - \lambda b & - I \\ b^{+} + \lambda c & - \lambda I \end{vmatrix} = \frac{1}{|b|} \begin{vmatrix} - a - \lambda b & - I \\ b^{+}+(a+c)\lambda+\lambda^2 b & 0 \end{vmatrix} ,$$

where the second determinant results from the first one if the first two lines are multiplied by λ, and subtracted from the last two lines. At this point, the 4 by 4 determinant has been effectively reduced to a 2 by 2 determinant, and one can write

$$F(\lambda) = \det | b^{+} + (a + c)\lambda + b\lambda^2 | / \det | b | , \quad (6.11)$$

which is the desired result of this calculation.

The first corollary of (6.11) states that $F(0) = 1$, i.e., the determinant of the linear transformation (6.9) is 1; the transformation conserves the four-dimensional volume of the submanifold $(q_1 = \bar{q}, p_1 = \bar{p})$ in the neighborhood of the periodic orbit. This proposition will be generalized in the next chapter. The present purpose is to find out more about the eigenvalues of the transformation (6.9), i.e., the solutions of the equation $F(\lambda) = 0$.

Since the matrix elements in a, b, and c are real, the solutions of this algebraic equation either are real, or come in complex conjugate pairs.

Moreover, the expression (6.11) shows that, if λ is any zero of F, then so is $1/\lambda$. Thus, we have the following four possibilities which carry the classical designations:

(i) *elliptic* for $\lambda = \exp(i\chi), \exp(-i\chi)$ with χ real;
(ii) *direct parabolic* for $\lambda = +1$ or *inverse parabolic* for $\lambda = -1$;
(iii) *direct hyperbolic* for $\lambda = \exp(\pm\chi)$, or *inverse hyperbolic* for $\lambda = -\exp(\pm\chi)$ with real χ;
(iv) *loxodromic* for $\lambda = \exp(\pm u \pm iv)$ with independent signs and real values of u and v.

The first three cases occur in a system with two degrees of freedom, but the fourth case requires at least three degrees of freedom.

The values of $F(1)$ for the three first cases will come up in the trace formula (cf. Chapter 17); they are given by the expressions: $-4\,(\sinh(\chi/2))^2$ for direct hyperbolic, 0 for direct parabolic, $4\,(\sin(\chi/2))^2$ for elliptic, $+4$ for inverse parabolic, and $4\,(\cosh(\chi/2))^2$ for inverse hyperbolic orbits. Notice that these values of $F(1)$ span the full range from $-\infty$ to $+\infty$; Greene (1979) gave the name *residue* to this quantity $F(1)$ which characterizes the neighborhood behavior of the periodic orbit.

Since the first three cases describe transformations of a two-dimensional manifold in the neighborhood of the periodic orbit, they can be represented schematically as shown in Figure 11. In order to get from the initial point P' to the final point P'' in the transformation (6.9), the point P has to slide by a certain amount along either ellipses, parallel lines, or hyperbolas, as indicated by the arrows.

The situation in an integrable system (see Figure 11d) can only be reconciled with the parabolic case. When a mechanical system is integrable, the transformation (6.9) in the neighborhood of a periodic orbit has only the eigenvalues ± 1. Since periodic orbits are dense, this special feature has to be true everywhere. For systems with two degrees of freedom in particular, the 2 by 2 matrices in (6.9) reduce to real numbers; the condition for parabolic behavior of the periodic orbit becomes $a + c = \pm 2b$.

Integrable systems are, therefore, non-generic; they are exceptional, unless some principle in nature exists which gives preference to them over all the others; but none has been found so far. If a mechanical system with two degrees of freedom is generic, the neighborhood of a periodic orbit has to be elliptic or hyperbolic. It becomes then very difficult to reconcile a collection of the local behavior as depicted in Figure 11 with some simple overall pattern such as in Figure 11d. This is the reason why chaotic systems are unavoidable in mechanics; in most cases there is an intimate mixture of elliptic and

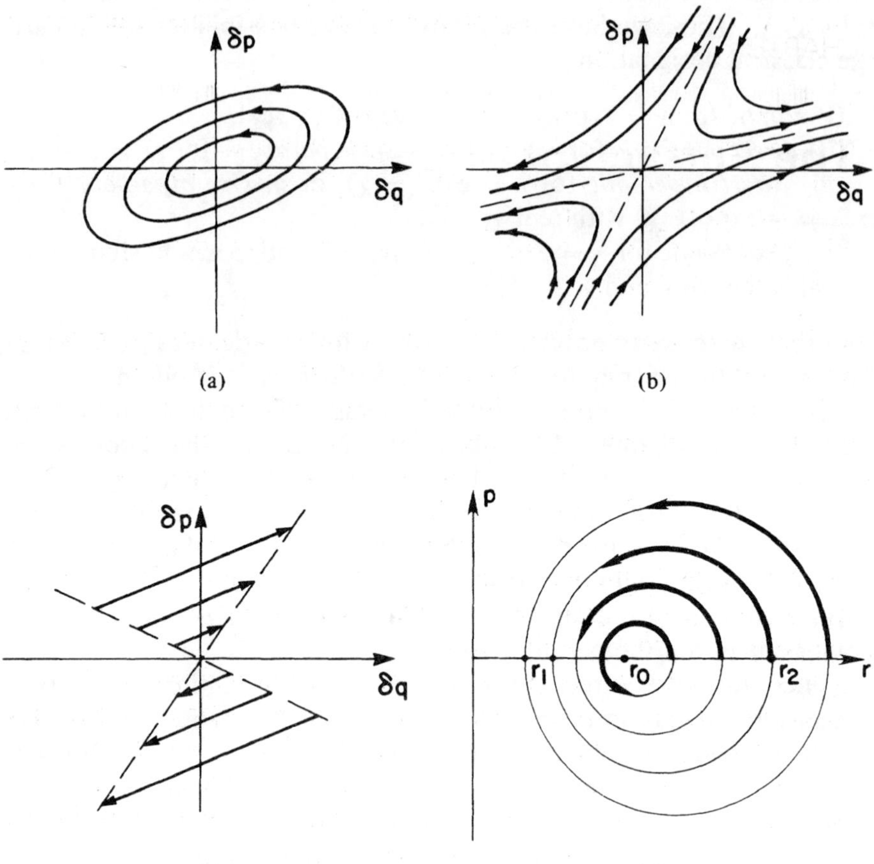

Figure 11 Area-preserving maps in the neighborhood of an elliptic (a), hyperbolic (b), parabolic (c) fixpoint, and the surface of section (d) of an integrable system ($\phi = 0$ at the end of section 6.1).

hyperbolic behavior which we shall call *soft chaos;* Markus and Meyer (1974) provide a mathematical version of this statement. Phase space might look integrable at some rough scale, although the fine details are much more complicated.

If the behavior is hyperbolic throughout, however, a relatively simple overall pattern can usually be established in spite of the prevailing chaos. In fact, the chaotic features in such a system turn out to be particularly striking, a condition we shall therefore call *hard chaos.* In our view, classical mechanical systems cover a wide range of chaotic behavior, with integrability as a rather exceptional situation on one end, and hard chaos as a structurally stable, i.e., immune to small perturbations, but not the most general situation at the other end.

CHAPTER 7

The Surface of Section

The trajectories in phase space often have very involved global features which are in apparent contradiction to the locally smooth flow. Poincaré proposed to deal with this problem by intercepting the flow at discrete times, rather than by following up on every little shift and displacement. His method, known as the surface of section, has become the main tool for studying chaotic systems, and will be discussed in this chapter for the special case of Hamiltonian systems. The main ideas are all of geometrical origin and will be presented in this light.

7.1 The Invariant Two-Form

The adjectives 'dynamical' and 'Hamiltonian' have been used quite loosely in the preceding chapters; it is now important, however, to follow the common usage. A *dynamical system* is a smooth vector field in phase space, and its trajectories are obtained by joining its little local arrows into continuous curves. A *Hamiltonian system* is the special case of a dynamical system where the vector field is defined by the formulas (2.2) in terms of the Hamiltonian $H(p, q)$.

This book deals exclusively with Hamiltonian systems; they turn out to be a very particular subset somewhere at the outer boundary of the large set of dynamical systems. Nature has decided to be Hamiltonian at its most basic level; non-Hamiltonian dynamical systems come up in physics only as phenomenological models for the more complicated underlying processes. The characteristic differences between the two types of systems shows up most clearly when discussing the surfaces of section.

The first step in our approach is the construction of the *canonical two-form* Ω. For this purpose, we choose an arbitrary point (p, q) in phase space and two more points in its neighborhood. The latter are defined by giving their displacement from (p, q), i.e., a vector $(\delta p_1, \ldots, \delta p_n, \delta q_1, \ldots, \delta q_n)$ for the first, and a vector $(\Delta p_1, \ldots, \Delta p_n, \Delta q_1, \ldots, \Delta q_n)$ for the second displacement. These two vectors span a parallelogram whose area is defined as

$$\Omega(\delta, \Delta) = \sum_{i=1}^{n} (\delta p_i \, \Delta q_i - \delta q_i \, \Delta p_i) . \tag{7.1}$$

The canonical two-form Ω is *antisymmetric,* in its two arguments, which means that $\Omega(\delta, \Delta) = -\Omega(\Delta, \delta)$.

The three points in phase space (p, q), $(p + \delta p, q + \delta q)$, and $(p + \Delta p, q + \Delta q)$ are now taken as the starting points for three different trajectories, each a solution of (2.2) with a well-defined Hamiltonian $H(p, q)$, which will remain the same throughout this chapter. Therefore, $p, q, \delta p, \delta q, \Delta p$, and Δq become functions of the time t. Since the displacements from (p, q) are assumed to be small, however, the equations of motion (2.2) can be simplified by expanding around the central trajectory $p(t), q(t)$. Therefore,

$$\begin{aligned} \frac{d\,\delta p_j}{dt} &= -\frac{\partial^2 H}{\partial q_j \partial p_k}\,\delta p_k - \frac{\partial^2 H}{\partial q_j \partial q_k}\,\delta q_k , \\ \frac{d\,\delta q_j}{dt} &= \frac{\partial^2 H}{\partial p_j \partial p_k}\,\delta p_k + \frac{\partial^2 H}{\partial p_j \partial q_k}\,\delta q_k , \end{aligned} \tag{7.2}$$

where one inserts the functions $p(t)$ and $q(t)$ into the second derivatives of H. The vector $(\Delta p, \Delta q)$ satisfies the same set of linear, first-order, ordinary differential equations. Equations (5.16) for the displacements from Hill's periodic orbit in the lunar problem are a special case of (7.2).

It is now a trivial exercise to check that

$$\frac{d\,\Omega(\delta, \Delta)}{dt} = 0 , \tag{7.3}$$

whatever the initial values of $p, q, \delta p, \delta q, \Delta p$, and Δq. In words, the value of the area Ω for the little parallelogram remains the same as the three neighboring trajectories proceed, each along its own course.

A more striking formulation of this result starts from an arbitrary, simply connected, two-dimensional surface W bounded by a curve C. Such a surface is parametrized by two real variables (u, v) with the help of functions $p(u, v)$ and $q(u, v)$ where (u, v) is limited to a simple domain. This domain is divided into small rectangles $(\delta u, \Delta v)$ by the increments δu and Δv, each of which gives rise to a pair of displacements $(\delta p, \delta q)$ and $(\Delta p, \Delta q)$. If the corresponding areas $\Omega(\delta, \Delta)$ are summed over the domain in (u, v), the surface W in phase space gets a value A for its total area.

Each combination (u, v) defines the starting point for a trajectory; there is now a two-parameter family of trajectories in terms of the functions $p(u, v, t)$ and $q(u, v, t)$. The whole surface W is thereby made to move as time t changes; we can speak of a surface $W(t)$ with an area $A(t)$. By virtue of the equation (7.3), the area A is constant, i.e., does not vary with time. This fact is expressed sometimes by saying that A is an *integral invariant* of the Hamiltonian system.

Notice that the constancy of A does not depend at all on the particular Hamiltonian that describes the system; on the contrary, the invariance of the integral A depends only on the peculiar structure of the equations of motion (2.2). The integral invariant Ω says something rather profound of a purely geometric nature concerning mechanical systems in general. Anticipating again some results to be discussed much later, the integral A is limited by quantum mechanics to simple multiples of Planck's quantum. This may be another instance where quantum mechanics lurks in the background, although neither Poincaré nor we could suspect such a relation at this point of the discussion.

7.2 Integral Invariants and Liouville's Theorem

The constancy of Ω is the generalization of a theorem of Liouville which has been known since the middle of the last century, and which plays a central role in statistical mechanics. The derivation of this theorem from (7.3) leads to further theorems that are needed for the work in the next section.

Let us consider a total of four neighboring points for (p, q), or equivalently, four vectors, the previous $(\delta p, \delta q)$ and $(\Delta p, \Delta q)$, and the new (dp, dq) and (Dp, Dq), which give the displacements from (p, q). Together they span a four-dimensional parallelepiped to which can be assigned the volume $\Omega_4(\delta, \Delta, d, D)$, which is defined as

$$\Omega(\delta, \Delta)\, \Omega(d, D) \;-\; \Omega(\delta, d)\, \Omega(\Delta, D) \;+\; \Omega(\delta, D)\, \Omega(\Delta, d)\,. \quad (7.4)$$

This combination is the only one that is antisymmetric under any exchange of two arguments, as for example $\Omega_4(\delta, \Delta, d, D) = -\,\Omega_4(d, \Delta, \delta, D)$, apart from an arbitrary constant factor.

This four-dimensional volume Ω_4 is also called a *four-form;* and since it is entirely based on the two-form Ω, it is often written as $\Omega\wedge\Omega$. The symbol $\wedge$ indicates the formation of an antisymmetric product like (7.4), also called the *exterior product* of Ω with itself.

The base point (p, q) and its four neighbors can again be taken as the initial conditions for one central and four neighboring trajectories, by solving (2.2) and (7.2). The four vectors δ, Δ, d, D are then functions of the time t, and so is the volume Ω_4. Equation (7.3) now shows that the value of Ω_4 does not depend on time, because in taking the time derivative of (7.4) each factor in the three terms is constant.

The exterior product of Ω with itself can be formed as many times as there are degrees of freedom. The antisymmetrizing operation like (7.4) yields zero when the vectors are not linearly independent, and only $2n$ vectors at most can be independent in a $2n$-dimensional space. The $2n$ – *form* Ω_{2n} is the ordinary (Cartesian) $2n$-dimensional volume of phase space.

As in the preceding section, one can now consider a simply connected piece W of 4, ... , or $2n$ dimensions in phase space. This piece W has a well-defined volume in terms of the exterior products of Ω with itself. Also, W can serve as the base for an ensemble of trajectories that move simultaneously through phase space. At each moment, this moving piece $W(t)$ has a volume that is always computed with the help of Ω and its exterior products. Therefore, *the volume of W is constant in time.*

This last statement for a $2n$-dimensional piece of phase space is Liouville's theorem. The importance of this theorem lies in its implications for the long-time behavior of the generic trajectory in a typical Hamiltonian system. The word *generic* is meant to exclude trajectories with special properties such as the periodic orbits, and the word *typical* is designed to remove special Hamiltonians from our discussion, such as the integrable ones. Non-generic trajectories have initial conditions whose measure in terms of Ω is zero; in a more vague sense, atypical Hamiltonians have near them an overwhelming number of typical ones.

In most Hamiltonian systems an arbitrarily chosen piece W of phase space with an initially simple shape gets progressively deformed into a grotesque tangle of interwoven arms and branches, while staying simply connected and keeping the same volume (cf. Figure 24). Eventually every piece of phase space will contain some part of this tangle $W(t)$, and the available volume will be distributed equally; i.e., the volume of $W(t)$ contained in some particular piece is proportional to the volume of that piece; it constitutes the same fraction everywhere. The Hamiltonian system is then called *ergodic,* which is the normal situation, although it is not true in integrable systems.

The word 'ergodic' needs an explanation: of its two Greek components, 'erg' refers to energy, and 'odos' to the trajectory. In a conservative Hamiltonian system, the energy stays constant so that each trajectory stays on a $(2n - 1)$-dimensional submanifold $H(p, q) = E$. The description of the increasing tangle in the preceding paragraph has to be restricted to such a surface of constant energy. The invariant volume on a constant energy surface, however, is the result of the more general invariant volume for the whole phase space.

When one starts with some simply connected piece of the constant energy surface $H(p, q) = E$, it is necessary to enlarge its content by including the whole slab of phase space between the energies E and $E + \varepsilon$ where ε can be an arbitrarily small increment. Liouville's theorem can only be applied to the slab. The thickness of the slab in terms of the coordinates (p, q) can vary from place to place. If one has defined some measure of $(2n - 1)$-dimensional volume on the constant energy surface (e.g., by using the $2n$ components of p and q like Cartesian coordinates), then it is essential to divide this measure by the absolute value of the gradient of H with respect to p and q. Otherwise, Liouville's theorem does not hold on the surface of constant energy.

7.3 Area Conservation on the Surface of Section

The flow of trajectories on a surface of constant energy $H(p, q) = E$ will now be studied in more detail. It would be difficult to follow all the many curves as they wind around this $(2n - 1)$-dimensional submanifold of phase space. Instead, we will concentrate on a $(2n - 2)$-dimensional submanifold of the constant energy surface, called a *surface of section* Σ, which is transverse to the flow; i.e., Σ cuts the flow, or put differently, Σ is never tangent to the flow. Apart from this requirement, the only other condition to be satisfied is that every trajectory actually does intersect Σ.

The points on this surface of section Σ can serve as initial conditions for the trajectories at the given energy E. Let us pick one such point $(p^{(0)}, q^{(0)})$ and start a trajectory there at the time $t = 0$. Unless we have chosen a bad surface of section, or a poor point on it, this trajectory will intersect Σ again at some later time t_1 in the point $(p^{(1)}, q^{(1)})$, but not at any time between 0 and t_1. In this manner, the point $(p^{(0)}, q^{(0)})$ is mapped into $(p^{(1)}, q^{(1)})$; Σ is mapped into itself. This transformation of Σ into itself gives a somewhat simplified picture of the flow of trajectories on the surface of constant energy.

The integral invariants for the flow of trajectories in phase space are now carried over into this transformation. Continuous time as the governing parameter is translated into some kind of discrete stepping mechanism. The basic geometric features of mechanics up to now were always tied to the coordinates (p, q) being continuous functions of the time t ; but when we use the surface of section, we look at each trajectory only as it crosses Σ, i.e., at the times $t = 0$, t_1, and at any of the subsequent times, t_2, t_3, and so on. To make things even more strange, the exact values for t_1, t_2, etc. depend on the initial point $(p^{(0)}, q^{(0)})$. It is very often neither desirable nor even possible to find Σ such that the times of intersection t_1, t_2, etc. are the same for all the points on Σ.

As in the previous two sections, let us look at neighboring points to $(p^{(0)}, q^{(0)})$, but this time located in Σ. The vectors $(\delta p, \delta q)$ and $(\Delta p, \Delta q)$ are, therefore, tangent to Σ. Since the neighboring points belong to the same constant energy surface and satisfy $H(p, q) = E$, we have

$$\frac{\partial H}{\partial p_k}\delta p_k + \frac{\partial H}{\partial q_k}\delta q_k = 0 , \tag{7.5}$$

with a similar equation for $(\Delta p, \Delta q)$; the derivatives of H are taken in $(p^{(0)}, q^{(0)})$. This relation can be interpreted differently if we introduce the vector $\tau = (\dot{p}, \dot{q})$, i.e., the tangent to the trajectory at $(p^{(0)}, q^{(0)})$. With the help of the equations of motion (2.2) and the definition (7.1) of the two-form Ω, one can write (7.5) as $\Omega(\delta, \tau) = 0$, and $\Omega(\Delta, \tau) = 0$. Of course, one has trivially $\Omega(\tau, \tau) = 0$.

The two-form $\Omega(\delta, \Delta)$ is now transported along the trajectory that starts in $(p^{(0)}, q^{(0)})$, until the time t_1 when this trajectory intersects Σ again. The value of $\Omega(\delta, \Delta)$ does not change; but the new vectors $(\delta p^{(1)}, \delta q^{(1)})$ and $(\Delta p^{(1)}, \Delta q^{(1)})$ are not necessarily tangent to Σ any longer, because the two neighboring points move along trajectories that differ from the one through $(p^{(0)}, q^{(0)})$, and hit Σ at a time that differs slightly from t_1. This complication can be removed, however, as will be shown in the next paragraph.

To the first order in the displacements, a change in the time t_1 adds to $(\delta p^{(1)}, \delta q^{(1)})$ a multiple of the vector τ_1, which is the tangent at time $t = t_1$ to the original trajectory through $(p^{(0)}, q^{(0)})$. If we continue the neighboring trajectory until it intersects Σ, we get a vector δ_1, which is tangent to Σ in $(p^{(1)}, q^{(1)})$, but which differs from $(\delta p^{(1)}, \delta q^{(1)})$ by a multiple of τ_1. Since $\Omega(\delta, \tau) = \Omega(\Delta, \tau) = 0$ for all times, such corrections do not change the value of $\Omega(\delta, \Delta)$ at t_1.

If one now looks at the map of Σ into itself, the vectors δ or Δ that join two neighboring points get mapped into the vectors δ_1 or Δ_1 joining the corresponding transformed points. The above arguments show that

$$\Omega(\delta, \Delta) = \Omega(\delta_1, \Delta_1) . \tag{7.6}$$

The discontinuous transformation of Σ into itself has, therefore, the same geometrical properties as the continuous map of the whole phase space as time increases. In particular, Liouville's theorem, the conservation of volume in Σ, follows by the same reasoning as in the preceding section.

7.4 The Theorem of Darboux

The surface of section is especially useful in systems with two degrees of freedom, because Σ then has two dimensions, and the map of Σ into itself can be visualized rather easily. Also the choice of Σ is often quite straightforward; e.g., if we start with the pairs of canonically conjugate coordinates (p_1, q_1) and (p_2, q_2), we define Σ as the submanifold $q_1 = 0$ in the energy surface $H(p, q) = E$. The second pair (p_2, q_2) is then the natural coordinate system in Σ, and the invariant element of area is simply $dp_2 dq_2$.

While almost all the examples in this book have no more than two degrees of freedom, and the above remarks are sufficient for their discussion, one would like eventually to advance into the vast arena of Hamiltonian systems with three degrees of freedom. The surface of section Σ then has four dimensions, however, and only very few pioneers (e.g., Froeschlé 1970) have had the courage to penetrate into this jungle. For example, the restricted three-body problem where the two primary bodies, such as the Sun and the Earth for the Moon, or the Sun and Jupiter for the asteroids, move in a fixed circle, has three degrees of freedom, and requires a four-dimensional surface of section. The neighborhood of a periodic orbit was already studied in Section 6.3 with the help of this more general setting.

Among the questions which arise is the following: Is it always possible to make Σ part of a canonical coordinate system such that the last two pairs (p_2, q_2) and (p_3, q_3) are the coordinates for Σ, while $p_1 = E$ represents the energy, and $q_1 = t$ is the time, at least in the neighborhood of Σ ? The general answer to this question is affirmative, and the construction of the required canonical coordinates is based on a theorem by Darboux. The whole discussion will be carried out for a system with three degrees of freedom because all the essential elements of the argument are present there already.

The original system is described by the usual canonical coordinates $(p_1, p_2, p_3, q_1, q_2, q_3)$ in phase space with the Hamiltonian $H(p, q)$. The surface of section Σ is defined in terms of four variables $z_1, \ldots, z_4$; the coordinates (p, q) are known functions of the z's. The map of Σ into itself leaves the two-form Ω invariant, if one restricts the displacements

δ and Δ to be tangent to Σ. Varying any one of the z's generates such a displacement $\delta = (\delta p, \delta q)$. The two-form Ω that is originally defined for any displacements in phase space, becomes thereby a two-form Ω_Σ which is restricted to the displacements in Σ.

In terms of the variables z, the two-form Ω_Σ does not look as simple and concise as the definition of Ω in terms of p and q in (7.1). But the transition from Ω to Ω_Σ is easy to work out; the displacements δp, δq, Δp, and Δq in (7.1) have to be written in terms of δz and Δz. The result can be expressed with the help of the *Lagrange bracket* with respect to the two variables z and ζ,

$$\{z, \zeta\} = \sum_{i=1}^{i=3} \left(\frac{\partial p_i}{\partial z} \frac{\partial q_i}{\partial \zeta} - \frac{\partial p_i}{\partial \zeta} \frac{\partial q_i}{\partial z} \right) . \tag{7.7}$$

The resulting formula is

$$\Omega_\Sigma(\delta, \Delta) = \frac{1}{2} \sum_{k,\ell=1}^{4} \Phi_{k\ell} \, (\delta z_k \, \Delta z_\ell - \delta z_\ell \, \Delta z_k) . \tag{7.8}$$

The coefficients in this two-form, $\Phi_{k\ell} = \{z_k, z_\ell\}$, are functions of the parameters $z_1, \dots, z_4$, and are assumed to be known in Σ. These functions have the evident antisymmetry $\Phi_{k\ell} = -\Phi_{\ell k}$.

The question of finding the canonical coordinates in Σ now becomes: Are there functions f_2, f_3, g_2, g_3 of $z_1, \dots, z_4$ such that

$$\Omega_\Sigma(\delta, \Delta) = \delta f_2 \Delta g_2 - \delta g_2 \Delta f_2 + \delta f_3 \Delta g_3 - \delta g_3 \Delta f_3 , \tag{7.9}$$

where δf_2, Δf_2 etc. have to be expressed in terms of δz_1, Δz_1 and so on ?

The answer comes in the *theorem of Darboux:* Given any two-form (7.8) in Σ, the necessary and sufficient conditions for finding new coordinates $f_2, \dots, g_3$ such that (7.9) holds, are the equations

$$\varepsilon_{ijk\ell} \frac{\partial \Phi_{k\ell}}{\partial z_j} = 0 , \tag{7.10}$$

where the summation over the indices j, k, and ℓ from 1 to 4 is implied as usual by the Einstein convention concerning indices that occur twice in a product. The Kronecker symbol $\varepsilon_{ijk\ell}$ differs from 0 only when the four indices are all different; $\varepsilon = 1$ when $(i\,j\,k\,\ell)$ is an even permutation of (1234), $\varepsilon = -1$ when $(i\,j\,k\,\ell)$ is an odd permutation of (1234). It is a straightforward exercise in differentiation to show that the Lagrange brackets $\{z_k, z_\ell\}$ satisfy (7.10).

The necessity of (7.10) for the required existence of the new canonical coordinate pairs (f_2, g_2) and (f_3, g_3) is easy to prove because (7.10) is satisfied in the new coordinates as follows from the right-hand side of (7.9), and a straightforward calculation shows that (7.10) must hold in all coordinate systems if it holds in one of them. The difficulty with Darboux's theorem lies in demonstrating the sufficiency of (7.10). We shall not attempt even to give a glimpse of how this part of the proof can be carried out; but before sending the reader to the relevant sources, a few cautionary remarks seem appropriate.

The two modern references, Abraham and Marsden (1978) and Arnold (1978), rely completely on their discussion of *symplectic geometry,* which takes up dozens of pages before coming to Darboux's theorem. To work through this whole build-up of preliminary material is made difficult at best by the use of modern, mathematical terminology; at the end, it is almost impossible for the physicist to tell where exactly the new coordinate system was actually shown to exist, rather than just being defined and manipulated. The original paper by Darboux (1882) is quite readable, and gets right away into the main business of showing which ordinary differential equations have to be solved in order to find the new coordinates. It is written for a mathematical audience of the time, however, not for physicists, and ends up being very long because many different cases of similar nature are treated without regard to their relevance in mechanics.

Since mechanics is a branch of physics, the conditions (7.10) cannot be left standing without mentioning another area where they play a central role, *Maxwell's equations* of electromagnetism. The four variables z now describe our everyday space-time continuum, and the antisymmetric functions $\Phi_{k\ell}$ are the components of the magnetic and the electric fields in the notation of special relativity, $B_1 = \Phi_{23}$, $B_2 = \Phi_{31}$, $B_3 = \Phi_{12}$, $F_1 = \Phi_{14}$, $F_2 = \Phi_{24}$, $F_3 = \Phi_{34}$. The conditions (7.10) are the one half of Maxwell's equations which does not involve the electric charges and currents. They proclaim the absence of magnetic monopoles, and Faraday's laws of magnetic induction; they guarantee the existence of a four-component vector potential A from which the magnetic and the electric fields are obtained by calculating the curl in four dimensions. Although the theorem of Darboux is based on the same conditions, its conclusion in the form of the canonical coordinates f and g is quite different from their use in electromagnetism.

7.5 The Conjugation of Time and Energy in Phase Space

Now that the surface of section Σ has been endowed with a system of canonically conjugate coordinates (f, g), one would like to complete the job by adding one more pair, at least in the neighborhood of Σ, calling it (f_1, g_1). The two most obvious candidates are the energy E for f_1, and the time t for g_1. The success of this endeavor depends on calculating the relevant two-forms (7.1) between E, t, and the coordinates in Σ. The part concerned with t is easy, but E has to be treated with some care.

Since each point on Σ is the starting point for a trajectory, we will use the simple expedient of defining $g_1 = 0$ on Σ, and setting $g_1 = t$ along each trajectory since leaving Σ. We can write δt instead of δg_1; the displacement $(\delta p, \delta q)$ corresponding to δt is given by the vector $(\dot{p}, \dot{q})\delta t$, or in view of (2.2) by $(-\partial H/\partial q, \partial H/\partial p)\delta t$ which will be called δ for simplicity's sake. We can now calculate its two-form Ω according to (7.1) together with an arbitrary vector Δ in Σ. The result is the product $\delta t\,[\Delta p(\partial H/\partial p) + \Delta q(\partial H/\partial q)]$. Since Σ belongs to a surface of constant energy, this change of H with the displacement Δ vanishes, and $\Omega(\delta, \Delta) = 0$.

The surface of constant energy $H(p, q) = E$ is now endowed with the coordinates f_2, f_3, g_1, g_2, g_3, sufficiently close to Σ. A change in any one of these five coordinates leads to well-defined displacements inside Σ and along the trajectories of energy E. The sixth and last coordinate in phase space has the job of changing the value of $H(p, q)$ from E to $E + \delta E$. This new coordinate f_1 will be chosen such as to have a vanishing value of Ω with any displacement Δ inside Σ, whereas $\Omega(\delta f_1, \Delta g_1) = \delta E\,\Delta t$. The only new kind of displacement in phase space is in the direction $\theta = (\partial H/\partial p, \partial H/\partial q)$. The displacemnt corresponding to a change in f_1 is, therefore, a linear combination of θ with the five independent deviations inside the constant energy surface, so as to produce the correct values for Ω.

The details are tedious to work out, and the end result is not of interest in itself. Of the six coefficients in the linear combination, the one that goes with δg_1, i.e., with a vector in the direction $(-\partial H/\partial q, \partial H/\partial p)$, remains arbitrary; the antisymmetry (7.2) of Ω allows only for five independent conditions, whereas six would be required for a unique determination.

This ambiguity is natural as can be seen even in the original coordinates (p, q): the two-form (7.1) does not change, for example, if one replaces p_2 by $p_2 + q_2$ while leaving all the other coordinates as before. What does change, is the one-form $p\,\delta q$, and that would affect the value of the action integral S in (2.3). When the integral of the one-form is taken over a closed loop, however, for example, a periodic orbit, the ambiguity disappears.

It is important to emphasize that the addition of $f_1 = E$ and $g_1 = t$, as canonically conjugate variables to the coordinates in Σ, is generally feasible only in the neighborhood of Σ in phase space. As t increases, $\Sigma(t)$ sweeps out the energy surface, and some parts of $\Sigma(t)$ will eventually intersect $\Sigma(0)$; but at no time will $\Sigma(t)$ and $\Sigma(0)$ fully coincide.

This complication was already discussed in Section 7.3; it is all the more remarkable that the transformation of Σ into itself, after correcting for the different return times, still conserves the two- form Ω. The special coordinates (E, t) in the neighborhood of a particular trajectory played an important role in calculating the determinant $D(q'' q' E)$ in Section 2.4.

The construction in this section should be sharply distinguished from another one which also presents E and t as an additional pair of conjugate coordinates. The dimension of phase space now gets increased from $2n$ to $2(n + 1)$ where n is always the number of degrees of freedom. This idea is already present in the expressions (1.8) and (1.9) for the variation of the Lagrange action R, where the one-form $p\,\delta q$ gets the extra term $E\,\delta t$. A similar enlargement of phase space was used in Section 5.2 where the variable ℓ', the mean motion of the Sun acting as a time-dependent perturbation on the Moon-Earth-Sun system, was considered to be a new angle variable k with a corresponding new action K to complete the pair. The resulting enlarged system was thereby made conservative.

Including (energy, time) as a new pair in addition to (momentum, position) is, of course, essential in *special relativity,* and leads again to a conserved Hamiltonian system even though there might be time-dependent forces. As an example, let us consider a charged particle in an electromagnetic field, which is given by the *vector potential* (A_0, A_1, A_2, A_3) where each component is a given function of the time variable $x_0 = ct$, and the Cartesian coordinates x_1, x_2, x_3; c is the velocity of light. In order to find equations of motion that look like (2.2), one needs a conjugate variable p_0 for x_0, a new time-like variable τ, and a Hamiltonian function $M(p_0, \ldots, p_3, x_0, \ldots, x_3)$. The equations of motion have to yield the movement of an electric charge e in the *combined electric and magnetic fields* of the given vector-potential A, equivalent to Newton's equation of motion for a particle of rest-mass m_0 subject to the *Lorentz force* as in equations (18.1) and (18.2). If these terms are not understood at this point, the reader is requested to come back to this place after looking at Chapter 18 where the hydrogen atom in a magnetic field is discussed at length.

The most convincing form of the Hamiltonian M is

$$M = \left(p_0 - \frac{e}{c}A_0\right)^2 - \sum_{i=1}^{3}\left(p_i - \frac{e}{c}A_i\right)^2 , \qquad (7.11)$$

with a new variable τ to play the role of time. If this expression is used in (2.2) with M replacing H and x replacing q, and the result is compared with the usual equations of motion for an electric charge, such as (18.1), the following interpretation emerges: Since the system is conservative, i.e., M does not depend on τ, the value of M remains constant and its value is $m_0^2 c^2$. The variable τ is the proper time of the particle, i.e., the ordinary time measured in the instantaneous rest-frame, and $p_0 = E/c$ (cf. Garrod 1968).

In this form, the rest-mass m_0 stays constant because of the equations of motion in an electromagnetic field. It may change only if there are internal degrees of freedom; these work always in the proper time τ which is the natural time-like variable in the Hamiltonian M. As an example, one can think of *radioactive decay;* e.g., the 'heavy electron', alias μ-meson, decays into an ordinary electron with a probability which decreases with its speed, because the speed is measured in the time-frame of the laboratory while the internal clock always runs more slowly, as in the famous *twin paradox* of special relativity.

The end of this chapter may be the appropriate place to mention a bit of particle physics that has a direct bearing on the question of chaos in simple physical systems. The theoreticians of high-energy physics have arrived at a consensus in recent years, whereby one or the other of the *non-Abelian gauge fields* are able to explain the bewildering variety of elementary particles. These theories were first proposed by Yang and Mills in 1954 to generalize Maxwell's theory of electromagnetism which is an Abelian gauge field. But the new theories lead to non-linear field equations, although they are still amenable to the techniques of perturbation theory.

Unfortunately, the effective coupling constants are outside the domain where the perturbation expansion can be expected to converge, even if only asymptotically. A brute-force numerical assault on the problem, lattice gauge theory, seems to be the only way to extract quantitative results at this time; the work has been in progress for some years. It seems almost inevitable that chaotic behavior is present in these systems, and that no real understanding will be achieved until this aspect of the problem is taken into account.

The first steps in this direction have been made by Savvidy (1982 and 1983) together with Martinyan and Prochorenko (1988). They have studied the classical non-Abelian field equations in the simple approximation where they are constant in space and depend only on time. With three non-vanishing independent components of the field, they obtain a rather ordinary-looking Hamiltonian system with three degrees of freedom and a polynomial interaction. It can be studied by the standard methods involving the surface of section (cf. the next chapter), and is found to be chaotic.

CHAPTER 8

Models of the Galaxy and of Small Molecules

This chapter deals with some special models that were first discussed by the astronomers who tried to understand the motion of stars in the gravitational field of the Galaxy. The pioneering work was done by George Contopoulos (1960) and some of his students; they were the first to realize that the stellar trajectories in the gravitational potential of a typical galaxy can be either integrable or chaotic depending on their initial conditions. That discovery was crucial in understanding the observed velocity distribution of stars in our solar neighborhood.

The work of Michel Hénon and Carl Heiles from the early 1960s is based on a special case which finally brought the idea of chaos home to the physicists. As astronomers, these two authors, like their predecessors, were interested in understanding our galaxy, but they chose a particular mathematical model for ease of computation while keeping the essential features of the galactic environment. Their model has become the testing ground for various general methods in the study of chaotic dynamical systems. We shall discuss three of these: The Birkhoff-Gustavson normalization, the analysis of singularities in the complex-time plane, and the study of discrete algebraic transformations. The chemists have adopted the same model to describe the motion of the nuclei in a small molecule as shown in the spectrum of molecular vibrations as well as in the transformation of molecules and their reactions (cf. Brumer 1981).

8.1 Stellar Trajectories in the Galaxy

Poincaré reviewed the purely analytical approach to celestial mechanics in the nineteenth century, and came to very disturbing conclusions: the motion along conic sections, or even the more general movements in the neighborhood of a periodic orbit, could no longer serve as a starting point in many situations. It became necessary to face a great proliferation of periodic orbits, and to cope with trajectories which defied any kind of simple repetitive pattern.

The only way to gain new insight was to carry out numerical calculations. This task was first taken up by Sir George Darwin (son of Charles who had written *The Origin of Species* and *The Descent of Man*) and F. W. Moulton, before World War I, and was then continued primarily by Strömgren and his school in Copenhagen during the 1920s and 1930s. This work and much of what followed is surveyed in the *Theory of Orbits* by Szebehely (1967). Together with the work on stellar trajectories in galaxies, this field, now called *dynamical astronomy,* is the modern version of celestial mechanics; Contopoulos (1979) reviewed it for the benefit of the quantum-oriented physicists.

Since the gravitational force between stars decreases as the inverse square of the distance, stars can never really get away from one another; but they don't hit one another either. Quite in contrast, the molecules in a gas move freely most of the time when outside the short range of their interaction; but they get close enough to one another with reasonable frequency, so as to change their direction of motion very drastically in a collision between only two of them. Stars move in the integrated potential of the whole galaxy, and any close encounters are prevented by the very long-range nature of their mutual attraction. This situation is also found in a plasma, i.e., a fluid whose particles are electrically charged rather than neutral as in ordinary gas and fluids.

The manner in which mass is spread through the Galaxy can be inferred from the observed distribution of stars in the sky. The gravitational potential $V(x, y, z)$ is then found from solving *Poisson's equation*

$$\frac{\partial^2 V}{\partial x^2} + \frac{\partial^2 V}{\partial y^2} + \frac{\partial^2 V}{\partial z^2} = -4\pi G \mu(x, y, z) , \tag{8.1}$$

where μ is the mass density and G is the gravitational constant. Each star moves as if alone in the fixed potential V (cf. Ollongren 1965).

The exact shape of μ is not known; but it corresponds to a flat disk with known radius in which the stars are fairly evenly distributed. An important feature is the *cylindrical symmetry,* the fact that μ depends only on $\rho = (x^2 + y^2)^{1/2}$ and z. Therefore, also V depends only on ρ and z, and as a consequence the angular momentum M of a star with respect to the z-axis, through the center of the Galaxy and at right angles to its plane, is a constant of motion.

The equations of motion are thereby reduced to two degrees of freedom with the canonical pairs (p_ρ, ρ) and (p_z, z), and the Hamiltonian

$$H(p_\rho, p_z, \rho, z) = \frac{1}{2}(p_\rho^2 + p_z^2) + V(\rho, z) + \frac{M^2}{2\rho^2} . \tag{8.2}$$

Notice that the mass of the star does not appear in this Hamiltonian because it would multiply both the kinetic and the potential energy, and was divided out. The momenta p_ρ and p_z are, therefore, reduced to the corresponding velocities, and M is the areal velocity of Kepler's second law.

The nature of the stellar trajectories in the galactic potential V can be found by observing their velocities in the solar neighborhood. According to the principles of statistical mechanics, the distribution in phase space can depend only on the constants of motion. If there aren't any besides the Hamiltonian (8.2), and the angular momentum M, the stellar velocities in the meridian plane (ρ, z) have no preferred direction, contrary to experience. There is then bound to be a constant of motion F in addition to the energy H and the angular momentum M. Since it is not clear what it should be in terms of the variables p_ρ, p_z, M, ρ, z, the astronomers called it the *third integral* without knowing exactly its origin.

The problem of the third integral was approached systematically in the early 1960s, especially by Contopoulos and his students, using both analytical and numerical tools. A collection of the relevant work can be found in the volume edited by Contopoulos (1966). Many of the results are directly applicable to the study of such disparate fields as the design of particle accelerators, plasma physics, atoms in strong magnetic fields, and molecular vibrations. The work of Hénon and Heiles (1964) gives the most economical account of this large body of research.

8.2 The Hénon-Heiles Potential

The potential W in the preceding section has a basin or pit where most of the stars in the galaxy are trapped. A high barrier due to the centrifugal potential $M^2/2\rho^2$ keeps the stars away from the center of the galaxy; but since the gravitational attraction of the whole galaxy vanishes at large distances, stars with sufficient energy can escape into intergalactic space; a mathematical model for the combined potential should have these features. The cylindrical geometry, however, is not important any longer so that one might just as well replace (ρ, z) by the Cartesian (x, y).

The model potential $U(x, y)$ as a function of the position coordinates (x, y) should also be easy to evaluate, either analytically in a proof, or numerically in a computation. All these requirements leave us with a polynomial of third degree. Hénon and Heiles chose a potential whose value is constant along the sides of an equilateral triangle, on the straight lines, $x = (y - 1)/\sqrt{3}$, $x = -(y - 1)/\sqrt{3}$, and $y = -1/2$, so that $U(x, y)$ is given by

$$(y + \frac{1}{2})(x^2 - \frac{(y-1)^2}{3}) = x^2 y - \frac{y^3}{3} + \frac{x^2 + y^2}{2} - \frac{1}{6} . \quad (8.3)$$

The constant $-1/6$ is usually left out, so that the local minimum of U at the origin has the value 0, while $U = 1/6$ along the sides of the equilateral triangle. There are steep mountains beyond the three sides; but the vertices of the triangle are mountain passes on whose far side the potential decreases rapidly, and goes to $-\infty$ eventually, allowing the particle to escape.

If the results of Hénon and Heiles are applied to a real physical situation, the starting Hamiltonian would be

$$H = \frac{u^2 + v^2}{2m} + m\omega^2 \frac{x^2 + y^2}{2} + \lambda(x^2 y - \frac{y^3}{3}) . \quad (8.4)$$

Notice the three physical constants: the mass m, the frequency ω, and the length $a = m\omega^2/\lambda$ which make of this Hamiltonian a complete model for a non-linear system. The value of the parameter λ does not indicate the strength of the non-linear coupling between the motions in x and in y, but sets the length scale a. As pointed out in Section 3.6, however, Planck's quantum h has now a non-trivial numerical value in such a system if the three constants, m, ω, and a, are used as physical units.

The Hénon-Heiles system shows a remarkable similarity to the Toda lattice of Section 3.6. The simplest case is a tri-atomic molecule in one dimension where $n = 3$ in (3.12), and the lattice is periodic so that $q_4 = q_1$. Its configuration is decomposed into the center-of-mass coordinate $z = (q_1 + q_2 + q_3)/3$, and the two internal vibrational coordinates $x = (2q_1 - q_2 - q_3)/2\sqrt{3}$ and $y = (q_3 - q_1)/2$. The potential energy does not depend on z, so that the center-of-mass momentum becomes a constant of motion whose value has no effect on the internal degrees of freedom x and y. The kinetic energy is the sum of the squares of the momenta corresponding to x, y, z.

The three-particle Toda lattice is now explicitly reduced to two degrees of freedom, exactly as it was tested by Ford and coworkers (cf. Section 3.6). If one thinks of small-amplitude vibrations in a molecule, it is reasonable to expand the exponential functions in the potential energy of (3.12) in powers of x and y, and to stop the expansion at the first term which goes beyond the quadratic. The resulting Hamiltonian is the same as (8.3), which is the reason why the Toda lattice, coming after the work of Hénon and Heiles, was not expected to be integrable, but rather chaotic at moderate energies. The non-integrability of the truncated Toda-lattice was proven by Yoshida, Ramani, and Grammaticos (1988).

8.3 Numerical Investigations

It would be difficult to get even an approximate idea of the dynamical properties of the Hamiltonian (8.4) without looking first at many trajectories with the help of the surface of section. The symmetry of the equilateral triangle suggests the vertical $x = 0$ for this purpose. The equations of motion with the normalization $m = 1$, $\omega = 1$, and $\lambda = 1$ can be integrated with any simple routine such as Runge-Kutta to a sufficient accuracy and for many traversals.

An energy E is chosen, and a number of trajectories are computed. The points of the traversal are plotted in the domain $D(E)$ of the surface of section which is defined by the inequality $v^2 + 2U(0, y) < 2E$. Each trajectory is started at some arbitrary initial point $P_0 = (v_0, y_0)$ with $x = 0$ and $u > 0$ at the fixed energy E. The subsequent traversals $P_1 = (v_1, y_1), P_2, \ldots$ of $x = 0$ with $u > 0$ are obtained by numerical integration. Either they line up on an apparently smooth curve, or they scatter wildly throughout a portion of the domain $D(E)$.

If the traversals line up nicely, the corresponding trajectory lies on a torus in the energy surface; if the traversals cannot be accommodated easily on a smooth curve, there is no invariant torus in phase space to contain the trajectory. It would seem at first that the criterion for the existence of a smooth curve through a finite number of points in a plane is entirely a matter of individual taste. But in practice there is no doubt about which of the situations applies to any particular trajectory. The dichotomy seems always amazingly clear; nobody quibbles with the conclusions of Hénon and Heiles as drawn from the three Figures 12a, 13a, and 14a.

Figure 12a was calculated for $E = 1/12$; points belonging to the same trajectory are connected by a smooth line, drawn by hand as it were. Every trajectory at that energy seems to have a torus, although the nesting of all these tori is not trivial. Several isolated points in the surface of section are the centers of concentric closed loops that define a basin or a mountain as in a topographic map. Between these mountains and basins, however, are separating lines with self-intersections; they correspond in an ordinary pendulum to the motion that separates the libration from the rotation around the point of suspension. The approach to the point of self-intersection takes a very long time, or equivalently, very many traversals with the surface of section. The conclusion from Figure 12a is clearly that the system is integrable at $E = 1/12$.

Figure 13a is computed for $E = 1/8$ which is still well below the energy where escape from the inside of the equilateral triangle is possible. Yet, a large part of the surface of section is clearly ergodic; amazingly, the points that scatter all over a portion of the domain $D(E)$ come from a single trajectory. This ergodic region in $D(1/8)$ coincides to some extent with the portions of $D(1/12)$ where the basins and mountains are meeting. The remaining concentric closed loops in $D(1/8)$, generally called *islands,* correspond quite closely to the nested loops in $D(1/12)$.

Part of the energy surface $E = 1/8$ is obviously covered with invariant tori, whereas the remainder is not. The boundary seems rather sharp, or even smooth, although that is probably an illusion to be checked by more elaborate calculations. Nevertheless, the area of the ergodic portion in $D(1/8)$ is quite well defined, and can be assigned a numerical value by counting little squares in the figure.

Figure 14a presents the surface of section for the escape energy $E = 1/6$; but only the point $y = 1, \dot{y} = 0$ could lead to the system actually leaving the equilateral triangle. It is highly unlikely for a particular trajectory to go through that unique point. Almost all of the domain shows ergodic behavior; the exception consists of some tiny loops near the centers of the earlier structures of nested tori. These small islands can easily evade the notice of the investigator, because it is again unlikely that the correct initial condition be found in a random search. The element of area can serve as a probability measure for finding a particular island.

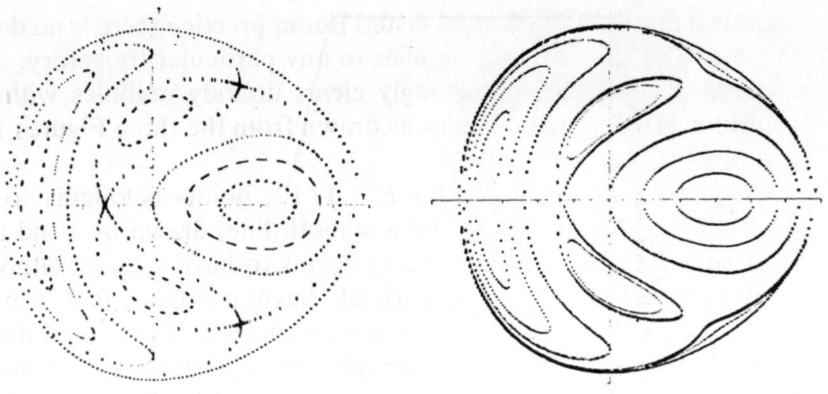

(a) (b)

Figures 12a and 12b Surfaces of section for the Hénon-Heiles potential at the energy 1/12, from numerical integration (a), and from Birkhoff-Gustavson renormalization (b) [from Gustavson (1966)].

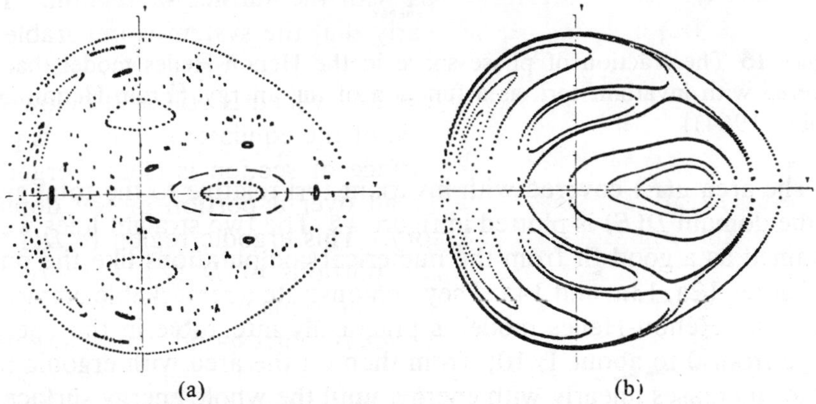

(a) (b)

Figures 13a and 13b Same as preceding figure for the energy 1/8.

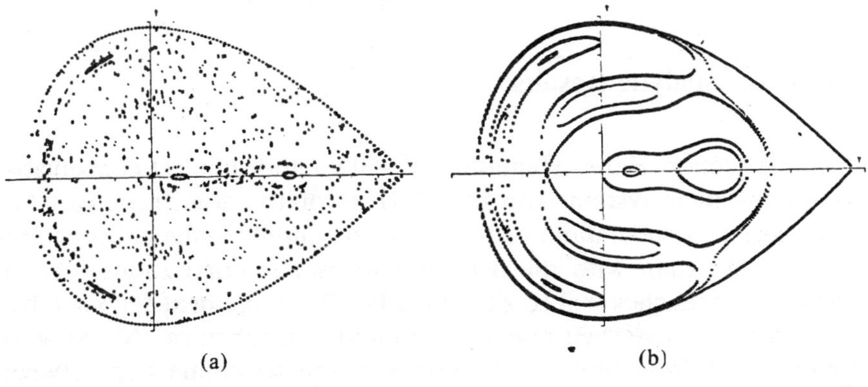

(a) (b)

Figures 14a and 14b Same as preceding two figures for the escape energy 1/6.

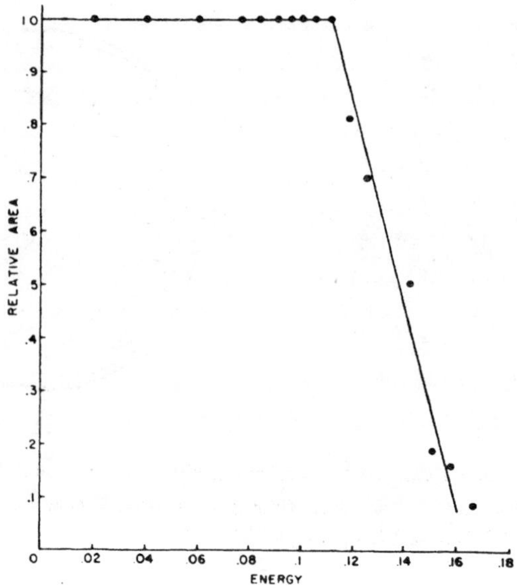

Figure 15 The fraction of phase space in the Hénon-Heiles model that is covered with invariant tori, as a function of the energy. [From Hénon and Heiles (1964)]

The area $\mu(E)$ covered with invariant tori relative to the total area of the domain $D(E)$ is plotted in figure 15. The two straight lines were obtained as a good fit from the numerical computations like the ones in Figures 12a, 13a, and 14a. They demonstrate a remarkable empirical fact: The Hénon-Heiles model is practically integrable in the energy range from 0 to about $1/10$; from then on the area with ergodic behavior increases linearly with energy, until the whole energy surface is ergodic at $E = 1/6$.

8.4 Some Analytic Results

Since the Hénon-Heiles Hamiltonian (8.4) is the simplest conceivable for a non-linear system with two degrees of freedom, no effort has been spared to understand its properties by purely analytical considerations rather than with the help of numerical computations. Three different approaches will be discussed briefly in the next sections, because they all have their analogs in quantum mechanics. Much work remains to be done, however, to establish a better connection between the classical and the quantal domain even in the limited confines of the Hénon-Heiles model.

The present section shows how a systematic perturbation theory has been devised by Gustavson (1966) on the basis of the earlier work by Birkhoff (1927). In the next section, we will present the results of an analysis by Chang, Tabor, and Weiss (1982), who examined for which choice of parameters a Hamiltonian like (8.4) becomes integrable. A third device was first invented by Hénon (1969) and has since become the basis for a large industry, both analytical and numerical. The ordinary differential equations of classical mechanics combined with the use of the Poincaré surface of section are replaced by a discrete, but still area-preserving map.

Birkhoff's method is designed to examine the motion of a dynamical system in the neighborhood of a point of stable equilibrium, like the center of the equilateral triangle in the Hénon-Heiles potential, or the equilibrium configuration of a large molecule. The Hamiltonian is expanded in the momentum and position coordinates; the lowest terms are quadratic, and positive definite since otherwise the equilibrium is unstable; they are diagonalized by a linear canonical transformation, and reduced to

$$H_2 = \sum_{j=1}^{n} \frac{\omega_j}{2} (p_j^2 + q_j^2) \ . \tag{8.5}$$

The stable equilibrium reduces in lowest order to a set of oscillators with the frequencies ω_j. The awkward form of the Hamiltonian results from the usual $(p^2 + m^2\omega^2 q^2)/2m$ by the canonical transformation $p \to p/(m\omega)^{1/2}$ and $q \to q(m\omega)^{1/2}$. Both p^2 and q^2 have the same dimension as Planck's quantum, and will eventually become multiples of it.

A sequence of canonical transformations (cf. Section 2.2) is now carried out with the help of the generating functions $W_3, \ldots, W_N$; the new Hamiltonian is expanded in a power series with respect to the new corrdinates p and q. The lowest term is kept exactly like (8.5), whereas the higher-order terms, up to a power N, are made to depend only on the combinations $\rho_j = p_j^2 + q_j^2$. The system is then called normalized to the order N. The remaining higher-order terms beyond N are assumed to be of no importance for understanding the properties of the dynamical system.

The paper by Gustavson in *The Astronomical Journal* (1966) gives a good explanation of the necessary algebraic manipulations. The principle is not difficult, although some of the details are tricky. The method will be further explained in Section 14.3.

The normalization procedure, and particularly the existence of additional constants of motion, depends on the numerical relations between the frequencies ω_j. If there exist no integers k_j such that the scalar product $(k, \omega) = 0$, the dynamical system has the n constants of motion $\rho_j = p_j^2 + q_j^2$ in terms of the new coordinates. If the condition $(k, \omega) = 0$ can be satisfied, however, and if, moreover, this can be done by ℓ independent vectors k, then there are $n - \ell$ independent combinations of the ρ_j that are constants of motion, in addition to the Hamiltonian itself.

Since $\omega_1 - \omega_2 = 0$ for the Hamiltonian (8.4), there is one more integral of motion F besides the Hamiltonian. F has only quadratic terms in the normalized coordinates; it can be written as a power series expansion in the original coordinates by inverting the canonical transformations generated by $W_3, ..., W_N$. The normalization and the construction of F was carried out by Gustavson for the Hénon-Heiles Hamiltonian (8.4), up to and including the eighth order in the u, v, x, y. The coefficients for the expansion of the normalized Hamiltonian, for the generating functions, and for the additional constant of motion F in the original coordinates are listed. If the Hénon-Heiles model were integrable, the expansion for F could be expected to converge; but a look at the coefficients says otherwise; several of them in eighth order are between 10 and 100, whereas in fourth order the largest coefficient is 5/3, and all the others are below 1 in absolute value.

Even when all the terms above the eighth are truncated, however, the expression for F gives a good account of the surface of section $x = 0$ wherever there are invariant tori. The Figures 12b, 13b, and 14b give the level lines, $F(u, v, x = 0, y) =$ constant, in the (v, y) plane with the energy held at $E = 1/12$, $1/8$, and$1/6$ as in the Figures 12a, 13a, and 14a. The islands are in good correspondence; but nothing in the new figures yields any indication of the chaotic regions in the earlier ones. Remarkably, as will be dicussed later, the quantal Hénon-Heiles model suggests that the formal constant of motion F has still some validity in the classically chaotic regions of phase space.

The Birkhoff-Gustavson construction has been applied to other problems where it yields similar results, e.g., the hydrogen atom in a magnetic field (cf. Section 18.2). Like all the other schemes for removing perturbing terms from the Hamiltonian with the help of canonical transformations, the task is accomplished by an averaging process that necessarily leads to the infamous small denominators. Deprit and coworkers (1969) have made the method particularly transparent; their work also casts the whole procedure into a recursive algorithm that is free of the coefficient-matching of Birkhoff and Gustavson.

8.5 Searching for Integrability with Kowalevskaya and Painlevé

The numerical evidence of the Hénon-Heiles calculations clearly indicates that the Hamiltonian (8.4) does not have a global third integral. Could one have known this in advance? Or since we know the final outcome of such an investigation is it possible to look at Hamiltonians similar to (8.4), and single out the integrable ones? This question is completely analogous to the question of Sofia Kowalevskaya concerning the motion of a *gyroscope* held outside its center of mass, and subject to the Earth's gravitational field (cf. Section 4.1). Her method was applied to Hamiltonians of the Hénon-Heiles type by Chang, Tabor, and Weiss (1982). This section presents the bare outline, so as to give at least a hint of the ideas involved in this type of analysis.

The main idea is to investigate the equations of motion in the complex domain, i.e., to allow the components of the momentum and of the position as well as the time to have complex values, and then to find out where the singularities of a typical solution are located in the complex-time plane. As a general rule, integrability requires an exceptionally simple structure of singularities; the criterion is the so-called *Painlevé property:* only isolated poles of bounded order. The reader should be forewarned, however, not to expect any clear-cut and sweeping theorems.

Let us start with the equations of motion in their Hamiltonian form (2.2). If the trajectory starts in (p_0, q_0) at the time t_0, it is natural to expand the right-hand sides as power series in the components of $p - p_0, q - q_0$, as well as $t - t_0$, and to assume that each power series converges within a circle of non-vanishing radius in the complex plane. The solution is then uniquely determined as a convergent power series with respect to $(t - t_0)$ inside the intersection of all these circles of convergence.

The trajectory is now defined in the complex-time plane, and can be continued analytically until it collides with one or more of the singularities. In order of increasing seriousness, they can be poles, algebraic branch points like a square root, transcendental branch points like a logarithm, or finally an essential singularity such as $\exp(-1/t)$ whose expansion in powers of t has negative powers going to $-\infty$. The location of some of these singularities may depend on the initial values p_0, q_0, t_0 of the trajectory; if so, they are called *movable singularities,* and our main task is to find where and of what kind they are, and whether they have the Painlevé property.

The most general potential with second- and third-order terms, first investigated by Contopoulos, leads to the Hamiltonian,

$$H = \frac{1}{2}(u^2 + v^2 + \mu x^2 + y^2) + \lambda x^2 y - \frac{y^3}{3}, \qquad (8.6)$$

where the units of mass, length, and time have already been chosen so as to leave only the essential parameters λ and μ. The equations of motion are used in the Newtonian form,

$$\ddot{x} = -\mu x - 2\lambda xy, \qquad (8.7a)$$

$$\ddot{y} = -y - \lambda x^2 + y^2. \qquad (8.7b)$$

The next step is elementary, but it has to be carried out with great care. Suppose that our trajectory has a singularity at the complex time t_0. We try to find the expansion for x and y in powers of $\tau = t - t_0$. The most singular, i.e., the algebraically smallest, powers of τ are written as

$$x = \tau^\alpha(a + c\,\tau^\gamma), \quad y = \tau^\beta(b + d\,\tau^\gamma), \qquad (8.8)$$

where $\gamma > 0$. The a and b terms in (8.8) are called the *leading terms* because they are the most singular, while the c and d terms are called the *resonance terms* for somewhat obscure reasons.

First only the leading terms are inserted into (8.7), and the most singular contributions are matched on either side of the = sign. The linear terms on the right-hand sides of (8.7) cannot compete with the non-linear ones. Equation (8.7a) gives $\beta = -2$, but (8.7b) permits two cases. Case 1 is obviously $\alpha = \beta$, while case 2 allows $\alpha > \beta$. Matching the coefficients in front of the powers of τ leads in case 1 to the conditions $\lambda b = -3$, $\lambda a^2 = b^2 - 6b$, whereas in case 2 one has $\alpha(\alpha - 1) = -2\lambda b$, $b = \beta(\beta - 1) = 6$ with a remaining undetermined.

At this point one could pursue each case separately in order to find the power series expansions for x and y starting with the leading terms. If we want to have only poles in the solution, it is important to require that α in case 2 be an integer; the value of λ is thereby severely restricted, a first indication that not every Hamiltonian (8.6) is sufficiently well behaved. (Actually, this condition on α will be relaxed later on, but for the time being there is no harm in adopting it.) In this manner one ends up with one particular trajectory, whereas there ought to be four free parameters for a general solution of (8.7). The time t_0 represents only one parameter; the remaining three come from the resonance terms.

If the full expressions (8.8) are inserted into (8.7), the leading terms have already been chosen to cancel the most negative powers of τ. The next terms in (8.7) involve the time derivatives of the resonances on the left, and the products of the leading terms with the resonance terms on the right. These terms in (8.7) cancel, provided the coefficients c and d satisfy two linear equations; and that in turn requires the vanishing of the corresponding determinant. In this way, the necessary conditions for the exponent γ in the resonance terms are established. For each acceptable value of γ the linear equations for c and d can be solved, yielding one free parameter in the solution of the equations of motion.

Case 1 yields either $\gamma = 6$ or $(\gamma - 2)(\gamma - 3) = -6(1 + 1/\lambda)$. Both solutions of this quadratic equation for γ have to be positive integers in order to obtain the required three parameters. Thus, we find a set of possible negative values for λ in case 1. The analysis of case 2 is different: the coefficient a in (8.8) is already arbitrary, corresponding to $\gamma = 0$; there is again $\gamma = 6$; the third value of γ is such that $\alpha' = \alpha + \gamma$ is the second solution of $\alpha(\alpha - 1) = -12\lambda$, i.e., both solutions have to be > -2 and integers, leading to a restriction on λ.

To continue the argument to its successful conclusion, it is now necessary to construct the full expansion starting from the values for the exponents α, β, and γ in (8.8) which have just been found. There are many tricky details that the reader has to study in the original paper. The present discussion is meant to stimulate interest in a totally different and potentially revealing approach to the question of integrability in Hamiltonian systems.

The final conclusion shows the following four cases where the Hamiltonian (8.6) is integrable:

(1) $\lambda = 0$ decouples the motions in x and y;
(2) $\lambda = -1$ with $\mu = 1$ also becomes separable when the coordinates $x + y$ and $x - y$ are used;
(3) $\lambda = -1/6$ has the "third" constant of motion $F = x^4 + 4x^2y^2 - 4u(uy - vx) + 4\mu x^2 y + (4\mu - 1)(u^2 + \mu x^2)$;
(4) $\lambda = -1/16$ with $\mu = 1/16$ is also integrable, but the expansions in the complex-time plane require algebraic branch points.

It would have been close to impossible to find the last two cases without the analysis in the complex-time plane that was first used by Kowalevskaya. The reader will find more details in the work of Yoshida (1983), in the article by Newell, Tabor, and Zeng (1987), and in the recent monograph by Tabor (1989).

8.6 Discrete Area-Preserving Maps

Investigating the Poincaré surface of section for an arbitrary Hamiltonian requires the numerical integration of the trajectories. Since most of them are inherently unstable in a chaotic system, no computational scheme is able to provide more than a few dozen intersections with the surface of section, say $x = 0$ as in Section 8.3. The quality of any calculation can always be checked by running the trajectories backward in time: let $P_0 = (v_0, y_0)$, $P_1 = (v_1, y_1)$, ... , $P_n = (v_n, y_n)$ be the consecutive intersections; then one could start a trajectory at $P'_0 = (-v_n, y_n)$; the subsequent intersections are $P'_1 = (-v_{n-1}, y_{n-1})$, ..., $P'_n = (-v_0, y_0)$ to the precision of the round-off error.

Even without this technical difficulty, it is often prohibitively expensive to run the integration routine over thousands of initial points P_0, because each intersection may require several hundred integration steps in order to approximate the continuous variation of the coordinates along the trajectory. The fine structure of the Poincaré map cannot be displayed, unless the map is drastically simplified. Instead of the real trajectories that belong to a well-defined Hamiltonian such as (8.7), an artificial transformation from the (x, y) plane into itself is studied.

Froeschlé (1968) and Hénon (1969) were among the first to try out this strategy in order to overcome the computational limits of the earlier calculations concerning the Hamiltonian (8.6). The choice of a map $(x, y) \rightarrow (x_1, y_1)$ was dictated by the paramount requirement that the element of area remain invariant, or equivalently that the Jacobian $\partial(x_1, y_1)/\partial(x, y) = 1$. Moreover, the map had to look like a rotation in the neighborhood of the origin, imitating the Poincaré map in the neighborhood of a point of stable eqilibrium such as the center of the equilateral triangle in the Hénon-Heiles potential. Finally, the simplest non-linear term was added in order to achieve the most economical numerical procedure.

After eliminating all trivial parameters, the following transformation T is found to be the most general when no more than quadratic terms are allowed,

$$\begin{aligned} x_1 &= x \cos\alpha - (y - x^2)\sin\alpha\,, \\ y_1 &= x \sin\alpha + (y - x^2)\cos\alpha\,, \end{aligned} \tag{8.9}$$

where α is the only non-trivial parameter left. This map has the further advantage that its inverse is given by similar quadratic formulas, as one can check immediately. Moreover, the fixpoints of T, T^2, T^3, T^4 can be calculated explicitly by solving the corresponding algebraic equations.

The resulting structure of islands, i.e., smooth invariant curves, and of areas where the consecutive points of transformation $P_0, P_1, P_2, ...$ scatter chaotically, is shown in Figure 16, where $\cos\alpha = .24$. The resemblance to Figure 12 is quite striking; but certain fine details can

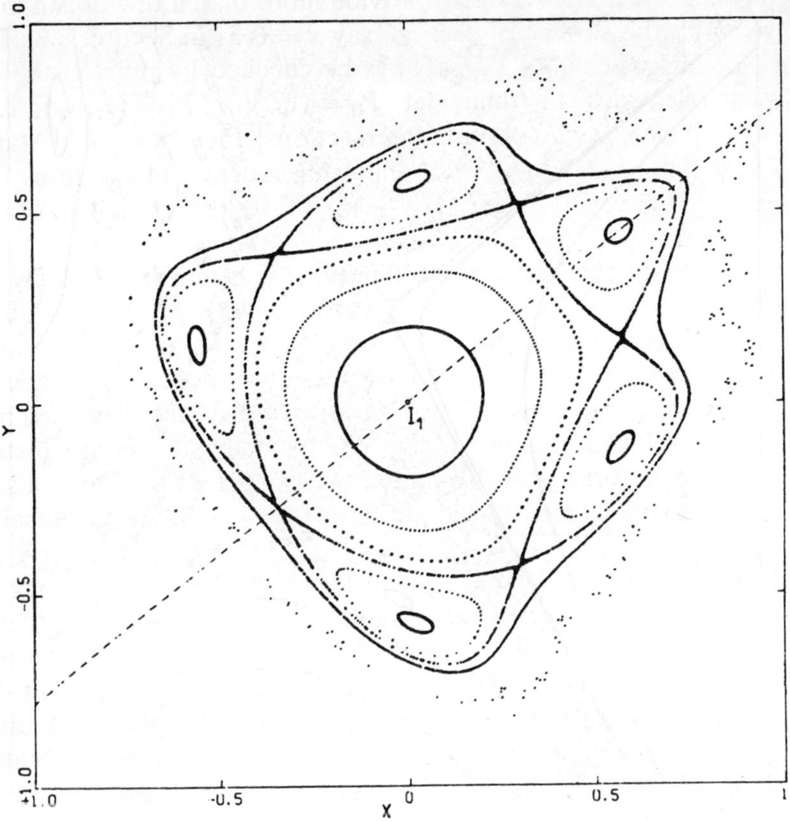

Figure 16 Consecutive points of the transformation (8.9) for $\alpha = 0.24$ [from Hénon (1969)].

now be explored, as demonstrated in Figure 17, where the neighborhood of an unstable fixpoint of T^5 is blown up. Notice the various sets of small subsidiary islands and the approach of the large island within .00001 of the fixpoint. The detailed structure could easily have gone unnoticed in a cruder calculation, but then quantum mechanics might ignore it knowingly, as we shall see later.

This map, sometimes called the *Hénon map,* has been investigated very intensively during the last decade in a slightly more general version which was also proposed by Hénon (1976). The transformation $(x, y) \rightarrow (x_1, y_1)$ now has the deceptively simple appearance,

$$x_1 = y\,, \quad y_1 = -\,\varepsilon x + \mu - y^2\,, \tag{8.10}$$

where ε and μ play the following role. The Jacobian $\partial(x_1, y_1)/\partial(x, y) = \varepsilon$ so that this map does not preserve the area unless $\varepsilon = 1$; if such is the case, however, the new map (8.10) is the same as

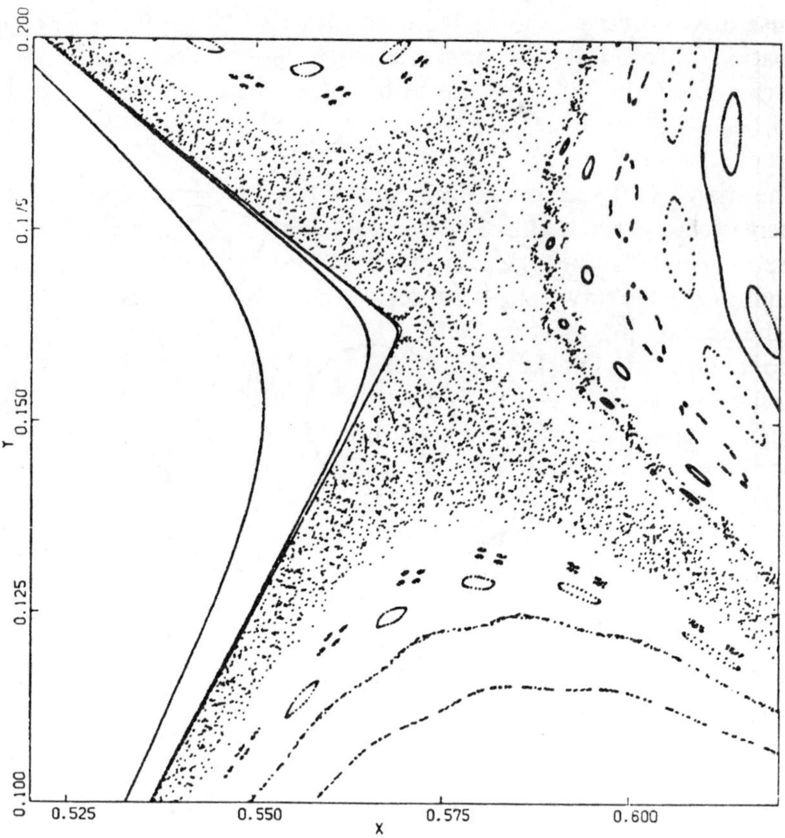

Figure 17 The neighborhood of an unstable fix-point in the preceeding figure is enlarged to show the intricate mingling of small islands with ergodic regions [from Hénon (1969)].

(8.9) up to a change of coordinates in the (x, y) plane. The rotation angle α in (8.9) is related to μ in (8.10) through $(1 + \mu)^{1/2} = 2 \sin^2 \alpha/2$.

When $\varepsilon < 1$, the map is a good model for a dynamical system with *dissipation*. As the the map is iterated, the original area of any portion of phase space becomes smaller by a factor ε at each step. The contraction, however, does not lead to a set of points whose diameter decreases indefinitely, but rather to a fractal set with dimension larger than 1. The consecutive points $P_0, P_1, \ldots.$ eventually get ever closer to this set, which has, therefore, been called a *strange attractor;* they move around this fractal without ever converging to a limit.

Since the volume in phase space is conserved, this kind of phenomenon is strictly excluded from Hamiltonian mechanics, which is the central theme of this book. The transformation (8.10) is of interest because area conservation appears as the formal limit of decreasing dissipation. Moreover, the opposite limit, $\varepsilon = 0$, is perhaps the best-known map of the straight line into self, the so-called *logistic map,* $y \to y_1 = \mu - y^2$; the variable x simply takes on the previous value of y, and can be ignored.

The map (8.9) has been studied in great detail and many of its mathematical properties have been established in the form of bona fide theorems which are not easy to prove at all. The reader has to work through a great many papers to get acquainted with the terminology and the techniques of this field. Recent surveys were written by Grebogi, Ott, and Yorke (1987), as well as Schuster (1988). One general conclusion, however, is worth mentioning: the transition from the logistic map to the Hénon map (8.9) is very complicated, and the former is not a very useful guide to the latter. The absence of an attractor in Hamiltonian systems, even a strange one, is responsible for some entirely different situations in phase space, and puts these systems into a class all by themselves.

As a final comment on the progressive reduction from conservative, to dissipative, to the logistic map, the ultimate link in this chain are maps of the logistic kind, but considered in the complex plane. This work goes back to the French mathematicians Fatou (1906) and Julia (1918) in whose honor these maps are called *Julia maps.* Their properties are astounding and have been studied in great computational and artistic detail, e.g., by Peitgen and Richter (1986). As in the analysis of Kowalewskaya and Painlevé, the recourse to the complex domain again reveals itself as fundamental for a genuine understanding. This conclusion becomes particularly evident when the logistic map ($\varepsilon = 0$ in 8.10) is investigated as a function of the complex parameter μ; its general behavior is best understood with the help of the *Mandelbrot set* (Mandelbrot 1980).

CHAPTER 9

Soft Chaos and the KAM Theorem

Chaotic behavior in a dynamical system is most easily understood as a breakdown of the invariant tori due to the perturbations. The KAM theorem deals with this process of disintegration and shows that it is gradual. The resulting situation in phase space, to be called soft chaos, is smooth wherever the tori are intact, but it has many rough spots that are associated with resonances or phase-locking. This phenomenon happens when two degrees of freedom get stuck with the ratio of their frequencies given by a rational number. Soft chaos can be explained by estimating the size of the domain of phase space where phase-lock occurs as a function of the perturbation strength.

9.1 The Origin of Soft Chaos

The key problems in mechanics are integrable: the motion of the planets around the Sun in astronomy, the symmetric heavy gyroscope in mechanics, the hydrogen atom in physics, and the hydrogen-molecule ion in chemistry. It is tempting to reduce the more complicated problems to these standard cases, and the method to carry out this reduction is perturbation theory. We have discussed the three-body problem of Moon-Earth-Sun in some detail in order to bring out the difficulties of such an approach and to suggest at least one alternate route, Hill's theory of the Moon. This chapter is meant to explain how perturbation theory breaks down and what is left of the old integrable structure of phase space.

The pictures for the surface of section in the Hénon-Heiles model show clearly how the smooth intersections of the invariant tori at low energies give way to isolated islands in a sea of chaotic behavior as the energy increases. Each island consists of nested simply closed curves with a single point at the center, which corresponds to a very special periodic orbit, comparable to a *soliton.* The motion in the neighborhood of this periodic orbit is stable; the intersections of a neighboring trajectory with the surface of section stay on one of the closed curves.

Stable periodic orbits are indigenous to mechanical systems; they are a generic feature as we argued at the end of Chapter 6, in contrast to integrability, which is highly exceptional. Stable periodic orbits are the best we can hope for as perturbations are allowed to destroy part of the invariant tori that characterize the integrable systems. The main question now concerns the extent of a particular island as it is saved from the floods of chaos.

Periodic orbits show up in the surface of section as a finite set of say n isolated points whose locations line up again on a simply closed curve. This last claim is, of course, moot, because any finite set of points in a plane can be lined up on a smooth curve; but a close look shows an ordered pattern of islands belonging to the same stable periodic orbit; it clearly arises from some underlying simple loop and is all that's left over from the original invariant torus. The periodic orbit jumps from one point in the finite set to the next in a well-defined manner, such as skipping every other point; thus, it runs around this simple loop a number of, say m, times before returning to the first point. Figure 16 shows this situation with $n = 5$ islands, which are visited by the periodic orbit in $m = 2$ turns around the underlying closed curve.

The physical interpretation of this phenomenon is made in terms of a *resonance* between the two degrees of freedom of the dynamical system. An integrable system is full of resonances since every pair of integers (m, n) yields a periodic orbit, as we saw in Section 6.2. Only few of them are effective, however, in capturing the neighboring trajectories the way it happens in Figure 16. Indeed, a whole area of the surface of section in Figure 16 imitates the same 5:2 repeating pattern as the periodic orbit. This phenomenon is called *phase-lock:* the phase angle of the first degree of freedom is locked in step with the phase of the second degree of freedom, and cannot be shaken loose by modifying the starting positions and momenta.

The capture of the trajectories in the neighborhood of a periodic orbit is the origin of chaos, because it is impossible to fit many bands of islands into a smooth overall *foliation of the energy surface,* i.e., a complete covering with non-intersecting manifolds of lower dimension such as the invariant tori. On the other hand, the very existence of the islands indicates a local kind of integrability whose usefulness depends on the extent of the island. The main purpose of this chapter is to estimate the size of a particular set of islands. If they do not cover the whole surface of section, one can assume that the remainder is chaotic. In a somewhat mysterious fashion, the old foliation by invariant tori is not completely destroyed; the mixed situation, where some invariant tori and whole islands are left over in phase space, will be called *soft chaos*.

This term is entirely devoid of mathematical precision, and its domain of applicability is only bounded on one side by the integrable systems and on the other side by hard chaos. The majority of Hamiltonian systems belongs to the category of soft chaos; the prime example is the hydrogen atom in a magnetic field: it is integrable both for a vanishing field and for very strong fields, so that it never quite detaches itself from integrability. Most research nowadays deals with softly chaotic dynamical systems, because they arise quite naturally when one starts with an integrable system like coupled linear oscillators and adds a sufficiently smooth perturbation such as the cubic potential in the Hénon-Heiles model.

In spite of all this work, soft chaos is understood only in some of its local features, while no more than superficial and general arguments can be given for the global characteristics. Just as one cannot say easily whether a particular dynamical system is integrable or not, so nobody has been able as yet to offer a good overall description of phase space with soft chaos. That may be the basic reason why the connection with quantum mechanics is still so poor; the failure is as much on the classical as on the inherently more subtle quantum side.

9.2 Resonances in Celestial Mechanics

By far the most striking resonance is known to everybody, although its explanation was only given by Lagrange (1764) in response to a prize question of the French Academy of Sciences; the 28-year-old author was awarded the prize, but he went back to the problem in 1780 to get better agreement with observations. Since the Moon turns always the same side toward the Earth, her rotation around her own axis is obviously synchronized with her motion around the Earth. But these two frequencies do not have the same value by accident, because otherwise they would differ by a small amount, and we would get to see the far side of the Moon after many years. The Moon is elongated toward the Earth, leading to spin-orbit coupling and to a complete phase-lock.

Another example of a resonance was well known to the observers before it was finally explained by Laplace. Its origin is a numerical coincidence between seemingly unrelated periods; the period of Jupiter around the Sun is 12 years while the period of Saturn is 30 years; the exact value of their ratio is .40268677, very close to 2:5. The orbits of both planets are very close to the ecliptic; they also carry most of the angular momentum in the solar system. If one ignores all the other planets, one ends up with another three-body problem, but of a quite different nature compared with the Moon-Earth-Sun system.

Molecular vibrations are full of such resonances; but the comparison with the Helium atom is more appropriate. The mass of Jupiter is about 1/1000, and the mass of Saturn is about 1/4000 of the solar mass, while the mass of the electron is about 1/8000 of the mass of the He-nucleus. As emphasized in Section 4.2, however, the main difference is the strength of interaction, which goes as the product of the masses in astronomy, rather than the product of the electric charges in physics and chemistry.

The eccentricities of the Kepler orbits are moderate by the standards of the solar system, 0.048 for Jupiter and 0.056 for Saturn. They play the crucial role in the resonance, however; if they were zero, the interaction between the two planets would depend only on the difference of the polar angles $\phi_J - \phi_S$, and the peculiar combination $\psi = 5\phi_S - 2\phi_J$ would never come up. The exact ratio of the frequencies ω_S/ω_J yields 880 years for ψ to change by 2π. The small perturbation with this angular dependence is able to build up more effectively than larger perturbations with shorter periods.

Let each planet be confined to the ecliptic; its coordinates with respect to the Sun are the osculating elements in the conjugate pairs (Λ, ℓ) for the energy and the mean anomaly and (M, g) for the angular momentum and the direction of the perihelion (cf. Section 4.4). The center-of-mass motion is eliminated, taking away two degrees of freedom; the total angular momentum $M_J + M_S$ is conserved, and the perturbation depends only on the difference $g_J - g_S$; only three degrees of freedom are left. Each planet gets one for moving in its own elliptic orbit, and the third degree of freedom allows for trading angular momentum back and forth, and thereby changing the individual eccentricities.

Formulas (4.3) and (4.4) have to be inserted into the gravitational potential between Jupiter and Saturn with the help of the expressions for Λ and M just following (4.7). The critical angle $\psi = 5\phi_S - 2\phi_J$ appears only in terms proportional to at least three powers of the eccentricities, and thus smaller by a factor $(1/20)^3 = 1/8000$ than the leading term in the perturbation. Calculating the positions requires two time integrations, or equivalently two divisions by the small denominator $5\omega_S - 2\omega_J$ whose period is 880 years compared with the basic periods of 12 and 30 years. Therefore, the perturbation gets boosted by a factor somewhere between 30^2 and 75^2.

The complete theory of the Jupiter-Saturn resonance is difficult because the ratio of their semimajor axes is more than 1/2 so that any expansion in this parameter converges poorly. The inequalities of 20′ for Jupiter and 48′ for Saturn were easily observed by the naked eyes of Brahe and Hevelius. The center of mass for Sun, Jupiter, and Saturn lies outside the Sun most of the time so that the Sun's motion is not negligible.

A famous example of resonance in the solar system may be more directly connected with the phenomenon of chaos. It was noted by Kirkwood in 1866 that the periods of *asteroids* in the belt between Mars and Jupiter are such as to avoid any resonance with Jupiter. These are the famous Kirkwood gaps of which the most glaring correspond to the ratios 2:1 and 3:1 for Jupiter's period to the asteroid's period (cf. Froeschlé and Scholl 1982). Somewhat less conspicuously missing are the ratios 5:2, 5:3, 7:2.

It is not clear whether a resonance always has a destabilizing effect. Our understanding of the Kirkwood gaps is incomplete at best; the situation is reminiscent of the gaps in the rings of Saturn, which are presumably due to the resonances with some of Saturn's satellites. The doubts in all these explanations have to do with phase-lock, and were first pointed out in 1812 by Gauss. The asteroid Pallas, one of only four known at that time, has a period ratio close to 18:7 with Jupiter. Gauss thought that this coincidence would tie down the motion of Pallas in the same way as the Moon's rotation is locked to its orbit around the Earth. Rather than make the motion more precarious, the resonance provides extra phase space in which to absorb small irregular noise-like perturbations.

9.3 The Analogy with the Ordinary Pendulum

Soft chaos will now be approached starting from an integrable system with two degrees of freedom. The action-angle coordinates for the unperturbed system are the canonical pairs (M, ϕ), (N, ψ). The perturbation does not depend explicitly on time, so that the Hamiltonian reduces to

$$H = H_0(M, N) + \frac{\varepsilon}{2} \sum_{m,n \neq 0,0} V_{m,n}(M, N) \exp(im\phi + in\psi) \;, \tag{9.1}$$

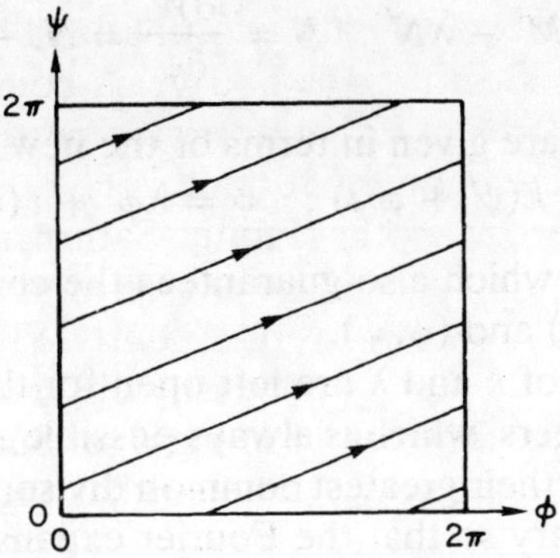

Figure 18 Trajectory in the (ϕ, ψ) plane of angle variables with a 2:5 resonance, using periodic boundaries.

where the coefficients $V_{m,n}$ are complex and satisfy the relation $V_{-m,-n} = V^+_{m,n}$ in order to make the second term on the right real-valued. The factor ε has been inserted to help us distinguish the various orders of perturbation.

Let us investigate the neighborhood of the particular unperturbed periodic orbit with frequency ratio k/ℓ, i.e., the neighborhood of the values $M=M_r$, $N=N_r$ where

$$\omega_1 = \frac{\partial H_0}{\partial M}\Big]_{M_rN_r} = k\omega_0\,, \quad \omega_2 = \frac{\partial H_0}{\partial N}\Big]_{M_rN_r} = \ell\omega_0\,. \tag{9.2}$$

The period of this orbit is $2\pi/\omega_0$ which would be 60 years in the case of the Jupiter-Saturn resonance where $k/\ell = 5/2$. The orbit in a ϕ versus ψ diagram can be represented in two ways: either in the square $0 \le \phi, \psi \le 2\pi$ with opposite sides identified as in Figure 18, or in the unlimited (ϕ, ψ) plane as in Figure 19; in both cases, one has a straight line at an inclination 2/5, not necessarily through the origin.

The first step in dealing with this resonance is a shift of the origin and a linear transformation of the axes in the action-angles variables. The origin of the (M, N) plane is moved into (M_r, N_r). The angles (ϕ, ψ) are transformed into (ϕ', ψ') so that the unperturbed periodic orbit is given by $\phi' =$ constant. Since ψ' becomes the coordinate along the periodic orbit and increases linearly with time at the rate ω_0, we will take the term $\omega_0 t$ out of ψ'. The canonical transformation is carried out with the help of the generating function

$$W = (M_r + \ell M' - \lambda N')\phi + (N_r - kM' + \kappa N')\psi - N'\omega_0\, t, \tag{9.3}$$

which has the same mix of the new actions (M', N') and old angles (ϕ, ψ) as W in the discussion of Section 5.3; the transformation is given by the same formulas (5.3).

The old actions are then given in terms of the new ones by

$$M = \frac{\partial W}{\partial \phi} = M_r + \ell M' - \lambda N' , \quad N = \frac{\partial W}{\partial \psi} = N_r - kM' + \kappa N' , \quad (9.4)$$

while the old angles are given in terms of the new ones by

$$\phi = \kappa\phi' + k(\psi' + \omega_0 t) , \quad \psi = \lambda\phi' + \ell(\psi' + \omega_0 t) , \qquad (9.5)$$

where $\kappa\ell - \lambda k = 1$, which also guarantees the conservation of area in both planes, (M , N) and (ϕ, ψ).

The exact values of κ and λ are left open for the moment, although they have to be integers, which is always possible as long as k and ℓ are relatively prime, i.e., their greatest common divisor is 1. The restriction to integers is necessary so that the Fourier expansion of the perturbation in (9.1) is still valid. Taking the time derivatives in (9.5) shows that $\dot{\phi}' = \ell\dot{\phi} - k\dot{\psi} = \ell\omega_1 - k\omega_2 = 0$, and $\dot{\psi}' = -\lambda\dot{\phi} + \kappa\dot{\psi} - \omega_0 = -\lambda\omega_1 + \kappa\omega_2 - \omega_0 = (-\lambda k + \kappa\ell)\omega_0 - \omega_0 = 0$. Therefore, both ϕ' and ψ' remain constant along the periodic orbit as long as there is no perturbation.

If the formulas (9.4) and (9.5) are inserted into (9.1), the exponent in the perturbation becomes $m\phi + n\psi = (m\kappa + n\lambda)\phi' + (mk + n\ell)(\psi' + \omega_0 t)$. The new integers $m' = m\kappa + n\lambda$ and $n' = mk + n\ell$ vary independently from $-\infty$ to $+\infty$. The coefficients in the Fourier expansion of the perturbation can be simplified by setting $V_{m,n}(M, N) \simeq V_{m,n}(M_r, N_r)$, since they vary slowly as functions of M and N, and are, moreover, small compared with H_0. They are also renumbered in terms of (m', n') rather than (m, n).

The first term in (9.1), the integrable part of the Hamiltonian H_0, is expanded around the resonance by assuming that the values of M' and N' are small. That requires taking first and second derivatives of H_0 with respect to M and N at the resonance (M_r, N_r); let these be called H_1, H_2, H_{11}, ... and so on. The new unperturbed Hamiltonian is given by (5.5); its constant term, $H_0(M_r , N_r)$, can be dropped; the linear terms cancel, as can be checked without difficulty; to terms of second order, the transformed Hamiltonian (9.1) becomes

$$AM^2 + 2BMN + CN^2 + \varepsilon \sum_{m,\, n \neq 0,0} V_{n,m} \exp(im\phi + in(\psi + \omega_0 t)) , \quad (9.6)$$

where A, B, C are defined by the matrix relations

$$\begin{pmatrix} A & B \\ B & C \end{pmatrix} = \begin{pmatrix} \ell & -k \\ -\lambda & \kappa \end{pmatrix} \begin{pmatrix} H_{11} H_{12} \\ H_{21} H_{22} \end{pmatrix} \begin{pmatrix} \ell & -\lambda \\ -k & \kappa \end{pmatrix} .$$

An overall factor 1/2 has been left out in (9.6); also, the primes are not attached to M, N, ϕ, ... so as to simplify the reading of (9.6) and the following paragraphs.

The only approximation so far is the expansion of $H_0(M, N)$ around the resonance at (M_r , N_r) to second order; it could have been carried

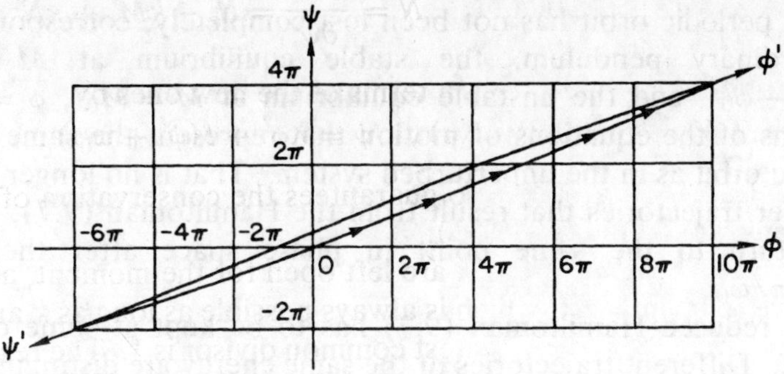

Figure 19 The transformation of the angle variables according as (9.5) from (ϕ, ψ) to (ϕ', ψ') with $(k, \ell) = (5,2)$ and $(\kappa, \lambda) = (3,1)$; the base vectors of the new unit-cell are extremely elongated.

further if necessary. Before indulging in more drastic simplifications, however, one has to ask: Is there still a periodic orbit in the neighborhood of the resonance when the perturbation is taken into account, and if so, where exactly is it located ?

The answer is by no means obvious. Proving the existence of a periodic orbit became an important part of Poincaré's work in celestial mechanics; his last scientific paper (Poincaré 1912) deals with this problem, and he shows in great detail how he failed to come to a mathematically rigorous result; his theorem was proven the following year by Birkhoff (1913). We will discuss an approximate solution for the trajectories of (9.6) without trying to justify the procedure by more than crude physical reasoning. Our purpose is to give a simple picture for the existence and size of islands around a periodic orbit.

The main argument is that any term in the perturbation with a non-vanishing variation in $\psi + \omega_0 t$ gets averaged out to zero; only the terms with $n = 0$ are important in lowest approximation. The summation over n in (9.6) is, therefore, left out at first, leaving us with the quadratic terms in M and N, and a perturbation that depends only on ϕ. Nothing essential is lost if only the terms $m = \pm 1$ are taken into account, and all the others are dropped. Thus, the potential energy in (9.6) becomes simply $\varepsilon \,|\, V_{1,0} \,|\, \cos(\phi + \phi_0)$, where ϕ_0 is the phase of the complex coefficient $V_{1,0}$.

The two degrees of freedom can now be separated in this simplified Hamiltonian. Since the angle ψ does not occur at all, N becomes a constant of motion. The conjugate pair (M, ϕ) is governed by the same Hamiltonian as the ordinary pendulum,

$$\frac{1}{2} A (M - M_0)^2 + \varepsilon \,|\, V_{1,0} \,|\, \cos(\phi + \phi_0) + \frac{AC - B^2}{2A} N^2 \,, \quad (9.7)$$

where $M_0 = - BN/A$.

The periodic orbit has not been lost completely; corresponding to the ordinary pendulum, the stable equilibrium at $M = M_0$, $\phi = \pi - \phi_0$, and the unstable equilibrium at $M = M_0$, $\phi = \phi_0$ are solutions of the equations of motion that represent the same kind of periodic orbit as in the unperturbed system. That is no longer true for the other trajectories that result from the Hamiltonian (9.7); they do not return to the same point in phase space after the period $T_0 = 2\pi/\omega_0$.

The reduced Hamiltonian (9.7) has to be kept at some constant value E. Different trajectories of the same energy are distinguished by different values of N. It is convenient, however, to discusss the Hamiltonian (9.7) as if the third term was missing, and the individual trajectories are characterized by the energy E. Thus, one arrives at Figure 20, which is the surface of section ψ=0 with the coordinates (ϕ, M); it is in complete analogy to the phase space of the ordinary one-dimensional pendulum.

Let us now construct a more realistic picture by taking the trajectories from the reduced Hamiltonian (9.7) back into the original (old, unprimed) coordinates, with the help of (9.4) and (9.5); the reduced (new) coordinates of (9.6) and (9.7) are written again with primes attached as in (9.3), (9.4), and (9.5). In particular, let us try to look at the surface of constant energy E by adding the old action N as a third dimension to the old angles ϕ and ψ in Figures 18 and 19. Since the stable periodic orbit becomes the straight line $\phi' = \ell\phi - k\psi = \pi - \phi_0$ with $N - N_r = -k\, M'_0 + \kappa N'$, we can imagine Figure 20 as sliding along this line! The representative point of some particular trajectory in Figure 20 now becomes a trajectory in (ϕ, ψ, N) space, as shown in Figure 21, looking like a square platter with lasagne al forno.

The surface of section $\phi = 0$, with the old coordinates ψ and N can now be understood. Figure 20 slides k times through the plane $\phi = 0$, before it arrives at its starting configuration; also, it is reduced in size appropriately to fill out the interval $0 \leq \psi < 2\pi$, yielding k copies of itself sitting next to one another. These k copies can finally be wrapped into a ring because the angle ψ is defined only modulo 2π. The conjugate variable N can then be viewed as a radial coordinate. In this way, one obtains a picture looking like Figure 16 in the Hénon-Heiles model. We will refer to Figure 16 as if it had been obtained from a Hamiltonian with a 5:2 resonance.

The discussion in this section shows how a perturbation at a resonance frequency replaces the nested invariant tori with a chain of islands. The centers of these islands form a stable periodic orbit which is surrounded by a new set of invariant tori. The islands are separated from each other by an unstable periodic orbit; they are bounded by the

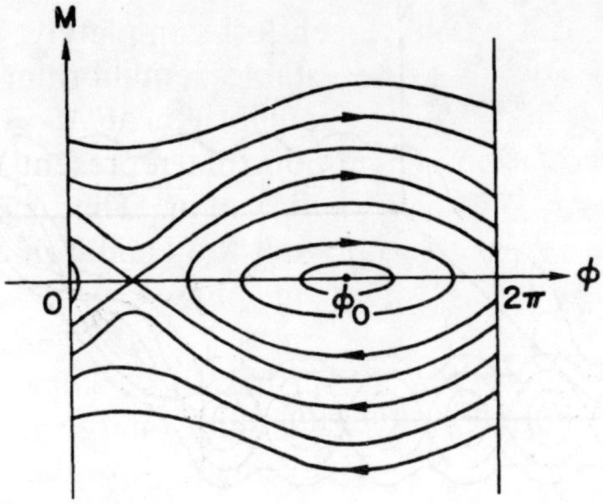

Figure 20 Surface of section for the Hamiltonian (9.7), which corresponds to the ordinary pendulum.

separatrix that forms the boundary between the vibrational and the rotational trajectories of the pendulum, and self-intersects in the unstable periodic orbit. The original layered structure of invariant tori continues to exist outside, and immediately adjacent, to the separatrix, at least in this primitive picture.

9.4 Islands of Stability and Overlapping Resonances

The main features of Figure 16 are largely independent of its precise construction; indeed, it could well have been obtained from a numerical computation of the trajectories in a circularly symmetric potential that is perturbed by a force with five-fold symmetry as in an organic molecule. A Poincaré section on a radial line, i.e., plotting the radial momentum p_ρ versus the radial variable ρ when the azimuthal angle $\theta = 0$, may lead to a figure just like 16 (cf. also Section 6.1).

Notice that the lines of constant energy in the equivalent pendulum of Figure 20 have an orientation: the closed loops go clockwise around the stable equilibrium; the open trajectories have decreasing (increasing) angles ϕ' according as $M' > (<) M'_0$. In Figure 16 the points in each island jump from one island to the another by skipping the one in between; thus there is a prevailing counter-clockwise drift for the whole pattern; but the trajectories outside the islands that correspond to the full rotations of the equivalent pendulum, inherit the increase (decrease) of the angle ϕ'; therefore the outermost lines in figure 16 turn counterclockwise more slowly than the innermost. This situation is consistent with the precession of the trajectories in a circularly sym-

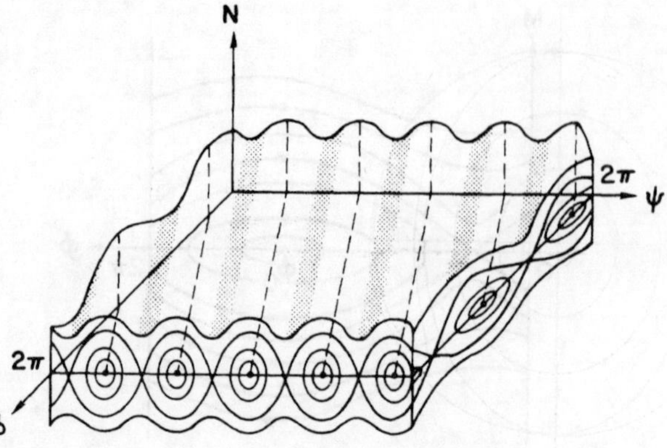

Figure 21 The foliation of the energy surface near a 5:2 resonance.

metric potential of Figure 11d, which is, therefore, maintained in spite of the perturbation.

If the prevailing motion of the islands, $4\pi/5$ per Poincaré section, is taken out of Figure 16, the basic scheme for a mapping of a circular annulus into itself appears. The outer and the inner rims turn in opposite directions, and, most importantly, the area of the map is conserved. Such a map is called a *twist map,* and plays a central role in modern treatises on celestial mechanics (cf. Siegel and Moser 1971; Moser 1973), as well as the study of classical chaos. It will be discussed further at the end of this chapter.

The perturbation disturbs the structure of nested loops of Figure 11d. Each term in the perturbation of (9.1) gives rise to its own set of islands, and difficulties are bound to arise when the islands from different resonances get into each other's way as the strength of the perturbation increases. They can no longer be separated from each other by a layer of smooth tori such as on the outer and the inner boundaries of the islands in Figure 16. The islands start to shrink and are surrounded by a chaotic region in phase space

This interpretation of Figure 16 is demonstrated by the calculations of Walker and Ford (1969) who investigated the Hamiltonian

$$H = H_0(M, N) + \alpha MN \cos(2\phi - 2\psi) + \beta M^{3/2} N \cos(2\phi - 3\psi)\ , \quad (9.8)$$

where H_0 is of second degree in (M, N), while α and β are arbitrary real coupling parameters. When $\beta = 0$ one finds Figure 22a with two islands surrounding the origin, while $\alpha = 0$ yields Figure 22b with a belt of three islands as expected from the analysis in the preceding section. With both coupling parameters small enough, the two sets of islands coexist, and are separated by a full invariant torus as shown in Figure 23a; everything looks as if the system were still integrable. When the coupling parameters α and β become fairly large as in Figure 23b, the

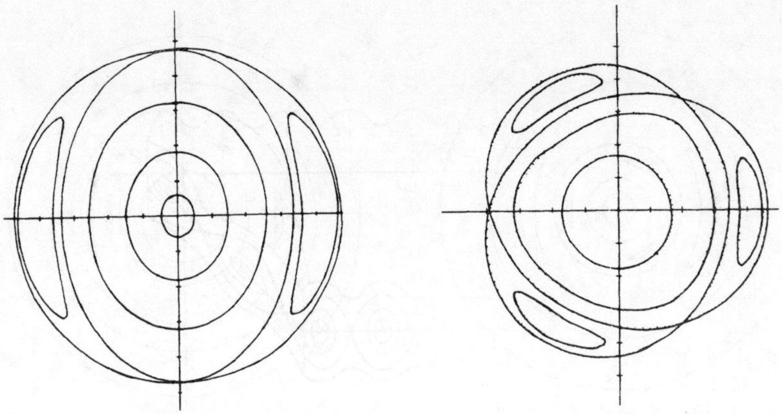

(a) (b)

Figures 22a and 22b Surface of section for the Hamiltonian (9.8) when (a) only the first resonance is turned on ($\alpha \neq 0$), and (b) only the second resonance is active ($\beta \neq 0$) [from Walker and Ford (1969)].

two belts of islands are no longer separated by an invariant torus, but each island is surrounded by a chaotic region.

The argument for the appearance of chaos because of the overlapping resonances can be made more specific with the help of the preceding section. The oscillatory region in the (M', ϕ') plane of Figure 20 becomes the set of islands in the (ψ, N) plane of Figure 16. The largest extent of this region is defined by the curve through the maximum of the potential energy in (9.7), and is given by $M' = \pm\, 2(\varepsilon \mid V \mid / A)^{1/2} \sin(\phi' - \phi_0)/2$. Since this curve separates the oscillatory and the rotatory regions of the phase space, it is called the *separatrix*.

The area contained inside the separatrix is $\int M' d\phi' = 8(\varepsilon \mid V \mid / A)^{1/2}$. The corresponding tube (torus) in phase space cuts the line $\phi = 0$ a total of k times so that the total area of the islands in the surface of section becomes $8k(\varepsilon \mid V \mid / A)^{1/2}$. This area depends only on the ratio k/ℓ because A is a quadratic function of (k, ℓ). According to the definition of A in (9.6), the quantity A/k^2 gives the change in frequency $\omega(M, N)$ with a change of the actions in the unperturbed system as one moves away from the resonance. The size of the islands increases, therefore, with the strength of the perturbation, and depends inversely on the derivative of the frequency with respect to the action, $\partial\omega/\partial I$.

Another way of judging the effectiveness of the perturbation in creating islands, emerges from studying the frequency ratio ω_1/ω_2. This ratio is basically the same for the whole island, because all the

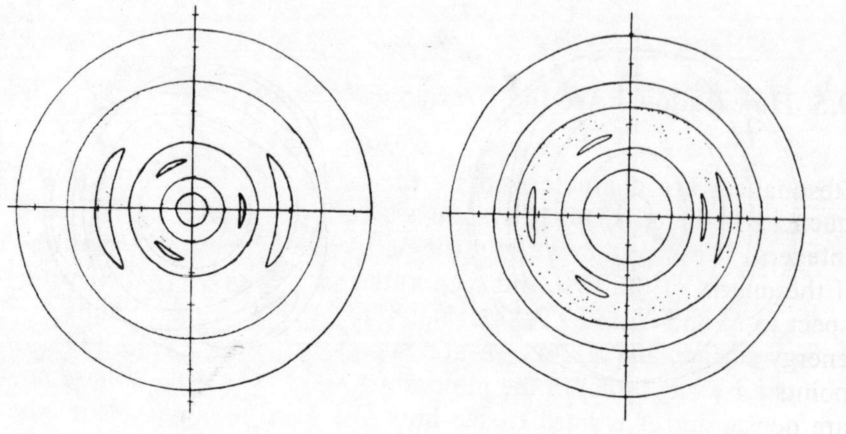
(a) (b)

Figures 23a and 23b Surface of section for the Hamiltonian (9.8) when both resonance terms are turned on; (a) with small coupling parameters α and β, and (b) with large coupling parameters [from Walker and Ford (1969)].

points in it go through the same k/ℓ cycle of the resonance; this is one form of phase-lock. The value of M' can be spread in the perturbed system as much as $2(\varepsilon\,|\,V|/A)^{1/2}$ around the resonance values (M_r, N_r) without changing the frequency ratio, or equivalently, the original actions can be changed by as much as $(\delta M, \delta N) = 4(\ell, -k)M' = (\ell, -k)(\varepsilon\,|\,V|/A)^{1/2}$. If we go back to (3.3), we can now calculate the change in the frequency ratio that such a change in the actions would have caused in the unperturbed system,

$$\frac{\omega_2}{\omega_1}\,\delta\!\left(\frac{\omega_1}{\omega_2}\right) = \frac{4\sqrt{\varepsilon\,|\,V_{\ell,-k}\,|}}{H_1 H_2}\begin{vmatrix} H_{11} H_{12} H_1 \\ H_{21} H_{22} H_2 \\ H_1 \; H_2 \; 0 \end{vmatrix}^{1/2}, \quad (9.9)$$

where the integers (k, ℓ) appear only as indices in the strength of perturbation $V_{\ell,-k}$. The determinant of first and second derivatives measures the non-linearity of the Hamiltonian $H_0(M, N)$ transverse to its gradient, i.e., in the direction where the frequency ratio changes.

If there are two active resonances (ω_1', ω_2') and (ω_1'', ω_2'') belonging to (k', ℓ') and (k'', ℓ''), each is embedded in an interval of width given by (9.9). Chaos arises when these two intervals overlap; this explanation of soft chaos has been pursued by Chirikov and his school, and is generally called the *resonance-overlap criterion* (cf. the reviews Zaslavskii and Chirikov 1972, and of Chirikov 1979). It allows to estimate qualitatively how strongly a system can be perturbed before it becomes softly chaotic. In practice, the overlap becomes critical already when intervals of only half the width (9.9) touch.

9.5 How Rational Are the Irrational Numbers?

Resonances are characterized by the fact that the ratio of two frequencies in the unpertubed system is a *rational number,* the ratio of two integers. If the system without the perturbation is not degenerate, i.e., if the matrix of the first and second derivatives of $H_0(M, N)$ with respect to M and N in (9.9) does not vanish, the resonant points on the energy surface $H_0(M, N) = E$ are exceptional, exactly as the rational points are exceptional in the interval (0,1). Nevertheless, these points are dense, and it is hard to see how one ever gets away from them; equivalently it is not clear how one ever escapes the dire consequences of even small perturbations $V_{k,\ell}$ in (9.1) if they don't vanish for any pair (k, ℓ).

The salvation can only be sought in the *non-rational,* alias *irrational, numbers* of which there are, of course, many more than rational ones. One has to be at a safe distance from the rational numbers, however, so as to avoid their destructive influence. In order to define such a distance, the manner in which an irrational number can be approximated by a rational one has to be studied. Humanity has pondered this problem ever since anybody tried to measure the diagonal of a square or the circumferene of a circle. In the first case the nature of $\sqrt{2}$ is at stake, and in the second the nature of the number π. Although a knowledge of *continued fractions* is not absolutely necessary for the discussion of resonances, even a cursory discussion of this beautiful subject will make everything better understood and will be useful in the study of hard chaos of Section 20.2. For a detailed presentation cf. textbooks on elementary number theory, e.g., Khintchine 1963; Drobot 1964; Stark 1970.

Every real positive number α can be expanded in the form

$$\alpha = a_0 + \cfrac{1}{a_1 + \cfrac{1}{a_2 + \cfrac{1}{a_3 + \dots}}} = \{a_0; a_1, a_2, a_3, \dots\}, \quad (9.10)$$

where $a_0, a_1, a_2, a_3, \dots$ are positive integers with the exception of a_0, which can also be 0. The continued fraction for α is obtained by the following algorithm: if $x > 0$, call $[x]$ the largest integer that does not exceed x; let $a_0 = [\alpha]$, $\alpha_1 = \alpha - a_0$, and $a_1 = [1/\alpha_1]$; now work out recursively $\alpha_{n+1} = \alpha_n - a_n$ and $a_{n+1} = [1/\alpha_{n+1}]$.

The reader will be immediately captured if not obsessed by continued fractions after trying out a few examples on a hand-held computer.

The continued fraction for a rational number is finite. For the diagonal of the unit-square one finds that $\sqrt{2} = \{1; 2, 2, 2, 2, ...\}$; one of Lagrange's great achievements was to show that the necessary and sufficient condition for a continued fraction to be periodic is for α to be a quadratic (irrational) number, i.e., the solution of a quadratic equation with integer coefficients like $\alpha^2 = 2$.

The next example takes Euler, both to figure out numerically and then to prove rigorously: for the base of the natural logarithms, he found that

$$e = 2.71828\,18284 = \{2; 1, 2, 1, 1, 4, 1, 1, 6, 1, 1, 8, 1, ...\}.$$

The number π seems to defy any simple rule since it yields $\{3; 7, 15, 1, 292, 1, 1, 1, 2, 1, 3, 1, 14, 2, 1, 1, 2, 2, 2, 2, 1, 84, 2, ...\}$; in case the reader should be worried by the many occurrences of 1, it was shown by Gauss that if α is chosen at random in the interval $(0, 1)$, then the probability for α_n to be in the interval $(x, x + dx)$ is given by $dx/((1 + x) \log 2)$ for large n; the domain $1/2 < \alpha_n < 1$ leading to $a_{n+1} = 1$ has a probability of 42%.

Celestial mechanics provides the oldest and most significant practical applications of continued fractions. At issue are the ratios between the solar, lunar, and planetary periods; there is no reason why they should be simple; but the authors of antique texts have always expressed these ratios in rational numbers, for both technical as well as philosophical reasons. Nowadays, one looks up the relevant periods as obtained from the best observations, given in mean solar days, and calculates the continued fraction.

The *Astronomical Almanac* for 1987 gives 365.242191 days for the *tropical year* (equinox to equinox) and 29.530589 days for the synodic month (new moon to new moon) for a ratio of $12.3682663 = \{12; 2, 1, 2, 1, 1, 17, 3\}$. For the truncated continued fraction $\{12; 2, 1, 2, 1, 1\} = 12 + 7/19 = 235/19 = 12.368421$, one can say that after one *Metonic cycle* of 19 years = 6939.601 days the new moon is about 2 hours late because 235 months take 6939.688 days. The *Islamic year* is defined as 12 synodic months so that the Islamic calendar is already ahead of the Christian calendar by 7 months after 19 Christian years; the Islamic calendar gains about 3 years for every Christian century. The *Jewish calendar* makes a compromise by decreeing a cycle of 12 short years interspersed with 7 long ones. The full length of the continued fraction yields $12 + 376/1021 = 12.3682664$ which is already the full accuracy of 7 decimals; a total of (only) 1021 years is required to bring about the near complete coincidence of the two calendars!

When the continued fraction of an irrational number α is truncated at the n-th term, a rational number p_n/q_n is obtained which is a particularly good approximation, and is therefore, called the *n-th convergent* of α. The integers p_n and q_n are calculated by the recursion formulas

$$p_{n+1} = a_{n+1}p_n + p_{n-1}\,, \quad q_{n+1} = a_{n+1}q_n + q_{n-1}\,, \qquad (9.11)$$

with the initial values $p_0 = a_0$, $q_0 = 1$, $p_1 = a_1a_0 + 1$, $q_1 = a_1$. The even-numbered convergents form a monotonically increasing sequence that converges to α from below, while the odd-numbered convergents converge monotonically to α from above. Compared with any other rational p/q the convergents are distinguished because they satisfy the inequality

$$|\,\alpha - p/q\,| < 1/2q^2\,. \qquad (9.12)$$

In contrast to this result, notice that if the integer q is fixed, it is generally not possible to get closer to α than $1/2q$.

All these propositions are easy to prove, as is the more precise estimate $|\alpha - p_n/q_n| < 1/q_nq_{n+1} < 1/a_{n+1}q_n^2$, where we have used $q_{n+1} > a_{n+1}q_n$. The n-th convergent is, therefore, particularly good when a_{n+1} is large; e.g., in the ratio (tropical year)/(synodic month) of the preceding paragraph one has $a_6 = 17$ so that $p_5/q_5 = 235/19$ is within $1/(17 \times 19 \times 19) \simeq .00016$ of the correct value. A similar case is π where $a_2 = 15$ so that $p_1/q_1 = 22/7$ is good to $1/(15 \times 7 \times 7) \simeq .00136$; and even more unexpected is $a_4 = 292$ so that $p_3/q_3 = 355/113$ is within .00000027 of π.

The additional occurrence of $a_{12} = 14$ and of $a_{21} = 84$ in the continued fraction for π makes one suspicious that π may be so close to the rational numbers that a torus with this frequency ratio might be easily destroyed by the nearby resonances. In fact, in order to guard against resonances, one would like to find numbers α where the difference $|\,\alpha - p/q\,|$ never falls much below the value $1/2q^2$, which is always realized by the convergents for every irrational number.

Liouville gave a simple and constructive proof to show: If α is the root of the equation $c_0x^N + c_1x^{N-1} + ... + c_N = 0$ where $c_0, c_1, ..., c_N$ are integers, and $c_0 \neq 0$, there exists a number δ such that $|\,\alpha - p/q\,| > \delta/q^N$ whatever integers p and q are chosen. E.g., if α is a quadratic number such as $\sqrt{2}$, no rational p/q will get close to α within less than $2/3(c_1^2 - 4c_0c_2)^{1/2}q^2$.

This theorem was used by Liouville to construct the first *transcendental number,* i.e., a number α_L that is not the solution of an algebraic equation with integer coefficients. Let α_L have the continued fraction with $a_n = 10^{n!}$; then $a_{n+1} = (a_n)^{n+1}$, and one shows easily that $q_n < 2\;10^{1!+2!+\dots+n!} < 10^{2(n!)} = a_n^2$; our estimate then shows that $|\,\alpha_L - p_n/q_n| < 1/a_{n+1}q_n^2 < 1/(q_n)^{(n+5)/2}$. If α_L were algebraic of degree N, its $2N$-th convergent would already beat Liouville's lower limit.

Transcendental numbers that beat the Liouville conditions are called *Liouville numbers*. Not all transcendental numbers are Liouville; in fact, there is an ongoing competition between mathematicians to 'lower the boom' on transcendental numbers. The present record for π is held by Gregory Choodnovsky (1979) who showed that $|\pi - p/q| > \delta/q^7$. Algebraic numbers, too, are much less rational than the Liouville conditions would suggest. Indeed, a difficult and non-constructive argument by Roth shows that for an *algebraic number* $|\alpha - p/q| > \delta/q^{2+\varepsilon}$ with ε positive and arbitrarily small, but not 0. These modern results on irrational numbers are quite subtle; they require a large formal apparatus and serve as a warning against the temptation to oversimplify the mathematics of soft chaos.

9.6 The KAM Theorem

The explanations of the first four sections of this chapter about the effect of small perturbations on an integrable system can be made mathematically precise. The necessary formal apparatus is formidable, however, and the results are disappointing in the sense that the rigorous limits on the relevant propositions are much more stringent than the numerical examples indicate. Nevertheless, it is important to be aware of the minimum that can be guaranteed by mathematical proofs, and to be reassured that the somewhat superficial discussion of this chapter has some validity.

The central KAM theorem and its proof were first suggested by Kolmogoroff (1954), but the details were worked out by Arnold (1963), and the same conclusions were then established under much broader assumptions by Moser (1962). Benettin, Galgani, Giorgilli, and Strelcyn (1984) have given a self-contained presentation of the proof along Kolmogoroff's original scheme.

The main object is to overcome the small denominators which appeared in the discussion of canonical transformations of Section 5.3, particularly in formula (5.8) and which cannot be avoided. One would like to be sure that the second, non-trivial term in the generating function (5.6) is small in spite of the denominator $\omega_m = (m, \omega) = m_0\omega_0 + m_1\omega_1 + m_2\omega_2 + m_3\omega_3$ where we abbreviate $m = (m_0, m_1, m_2, m_3)$. This condition requires that the corresponding perturbation V_m become small with increasing $|m| = |m_0| + |m_1| + |m_2| + |m_3|$ at a faster rate than ω_m. Our main discussion in this section will center on various ways in which this condition can be realized.

Satisfying this requirement is not the whole story, however, because even if one particular canonical transformation as given in Section 5.3 succeeds in overcoming its small denominator, all the other terms of the perturbation (3.5) have to be transformed before going on to the next step. The next term to be treated in the new series (3.5) is characterized by the integers m', and the new coefficient $V_{m'}$ may be quite different after the transformation compared with its value in the series (3.5) before the transformation. If the old coefficient was small for its associated frequency $\omega_{m'}$, there is *a priori* no assurance that this will still hold for the new coefficient.

Another difficulty comes from the modifications in H_0 at each step in the perturbation procedure. The expressions (5.3) for the old actions I_i in terms of the new ones J_j have to be inserted into $H_0(I_1, I_2, I_3)$ and the result has to be expanded in powers of ε. The even powers contain terms, like the square of (5.7), that yield a correction to H_0 because $\sin^2(...) = 1/2 - \cos 2(...)/2$, of which the first part adds to the new H_0. The zero-order Hamiltonian for the next step, $H'_0(J_1, J_2, J_3)$ differs, therefore, from $H_0(I_1, I_2, I_3)$ by terms of order ε^2. If the same values are assigned to J_i as to I_i, the new frequencies differ from the old ones; unforeseen resonances can arise.

An example is the near-resonance between the three main motions of the Moon in the combination $2\omega_{\text{node}} + \omega_{\text{perigee}} - 3\omega_{\text{sidereal}}$, which was mentioned in Section 5.4. The expansion in powers of the small parameter the $n'/n \simeq 1/13$ has to be carried to the third power before this resonance appears; in second order, one would believe that $\omega_{\text{node}} + \omega_{\text{perigee}} = 0$ as formulas (5.9) and (5.10) show to Newton's great distress. Hill's theory of the Moon circumvents this problem by obtaining the complete expansion in n'/n before treating any of the other small parameters.

The purpose of the KAM theorem is to demonstrate the continued existence of certain invariant tori as the perturbation parameter ε increases from 0. The frequencies ω_m are kept away from any resonance by requiring that

$$|\omega_m| = |(m, \omega)| > \delta\, |m|^{-\nu} = \delta\, (\Sigma\ |m_i|)^{-\nu} \qquad (9.13)$$

at each step, where δ and ν are independent of m. In a system with two degrees of freedom, the results of the preceding section can be used to insure this condition. If the frequency ratio $\alpha = \omega_1/\omega_2$ is irrational, then $|q\omega_1 - p\omega_2| = q\omega_2 |\alpha - p/q| \neq 0$ as long as both ω_2 and q differ from 0. Therefore, if $|\alpha - p/q| > \delta/q^{\nu+1}$, condition (9.13) holds. With more than two degrees of freedom, however, the question of the preceeding section becomes how to approximate simultaneously two or more ratios of frequencies, α, β, etc. by rational numbers; the beautiful theory of continued fractions is no longer available, and very little is known. Nevertheless, condition (9.13) is the central assumption.

Two more ideas are important to complete the proof of the KAM theorem. Instead of the stepwise expansion in powers of ε where the remainder is smaller than the last term by only one factor ε, a construction is used where the remainder is of order ε^{2k} after the terms of order ε^k have been treated. The elementary example of this superconvergence is found in Newton's method for solving equations such as $f(x) = 0$; if x_0 is a first guess, the first approximation x_1 is obtained from solving the linear equation $f(x_0) + f'(x_0)(x - x_0) = 0$; the second approximation x_2 follows then from the linear equation $f(x_1) + f'(x_1)(x - x_1) = 0$, and so on. When solving for x_2, it is no longer admissible to calculate $f(x_1)$ and $f'(x_1)$ by expanding in powers of the small difference $(x_1 - x_0)$. Also, this 'superconvergence' is not applied to the full perturbation, but only to a properly smoothed version, not unlike singling out a particular, slowly varying term in explaining the effects of a resonance. These sketchy indications are intended to show that the KAM theorem cannot be proved without techniques outside conventional perturbation theory.

The KAM theorem will now be quoted from Arnold's book (1978):

> *If an undisturbed system is non-degenerate, then for sufficiently small conservative Hamiltonian perturbations, most non-resonant invariant tori do not vanish, but are only slightly deformed, so that in the phase space of the perturbed system, too, there are invariant tori densely filled with phase curves winding around them conditionally-periodically, with a number of independent frequencies equal to the number of degrees of freedom. These invariant tori form the majority in the sense that the measure of the complement of their union is small when the perturbation is small.*

(A conditionally-periodic motion is the same as our multiperiodic motion in Sections 3.2 and 3.3.)

The resonant tori of the undisturbed system remove from phase space a layer whose thickness in the space of the frequency ratios α is essentially given by (9.9). The KAM theorem is made plausible if the widths of these layers for all rational frequency ratios are added up, and their sum is found to be finite; the strength ε can then be chosen to leave enough space for the frequencies which satisfy the criterion (9.13). If the perturbation is analytic as in the Kolmogoroff-Arnold formulation, the expansion coefficients V_m in (3.5) decay exponentially with $|m|$, and the sum over all vectors of integers m converges. If the dynamical system has n degrees of freedom, and the perturbation is only required to have μ continuous derivatives, as in Moser's version, the coefficients V_m decay as $|m|^{-\mu-2}$. Since the width of the resonance (9.9) goes with $|V_m|^{1/2}$, and there are $\simeq |m|^{n-1}$ terms for a given $|m|$, convergence requires at the least that $(\mu+2)/2 - (n-1) > 1$, or $\mu > 2n - 2$. Moser gave the sufficient condition $\mu \geq 2n + 2$; the reader will find more details of this type in the monograph of Lichtenberg and Liebermann (1981).

A similar argument applies directly to the criterion (9.13): the frequencies ω_m which violate (9.13) for some fixed m fill a layer of thickness $\delta/|m|^{\nu+1}$ in frequency space. Adding up all these excluded layers leads to the sum of $\delta/|m|^{\nu-n+2}$ over $|m|$, which converges provided $\nu > n - 1$. In a system with two degrees of freedom, therefore, $\nu > 1$, which excludes all the quadratic numbers, and probably most of the algebraic ones, too, according to Roth's result. In the most common form, the KAM theorem is based on $\nu = 3/2$. While this kind of argument yields reasonable inequalities for μ and ν, it deals only with the frequencies whose neighborhood gets transformed into soft chaos. By contrast, the KAM theorem insures the survival of the invariant tori, and tells us something about the set in frequency space that is complementary to the incipient chaos.

9.7 Homoclinic Points

The destruction of the invariant tori can be seen as a catastrophe that overtakes the dynamical system as it is subjected to an ever stronger perturbation. A lot of effort, both numerical and analytical, has been spent on trying to understand the detailed mechanism by which the orderly structure of phase space gets lost, and to get a glimpse of what happens after the disaster has occurred. The reader will have to work through some of the vast literature to get an adequate picture of what is known so far (cf. Arnold and Avez 1967; Devaney 1985; Guckenheimer and Holmes 1983; as well as the survey by Grebogi, Ott, and Yorke 1987). This and the next section will briefly mention two ideas that seem to dominate much of this work.

The discusssion in Section 9.3 showed how the layered structure of the nested tori (cf. Figure 11d) gets broken up so as to resemble the phase space of a pendulum (cf. Figure 21). The immediate cause is a resonance with a sufficiently strong perturbation; but no great harm is done as long as this resonance stays isolated. The phase-lock region looks like the vibrational motions of the pendulum, now represented by a set of islands as in Figure 16; these are smoothly bounded by the separatrix, beyond which the original layered structure again takes over, just as the rotational motions of the pendulum; the critical element in this situation is the smooth nature of the separatrix.

A first general problem, therefore, concerns the loss of the separatrix in one single chain of islands. The discussion of Sections 9.3 and 9.4 gives no indication why the points $P_1, P_2, \ldots$ of unstable equilibrium between the islands should not remain the intersections of two smooth torus-like surfaces, the two branches of the separatrix as it were; the transition from Figure 23a to Figure 23b, however, indicates something quite different.

The points $P_1, P_2, \ldots$ continue to be part of a periodic orbit as the perturbation increases; but the linear map in their neighborhood acquires a trace larger than 2. As a consequence, there are two well-defined directions through each of them: the stable one for the trajectories approaching ever more closely at each pass through the surface of section, and the unstable one for the trajectories getting further and further away, as in the typical hyperbolic periodic orbit of Figure 11b.

For sufficiently weak perturbations, these two directions, say the stable one at P_1 and the unstable one at P_2, are connected by the separatrix so that a trajectory that starts near P_1 on the unstable manifold eventually ends up near P_2, typically after a very long time. As the perturbation becomes strong enough, the stable as well as the unstable directions at $P_1, P_2, \ldots$ continue to define smooth curves in the surface of section, namely the intersections of the stable and the unstable manifolds with the surface of section. The existence of these rather smooth manifolds, and their smooth lines of intersection, when everything else in phase space seems to break down, is something of a mathematical miracle. The trajectories on the stable manifold are characterized by their tendency to approach ever more closely to P_1 or P_2 at each pass in the future; trajectories on the unstable maniflold approach in the same way when going backward in time. But the unstable line of P_1 does no longer join up smoothly with the stable line of P_2, or any other stable line in the surface of section, to form a separatrix; and similarly for P_2.

On the contrary, lines originating in different points P_i will intersect transversally in a well-defined set of points. They are called *heteroclinic points,* if the intersection occurs between the stable and unstable lines belonging to different points P_i and P_j ; if the stable and unstable lines come out of the same point corresponding to an unstable periodic orbit, their intersections are called *homoclinic points.* The existence of these points causes a profound reorganization in the neighborhood of the unstable equilibria of the pendulum. Although this description refers only to one isolated chain of islands, it is the presence of a neighboring resonance which is responsible for the emerging chaos.

The conservation of area in the surface of section is now the main culprit in making things so complicated, together with the fact that the stable or the unstable lines cannot intersect themselves. The relevant arguments were first presented in detail by Birkhoff (1935); but already Poincaré (1899, last chapter) had understood their full impact. Since a serious discussion would take too much space, we will try to give the reader at least a glimpse of classical chaos in the making due to an homoclinic point; Figure 24 shows schematically how the stable and unstable lines are bound to form very complicated folding patterns.

Let the stable line ω out of the unstable periodic orbit through P, intersect the unstable line α out of the same point P in the homoclinic point Q_0. The name α was chosen to remind the reader that the pre-images of Q_0 in the surface of section, $Q_{-1}, Q_{-2}, \ldots$, lie on α, and form a sequence of points converging to P in the past; similarly, the images of Q_0, called $Q_1, Q_2, \ldots$, lie on ω where they form a sequence converging to P in the future. Thus, lines α and ω keep intersecting each other and forming new homoclinic points.

Between any two consecutive intersections, Q_j and Q_{j+1}, the segments along α and ω define an open domain D_j; all these domains enclose an area A which is the same, independent of j. If self-intersections of either α or ω are to be avoided, these domains have to take on increasingly contorted shapes. They become thinner and longer, because one of the bounding segments gets short quickly with large $|j|$; e.g., for $j < 0$, the points Q_j lie on the initial portion of ω near P, and their distance obviously decreases exponentially.

Eventually, one of these domains doubles back on itself to the point where both ends run right across one of its predecessors; the cut of the boundaries is allowed only of α with ω. The abstract version of this situation is the famous *horseshoe* of Smale (1965); the original domain D of roughly rectangular shape is stretched in one direction, and then bent like a horseshoe big enough so that its two ends run across D. Part of D is mapped into itself, and this map can be repeated indefinitely, both forward and backward.

This process leads to a natural coding scheme in terms of binary sequences, because there is a choice between the two branches of the horseshoe at each step of the consecutive transformations of D. The long-term behavior of the trajectories is now characterized by the *symbolic dynamics* of the binary sequences; in particular, it is easy to see that there are infinitely many new periodic orbits, as well as trajectories with a seemingly random character, according as the binary sequence is periodic or random. The basic mechanism behind these

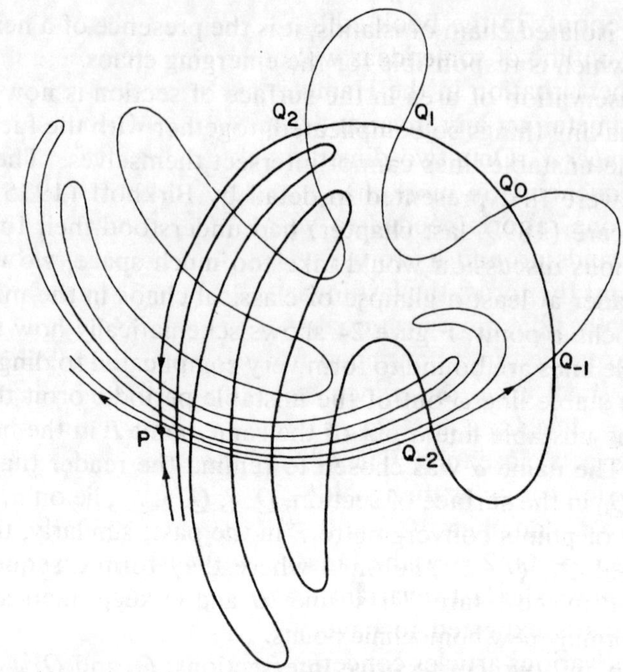

Figure 24 The formation of homoclinic points when the stable and the unstable manifolds of the same unstable periodic orbit intersect transversally.

unexpectedly complicated features in a simple dynamical system is the stretching and bending of an initially nice convex area in phase space, in agreement with Liouville's theorem.

The simple device of letting the stable and unstable lines in the surface of section intersect each other has produced the most extreme chaos, something akin to the flipping of a coin. While this picture is very generally valid, it does not yield the detailed and exhaustive information about the classical trajectories, in order to make the transition to quantum mechanics; the symbolic dynamics is applicable only to a very tiny subset in the surface of section. In contrast, we will find that all the trajectories in the Anisotropic Kepler Problem (Chapter 11), and the geodesics on a surface of negative curvature (Chapter 19), can be fully understood in terms of a simple coding scheme.

9.8 The Lore of the Golden Mean

A second general problem concerns the fusion of different chains of islands into a single region of chaotic character through the destruction of the invariant tori between them. The reader will find a useful selection of reprints in the book edited by MacKay and Meiss (1987); only a bare outline of some ideas will be given in this section.

As the perturbation in the Hamiltonian is allowed to increase, the critical parameter on any invariant torus in the original system is the *winding number* α. On a two-dimensional torus, it is the ratio between the two frequencies; in more than two degrees of freedom, however, more than one ratio is needed to characterize the way in which the trajectory winds around its torus. As a rule, the less rational α, the more resistant the torus; the last invariant torus to be destroyed by the perturbation has the most irrational frequency ratios.

If α is expanded in a continued fraction (9.10), and the values of a_n remain small, the difference between successive convergents, $1/q_n q_{n+1} = 1/q_n(a_{n+1} q_n + q_{n-1})$ remains large, and neither convergent ever gets close to α. The worst case is obviously $a_n = 1$ for all n; the recursion formulas then yield $p_n = f_{n-1}$, $q_n = f_n$, where f_n is the n-th Fibonacci number: $f_{-1} = 0, f_0 = 1, f_1 = 1, f_{n+1} = f_n + f_{n-1}$; and $\alpha = (\sqrt{5} - 1)/2 = \gamma$, the famous *golden ratio* γ of Greek geometry. Therefore, the last invariant torus to disappear as the perturbation increases is expected to have the winding number γ. A useful collection of various articles concerning this phenomenon has been put together by Cvitanovic (1984).

Even though more and more island chains are created, the remaining tori set up barriers which prevent the trajectories from drifting indiscriminately all over phase space. This conclusion holds rigorously for conservative systems with two degrees of freedom, because the two-dimensional tori effectively separate the three-dimensional energy surface into distinct open sets. With three degrees of freedom, however, a three-dimensional torus does not divide the five-dimensional energy surface into two disconnected open pieces; a trajectory that originates near one resonance eventually drifts toward some other resonance although there still exists an invariant torus at an intermediate set of winding numbers. This process, called *Arnold diffusion* (Arnold 1964) is actually very slow; it has been found to take a time $\simeq \exp(-1/\varepsilon^\sigma)$ with $\sigma \simeq 1/2$, while the diffusion of the trajectory inside its own original island chain takes a time $\simeq 1/\varepsilon^\tau$ with $\tau > 0$; ε is the strength of the perturbation (cf. Chirikov 1979). This further complication of the transition from integrable to chaotic behavior is still largely unexplored (cf. Piro and Feingold 1988), and we shall, therefore, return to the safety of systems with no more than two degrees of freedom.

The most closely studied systems are area-conserving maps, and among them in particular the so-called *standard map*,

$$I^{(n+1)} = I^{(n)} + K \sin w^{(n)}, \quad w^{(n+1)} = w^{(n)} + I^{(n+1)}. \tag{9.14}$$

The corresponding physical model, the *kicked rotator,* is a pendulum of length a and mass m, or equivalently, a rotator with moment of inertia ma^2 which is kicked with an impulse $F\,\Delta t\,\sin w$ at equal time intervals, Δt apart; w is the angular position at the time of the impulse; the motion is a uniform rotation between kicks. If one normalizes the angular momentum J by setting $I = J\Delta t/ma^2$, and calls $K = F(\Delta t)^2/ma$, the first equation (9.14) gives the increase in angular momentum from the n-th kick, and the second equation describes the increase of the angular position between the n-th and the $(n+1)$-th kick. The conservation of area can be checked by calculating the Jacobian $\partial(I^{(n+1)}, w^{(n+1)})/\partial(I^{(n)}, w^{(n)}) = 1$.

The ordinary pendulum arises in the limit $\Delta t \rightarrow 0$ which corresponds to $K \rightarrow 0$. The difference equations (9.14) become two ordinary first-order differential equations with the Hamiltonian $H_0 = (J^2/2ma^2) + F\cos w$. Near the stable equilibrium at $w = \pi$ the frequency of the motion is $\sqrt{F/ma^2}$; the separatrix including the unstable equilibrium at $w = 0$ is given by $H_0 = F$, where $I = \pm 2\sqrt{ma^2F}\,\sin(w/2)$, and the period is infinite. As Δt increases from 0, the trajectories tend to get stuck and form chains of islands around the periodic orbits of the difference equations (9.14); but for small K they are separated by what remains of the curves $H_0 = const.$, where the winding number α is irrational.

These are the *KAM tori,* so called because they are guaranteed to exist by the KAM theorem; they set up barriers for the classical trajectories, whose presence is critical for the understanding of chaos. For their practical calculation, the periodic orbits have to be found at a fixed K for a sequence of rational winding numbers that are the convergents for a given irrational number such as γ. As long as the corresponding KAM torus still exists, the trace for the local map near the periodic orbit will become 2 for the higher convergents. Greene (1979) calculated the residue (cf. section 6.4) for the relevant sequence of periodic orbits, and found their asymptotic behavior as a function of their order in the continued fraction expansion. He showed that the last KAM-torus for the standard map (9.14) belongs to the golden mean γ, and disappears when $K > .9716$, as shown in Figure 25.

According to Escande and Doveil (1981), however, it is not necessarily the winding number γ which determines the last KAM torus (cf. the review by Escande 1985). Nevertheless, the standard map gives a particularly simple example where the transition to global chaos in the phase space can be studied, even after K has become larger than its critical value. As a rule, the region of the last KAM torus acts as a barrier, on which the trajectories get stuck for a while even then, al-

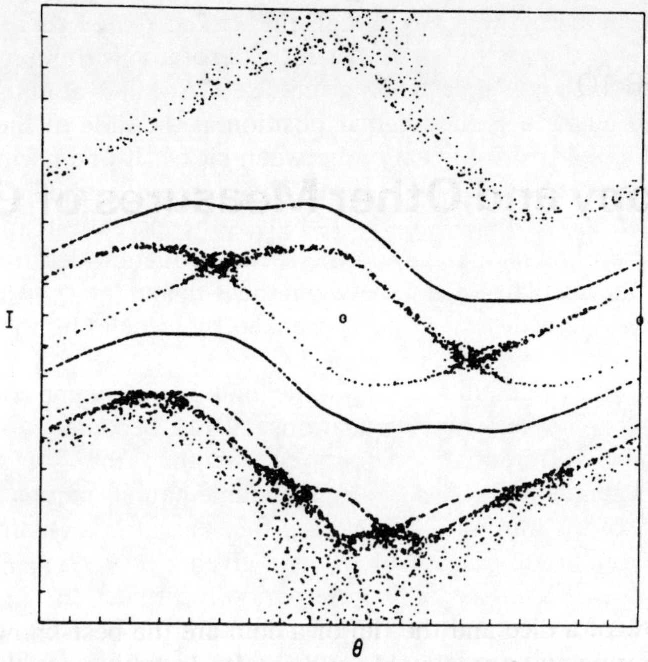

Figure 25 The last invariant torus separating different ergodic regions in the surface of section for standard map (9.12). [from Greene (1979)]

though they will eventually traverse it. It appears that this delay near a former KAM torus is crucial in quantum mechanics; a wave function is able to maintain itself in such a region, almost as if there was still a good invariant torus to settle on. Reinhardt (1985) has coined the phrase of 'vague' or 'fragmented' torus for this situation in classical phase space.

CHAPTER 10

Entropy and Other Measures of Chaos

The throw of a dice and the flip of a coin are the best-known devices for generating an unpredictable end result. In both cases the detailed sequence of events cannot be controlled at each step so as to guarantee a definite outcome. The main problem seems to lie with the starting conditions although relatively few parameters are involved: the position and the momentum of the center of mass, as well as the orientation in space and the angular momentum at a given instant in time. These quantities would have to be measured with an accuracy far exceeding any practical instrument; moreover, the drag from the air and the bouncing from surfaces would have to be investigated to a degree that is not ordinarily available. Nevertheless, there is a chain of cause and effect from beginning to end; nothing is inherently random in the elementary mechanics, but our ignorance makes it so.

Physicists have been aware of this paradoxical situation ever since Maxwell and Boltzmann laid the foundations of *statistical mechanics* at the end of the nineteenth century. But the resolution of the paradox depended on the large number of degrees of freedom, which obviates any effort to take into account the motions of all the particles involved. The mathematicians realized that similar problems of predictability arise in much simpler systems where the equations of motion can be written without much trouble and can be solved numerically on a (by modern standards) primitive computer. A number of very useful concepts were developed to define different types of such behavior; the most important notion is called *entropy*. This name was adopted from thermodynamics where it designates a measurable quantity that, according to Boltzmann's interpretation, indicates the prevailing degree of disorder.

While the idea of an entropy is of great help in understanding classical mechanical systems, nobody has been able to find its analog in quantum mechanics; therein lies the great unresolved mystery of quantum chaos. This chapter gives a brief introduction to the concept of entropy in classical mechanics, and some related notions, to make it quite clear that there is nothing imprecise about these quantities. For a rigorous and complete presentation, however, the reader will have to consult the mathematical literature; a particularly fine introduction is the book by Arnold and Avez (1967).

10.1 Abstract Dynamical Systems

In order to be on firm ground, the objects of investigation are defined in terms that are somewhat remote from physics. A *classical dynamical system,* also called a *flow,* is a collection of objects (M, μ, ϕ_t) where M is a differentiable manifold, alias a phase space, μ is a density defined on M, alias Liouville measure, and ϕ_t is a one-real-parameter group of diffeomorphisms (continuously differentiable map) of M into itself that preserves the measure based on μ, alias equations of motion. The restriction to a group rather than an arbitrary one-parameter family makes each time interval δt equivalent to any other of the same duration; that is exactly what happens in a conservative Hamiltonian system.

A particularly popular example is the *geodesic flow on a Riemannian surface N;* the manifold M is a set of points where each is the combination of a point in N and a tangent direction there; the measure μ is the product of the area (volume) of N and the (solid) angle of directions; the flow ϕ_t describes the motion of a particle that moves freely on N with unit speed for a time t.

As a purely mathematical model, one may instead study an *abstract dynamical system:* again a collection of objects (M, μ, ϕ_t) where now M is a measurable space with a measure μ and a group of automorphisms ϕ_t (maps of M into itself) that preserve the measure μ and where the variable t runs through the integers. In contrast to the preceding definition, the space M is only required to be endowed with a measure, and the automorphisms do no more than preserve this measure, so that sets of measure zero are allowed to do whatever they please. The maps on Poincaré sections belong to this kind of dynamical system, since the parameter for them is discrete; but there are now examples which seem at first far removed from ordinary mechanics.

The standard example is the *Bernoulli scheme* $B(p_1, \dots, p_m)$: Consider an alphabet of m letters which will be called $(1, \dots, m)$ for simplicity's sake; a point a in the space M is a sequence $a = (\dots, a_{-1}, a_0, a_1, a_2, \dots)$ where $a_i = 1, \dots, m$. In order to define a measure in this space, the sets $A_{ij} = \{a \,|\, a_i = j\}$ are given the measure $p_j > 0$ with $\Sigma p_j = 1$. The sets A_{ij} generate a so-called *σ-algebra* of measurable sets in M by taking all the possible intersections and unions with the obvious rules for the measure, such as $\mu(A_{i,j} \cap A_{k,\ell}) = p_j\, p_\ell$ provided $i \neq k$.

The group of automorphisms is defined as the integer powers of the *shift*, $\phi : a \to a'$ where $a'_i = a_{i+1}$ which translates the whole sequence a to the left by one place. In order to give a physical interpretation to the shift, we call the half-sequence $a_p = (\dots, a_{-2}, a_{-1}, a_0 \,|$ the *past*, and the half-sequence $a_f = |\, a_1, a_2, \dots)$ the *future*. After the event a_1 has occurred, the past becomes $a'_p = (\dots, a_{-1}, a_0, a_1 \,|$, and the future is $a'_f = |\, a_2, a_3, \dots)$.

The simplest among the Bernoulli schemes, $B(1/2, 1/2)$, is nothing but the coin toss: the experience of a particular person in this game is described by a binary sequence a of letters in the alphabet ($up = 1$, $down = 2$) which occur with equal probabilities $p_1 = p_2 = 1/2$. The probability for any special combination of sequences is the same as the measure of the subset in the space M that is made up of these sequences. Notice that no concept of closeness between the points a is required, only a notion of measure. Binary sequences will occur again in the discussion of the *Anisotropic Kepler Problem* (cf. next chapter); but they will be represented as points in a plane, and acquire thereby a notion of distance between different sequences, which is absent in a Bernoulli scheme.

A more involved example of an abstract dynamical system is the *subshift* where the space M consists again of the infinite sequences a of letters from a finite alphabet, and an automophism is obtained from the translation of a sequence by one place; but the measure μ on M is generated by the sets $A_{i,j,k} = \{a \,|\, a_i = j, a_{i+1} = k\}$ whose measure is given by a matrix of transition probabilities $p_{jk} \geq 0$ with the normalization $\Sigma\, p_{jk} = 1$ where, for each value of j, the sum is taken over k. Such a scheme has a memory built in, since the probability of the future event a_1 depends on the past event a_0. This is the basic model for a *Markoff process.* It is realized in the geodesic flow on a compact closed surface of constant negative curvature that will be discussed briefly in Section 20.3.

The Bernoulli schemes are the most unpredictable, and yet *deterministic* processes imaginable. The word deterministic conveys the basic assumption of classical mechanics: if the initial point a in the space M is known with infinite precision, then the 'trajectory' $a', a'', \ldots$ can be calculated for all later times, and even for all earlier times as well.

A mechanical system can be classified as being 'Bernoulli' although it may not look like that at all. Quite generally, two systems (M, μ, ϕ) and (M', μ', ϕ') are called *isomorphic* if there exists a measure-preserving bijection $f : M \rightarrow M'$ (one-to-one map) modulo sets of measure 0 where $\phi'(f(M)) = f(\phi(M))$, i.e., the order of f and ϕ does not matter, always up to sets of measure 0.

The two Bernoulli schemes $B(1/2, 1/8, 1/8, 1/8, 1/8)$ and $B(1/4, 1/4, 1/4, 1/4)$ are isomorphic; the reader may try to imagine how the bijection f can be established and will find it very difficult. Its existence depends on the equality of the entropies of the two schemes as will be discussed in Section 10.3.

Besides the isomorphism between two particular dynamical systems it is of interest to know whether a system is typical. The answer requires a concept of closeness (topology) between systems whereby each system is surrounded by a neighborhood of other systems, and the problem is to find the *generic* properties that are valid for a whole neighborhood even if the systems in it are not equivalent. Abstract dynamical systems have a better chance to be generic than the classical dynamical systems, which are described under more restrictive assumptions. In particular, since Hamiltonian systems, which are the topic of this book (and the basis of physics), have very few generic features, there is a tremendous variety of truly dissimilar types.

10.2 Ergodicity, Mixing, and K-Systems

The mathematicians have developed a finely tuned hierarchy of possible behavior in dynamical systems. We will mention the three most commonly used, and define them only with the degree of precision necessary to convey their basic differences.

(i) *Ergodicity:* For each integrable function $f : M \rightarrow R$, the spatial mean $\int f \, d\mu$ equals the temporal mean $(1/T)\int_0^T f(\phi_t x) \, dt$ when $T \rightarrow \infty$, and where x is any point in M with the exception of a set of measure 0. The integral over t is replaced by a sum when the parameter t varies in discrete steps. This definition corresponds to the original idea that Boltzmann invoked to explain the second law of thermodynamics. It turns out to be widely realized in dynamical systems; but it is not nearly strong enough to guarantee what Boltzmann intended to accomplish with it.

The repeated rotations around a point by an irrational angle are ergodic; this is the famous theorem that Weyl proved for the first time, although it must seem obvious to most non-mathematicians. The space M is the circumference of the unit-circle, the measure μ is the length along the circumference, and the automorphism is the rotation by the angle $2\pi\alpha$ where $0 < \alpha < 1$. Pandey, Bohigas and Giannoni (1989) have shown that this system has some very non-random features which make it unsuitable as a model in statistical mechanics; the angular distances between the points on the circle take values that are severely restricted rather than being arbitrary.

Ergodicity implies that the phase space M cannot be decomposed into subsets with non-vanishing measure each of which remains invariant. This idea is presented in many textbooks as the *ergodic hypothesis* which underlies thermodynamics. And yet, no explicit model for the basic phenomenon of heat conduction was known until Casati, Ford, Vivaldi and Visscher (1984) gave a very appealing demonstration: a few particles move inside a segment of finite length; when they reach the left end, they are reinjected with a kinetic energy corresponding to the lower temperature T_1, whereas they reappear with the higher temperature T_2 upon reaching the right end of the interval. During their travel back and forth, these particles hit massses that are suspended from a point like a pendulum and deprive the particles of their energy. When the pendulum comes back to its original position, the particle gets its energy back and can continue its trip down the segment; but the waiting time is longer for the higher kinetic energy, and this retardation of the fast particles accounts for an orderly transfer of energy at a finite rate that is proportional to the temperature difference as in Fourier's law.

(ii) *Mixing:* Given two subsets A and B of M, as $t \to \infty$ one has $\mu(\phi_t(A \cap B)) = \mu(A)\,\mu(B)$; each subset eventually gets spread out homogeneously. The rotations by an irrational angle do not satisfy this condition; two overlapping intervals of the circle remain exactly as they were at the start. On the other hand, it takes only relatively little stirring to mix coffee and milk quite thoroughly.

Neither ergodicity nor mixing gives us any understanding of a peculiar paradox in chaotic systems: on the one hand, they seem to obliterate any simple pattern that a trajectory might be designed to follow; on the other, they are full of the most rigid of such repetitive behavior, i.e., periodic orbits for isolated starting conditions. Indeed, the rotations by a fixed irrational angle $2\pi\alpha$ yield no periodic orbits whatsoever, although they are ergodic. The paradox appears when the scrambling of the space M proceeds in a much more pervasive manner, which was first studied by Kolmogorov.

Before giving the relevant definition, however, a number of technical terms have to be defined. A collection Ξ of subsets in M is called an *algebra,* if it is closed under the operations of taking the union for a denumerable family of subsets and of taking the complement of a subset (and, therefore, of taking the intersection between subsets). An algebra Ξ_0 is said to be contained in the algebra Ξ_1, if for every subset $A_0 \in \Xi_0$ there exists a subset $A_1 \in \Xi_1$ such that $\mu(A_0 \cup A_1 - A_0 \cap A_1) = 0$, i.e., up to sets of measure 0 the algebra Ξ_1 divides up the space M at least as well as the algebra Ξ_0. If a number of algebras Ξ_i are available, then there exists a maximal algebra $\cap\, \Xi_i$ which is contained in all of them, and a minimal algebra $\cup\, \Xi_i$ which contains them all. If the map ϕ is applied to the subsets of an algebra Ξ, the images of all the subsets of Ξ generate another algebra with the obvious name $\phi(\Xi)$.

(iii) *K-system:* there exists an algebra Ξ of measurable subsets in M which is contained in the algebra $\phi(\Xi)$ in such a way that $\cap\, \phi^n(\Xi) = \Omega$ and $\cup\, \phi^n(\Xi) = \Psi$, both taken over all integers n; the algebra Ω contains only sets of measure 0 or 1, whereas the algebra Ψ contains all the measurable subsets of M. Similar definitions apply to the classical dynamical systems where the parameter t is continuous.

The prime example of this construction are the Bernoulli schemes. The algebra Ξ is generated by the subsets $A_{i,j}$ with $i > 0$. Since $\phi(A_{i,j}) = A_{i-1,j}$, this algebra is further subdivided with each subsequent map ϕ. Every measurable subset with non-vanishing measure is caught eventually, with the possible exception of subsets with measure 0, or the whole space, because the complements always belong to an algebra.

10.3 The Metric Entropy

The concept of entropy in a dynamical system was formalized by Kolmogorov on the basis of the definition that Boltzmann had originally used in his theory of gases. Similar expressions are found throughout statistical mechanics and have become the essence of information theory. Again, an awesome degree of mathematical abstraction seems unavoidable, and we can only give a (possibly bitter) taste of the necessary details.

The space M has to be subdivided by a *decomposition* α, which is a collection of measurable subsets A_i where the index i now belongs to a collection I ; the union of these subsets is M, and all their intersections are void, up to sets of measure 0. The *entropy* $h(\alpha)$ *with respect to the decomposition* α is defined as

$$h(\alpha) = - \sum_{i \in I} \mu(A_i) \mathrm{Log}(\mu(A_i)) , \qquad (10.1)$$

where *Log* designates the logarithm with the base 2.

The next step in the construction requires the joining of two decompositions α and β in a straightforward manner: a decomposition $\gamma = \alpha \cup \beta$ consists of all the intersections of a subset $A_i \in \alpha$ with a subset $B_k \in \beta$. Now, the *entropy* $h(\alpha, \phi)$ *of the automorphism* ϕ *with respect to the decomposition* α becomes the limit

$$h(\alpha, \phi) = \lim_{n \to \infty} \frac{1}{n} h(\alpha \cup \phi^{-1}(\alpha) \cup \phi^{-2}(\alpha) \cup \cdots \cup \phi^{1-n}(\alpha)) \qquad (10.2)$$

A lot of hard work goes into showing that this limit exists.

The last step consists in getting rid of the reference to the decomposition α by looking for the supremum (the least upper bound) when all finite and measurable decompositions of α of M are considered. Thus, one finds the *metric entropy of the automorphism* ϕ

$$h(\phi) = \underset{\text{finite } \alpha}{\text{supremum}}\, h(\alpha, \phi) . \qquad (10.3)$$

More hard labor is required; but eventually the goal is reached in the form of two theorems.

The entropy $h(\phi)$ *is an invariant of the automorphism in the sense that two isomorphic automorphisms* ϕ *and* ϕ' *have the same metric entropy.* As an example, the two Bernoulli schemes at the end of Section 10.1 have the same value of the metric entropy because they are isomorphic; the equality of the entropies will be obvious as soon as the entropy for Bernoulli schemes is actually calculated in the next paragraph. Meanwhile another theorem is needed to allow us to make this calculation possible. It is natural to use a finite decomposition α to *generate* the algebra Ξ which goes into the definition of a K-system at the end of the preceding section, by joining α with $\phi^{-1}(\alpha)$, $\phi^{-2}(\alpha)$, and so on. If this can be done, α is called a generator with respect of ϕ. *If* α *is a generator for the automorphism* ϕ, *the metric entropy* $h(\phi) = h(\alpha, \phi)$. In other words, it is no longer necessary to find the least upper bound for all possible finite and measurable decompositions of M.

The generating decomposition α for the Bernoulli scheme $B(p_1, \ldots , p_m)$ is obtained from the sets $A_{11}, \ldots , A_{1m}$ as defined in Section 10.1. Since $\phi^{-k}(A_{1j}) = A_{k+1j}$, the subsets of the decomposition $\alpha \cup \phi^{-1}(\alpha) \cup \cdots \cup \phi^{1-n}(\alpha)$ are the intersections $A_{1j_1} \cap A_{2j_2} \cap \ldots$

$\cap A_{nj_n}$ whose measure is $p_{j_1} p_{j_2} \cdots p_{j_n}$. These values have to be inserted into (10.3) which leads back to (10.2) where the family of indices I now consists of $(j_1, j_2, \ldots, j_n)$ where each j varies independently from 1 to m. There is some minor rearrangement of the terms, which then leads to the metric entropy of the shift ϕ in $B(p_1, \ldots, p_m)$

$$h(\phi) = - \sum_{j=1}^{m} p_j \operatorname{Log}(p_j) \ . \tag{10.4}$$

The entropies for $B(1/2, 1/8, 1/8, 1/8, 1/8)$ and $B(1/4, 1/4, 1/4, 1/4)$ have the same value 2 Log 2 = 2, whereas the coin-toss $B(1/2, 1/2)$ has the entropy Log 2 = 1. Intuitively the scrambling of the space M is twice as effective at each step when the entropy is twice; that coincides with our impression of the two former Bernoulli schemes being equivalent to a double coin-toss.

Ornstein proved the remarkable theorem: *All Bernoulli schemes with the same entropy (10.4) are isomorphic.* This proposition is a model of simplicity; but its proof is very hard. It requires the construction of the isomorphism f, which is extremely difficult to find because it cannot be defined in a finite process. The development of this field, including some of the work to be discussed in the next sections, was recently reviewed by Adler (1987).

A favorite model of a strongly ergodic system is a point particle on a flat torus with a circular hole; the particle moves with constant velocity in a straight line until it hits the hole, where it undergoes an elastic reflection, maintaining its momentum parallel to the boundary, and reversing the momentum at right angles to the boundary. The motion gets badly defocused in this kind of billiard game, and Sinai (1968, 1970) has shown that this system has a non-vanishing entropy. A related model is the motion inside a stadium, i.e., two half-circles joined by two parallel lines; although its ergodic properties are not so obvious, Bunimovich (1974, 1979) was able to show that there is again a non-vanishing metric entropy (cf. Bunimovich and Sinai 1980). A recent update of these mathematically oriented developments was given by Katok and Strelcyn (1986).

10.4 The Automorphisms of the Torus

Since mechanics is closer to geometry than to algebra, one would like to have an example where the scrambling of phase space can be seen more intuitively. The automorphisms of the torus fulfill this role; they also show the relation between the metric entropy of the preceding section and the topological entropy of the next section.

Let L be an m by m matrix with integer elements such that det $|L| = 1$. The space M is the m-dimensional hypercube with opposite sides identified just like the m-dimensional torus of the angle variables in Section 3.3. Its points are given by the column vectors x with the components $(x_1, \ldots, x_m)$ where $0 \leq x_i < 1$. The measure is the Euclidean volume, which is preserved because det $|L| = 1$. The automorphism ϕ transforms x into the column vector y with the linear transformation

$$y = Lx \text{ (modulo 1)} . \tag{10.5}$$

The most famous example is *Arnold's cat map* with the 2 by 2 matrix (1, 1; 1, 2) in flattened notation. The pictures of the gradual distortion and disruption of a cat's head have become a compulsory item in every book on chaos.

The effect of the automorphism (10.5) on the m-dimensional cube can be understood by transforming the matrix L into diagonal form with real entries. This can be done for the cat map since L is symmetric; in general, however, the discussion is more complicated. The local neighborhoods get stretched in the direction where the eigenvalues λ of L are > 1, and get compressed in the directions where $|\lambda| < 1$. This argument has to be made more explicit in order to calculate the metric entropy; but the general result is again very simple:

$$h(L) = \sum_{|\lambda| > 1} \text{Log}|\lambda| . \tag{10.6}$$

This formula relates the metric entropy to the local distortion of the space M ; or equivalently, if one thinks of neighboring points as getting progressively further and further away from each other, the metric entropy gives the rate at which the two points diverge.

This last interpretation of the metric entropy is very important for the later applications. Quite generally, any two neighboring initial points can be followed through many successive automorphisms; their distance increases in an exponential manner with the number of automorphisms. The rate of this exponential drift follows from the linearized map and its eigenvalues exactly as for L; the expression on the right-hand side of (10.6) is called the *Lyapounoff number* for the automorphism in some particular neighborhood. Quite generally, the *metric entropy* is given by the average over the Lyapounoff numbers,

$$h(\phi) = < \sum_{|\lambda| > 1} \text{Log}|\lambda| > . \tag{10.7}$$

The exponential spreading of the trajectories was already mentioned in Section 1.6.

The linear automorphism shows very clearly the occurrence of periodic orbits. For this purpose, let us study the n-th power $\Lambda = L^n$ of the matrix L. A point x in the torus which returns to itself after n automorphisms satisfies the equation $\Lambda x = x$ *modulo* 1, or more explicitly, $\Lambda x = x + k$ where k is a column vector with m integer components. This condition can be written as $(\Lambda - I)x = k$, or in the form $x = (\Lambda - I)^{-1}k$, where I is the m-dimensional unit matrix. The transformation from the vector k to the vector x reduces the unit volume by a factor $K = \det |\Lambda - I|$. Therefore, if the vector k runs over the integer lattice points in m-dimensional space covering a volume equal to K, the corresponding points x are all located inside the m-dimensional unit cube.

The condition for a point x on the torus to have the period n is satisfied exactly K times. The integer

$$K = \prod_{j=1}^{m} (\lambda_j^n - 1) \simeq \prod_{|\lambda|>1} |\lambda_j|^n$$

in the limit of large n. If $N_n(L)$ designates the number of periodic orbits of 'length' n for the automorphism L, then in the limit of large n

$$\frac{1}{n} \operatorname{Log} N_n(L) \simeq \sum_{|\lambda|>1} \operatorname{Log}|\lambda| = h(L) , \qquad (10.8)$$

where the formula (10.6) for the metric entropy of L has been used. The metric entropy appears in a completely different context; the degree of scrambling that the automorphism L wreaks on the torus is expressed in the number of points that are mapped into themselves as the map is iterated. The arithmetical properties of linear maps on the 2-torus have been studied by Vivaldi (1987) and by Percival and Vivaldi (1987).

10.5 The Topological Entropy

The metric entropy tells us how fast the phase space M gets divided up by the repeated automorphisms ϕ. If the decomposition α consists of ν pieces, the decomposition $\alpha \cup \phi^{-1}(\alpha) \cup \cdots \cup \phi^{-n+1}(\alpha)$ can be expected to consist of ν^n pieces. The situation is more involved, however, because if one starts with a decomposition α of too many pieces, applying the automorphism ϕ^{-1} does not divide up every piece over and over again. On the other hand, if the starting decomposition has too few pieces, the repeated application of ϕ^{-1} does not keep pace with the scrambling of phase space.

Adler, Konheim, and McAndrew (1965) asked whether there is a minimum number ν for the generating decomposition α to calculate the metric entropy. In order to get away from the metric entropy, and define a new and independent quantity for the disorder that is created in the phase space by the automorphism ϕ, they went back to the original idea of the K-systems as explained in Section 10.2. They tried to count the number of subsets in the decomposition α and its offspring $\overset{n}{\cup}\phi^{-j}(\alpha)$ as n increases, assuming that the algebra Ξ of all measurable subsets for a K-system will be obtained in the limit of large n.

Instead of measuring the decreasing size of the pieces, their increasing number gives the same information, possibly even on a more fundamental level. This procedure can be viewed as setting up windows in the phase space M through which the trajectories have to pass; a trajectory can be characterized by the sequence of windows it passes as it is brought back by the succession of automorphisms. If the number of windows is too small, different trajectories end up with the same sequence of windows; whereas for too large a number of windows, some possible sequences of windows are not realized by any trajectory. This idea of associating with every trajectory a sequence of subsets by which it is characterized is reminiscent of the sequence of symbols that is the base for Bernoulli schemes, or *symbolic dynamics* in general.

Going back to the decompositions α and calling N the number of subsets, the *topological entropy* is defined as the infimum (largest lower bound) with respect to all decompositions α of the quantity

$$g(\phi) = \lim_{n\to\infty} \frac{1}{n} \operatorname{Log} N(\alpha \cup \phi^{-1}(\alpha) \cup \cdots \cup \phi^{1-n}(\alpha)) \quad . (10.9)$$

In this form the topological entropy seems quite remote because it seems difficult to count the different subsets, unless they are as simply constituted as in the Bernoulli schemes; indeed, there are m^n possible words of length n in an alphabet of m letters, so that $g = Log\ m$. In general, however, as a decomposition α is progressively cut into finer pieces by the successive automorphisms ϕ^{-1}, some cuts may not create new subsets because they duplicate earlier cuts. Another way of doing the same count may be, therefore, if not easier, then at least better defined.

The relevant idea appears to be discussed for the first time in a paper of Bowen; the connection with formula (10.9) can be grasped intuitively, but the mathematical intricacies are again discouraging for a physicist. As the decomposition α is refined by the automorphisms ϕ^{-1} , each piece can be characterized by a unique trajectory that returns to its beginning after n successive maps; in other words, by a periodic orbit. The number N in (10.9) is, therefore, also the number of periodic orbits of length n; *the topological entropy*, rather than giving the number of subsets in a generating decomposition, ***counts the number of periodic orbits.***

More generally, let us assume that there is a length s defined for the periodic orbits of a dynamical system; for a discrete automorphism, this may simply be the number of transformations to get back to the initial point; but for a Hamiltonian system, the length is either the period, or the action integral. The number of periodic orbits of length less than σ is then given by

$$N(s < \sigma) \simeq \exp(\ \sigma\ \tau\ \log 2)\ , \tag{10.10}$$

where τ is the topological entropy. This exponential is typical for chaotic systems, and is in sharp contrast with the count of periodic orbits in integrable systems, which was shown in Section 6.2 to be polynomial in the upper limit σ.

The number of periodic orbits for the linear automorphisms of the torus was worked out in the preceding section; indeed, formula (10.8) gives the topological entropy in this special case. The equality of the topological entropy with the metric entropy can be proven for certain classes of dynamical systems; the relevant statement is sometimes referred to as *Pesin's theorem* (cf. Pesin 1977). It does not hold for some scattering problems where the trajectories are more unstable than their increasing number would suggest, if the two entropies were the same.

The two different interpretations of the entropy, metric and topological, will be very important in the transition from classical to quantum mechanics in chaotic systems. This transition can be understood if some quantity of interest in quantum mechanics, such as the response to a external stimulus, or the probability for being deflected in a scattering experiment, can be expressed as a sum over all relevant classical trajectories. The convergence of this sum and the location of its singularities as a function of the energy E are the result of the competition between the number of terms and their absolute magnitude, i.e., between the kinds of entropy

A third element enters, however, in the guise of a phase factor (complex number of absolute value 1), where the phase angle is the classical action divided by Planck's quantum $\hbar$, either R from Chapter 1 or S from Chapter 2. The interplay among these three ingredients was discussed by the author (Gutzwiller 1986a). The statistical properties of the phases, $R/\hbar$ or $S/\hbar$, are sufficiently important that one is tempted to speak of *third entropy*.

10.6 Anosov Systems and Hard Chaos

Dynamical systems can be further analyzed than has been done so far, if the phase space M is endowed with a distance, indicated by the double bars $\| \quad \|$, in addition to the measure μ. Moreover, most applications of this new feature are made in spaces where functions can be differentiated, and the automorphisms are differentiable, so that one can speak of a tangent at a point, and so on. A Riemannian manifold is the most important example of such a space, and the reader might just as well fix her ideas on this case in the discussion of this section. The definitions can easily be extended from the discrete maps ϕ to the continuous flows ϕ_t. The map $\phi : M \rightarrow M$ is extended to the linear map $\phi^\bullet : TM_q \rightarrow TM_q$ where TM_q is the tangent space of M at the point $q \in M$.

The *Anosov systems* satisfy the following conditions: The tangent space at each point can be decomposed $TM_q = X_q \oplus Y_q$ so that for any positive integer n, there are constants a, b, λ that are independent of n and the lengths $\| \xi \|$, $\| \eta \|$ of the tangent vectors ξ and η, so that

$$\begin{array}{ll} \| (\phi^n)^\bullet \xi \| \geq a e^{n\lambda} \| \xi \| , \quad \| (\phi^{-n})^\bullet \xi \| \leq b e^{-n\lambda} \| \xi \| , & \text{if } \xi \in X_q ; \\ \| (\phi^n)^\bullet \eta \| \leq b e^{-n\lambda} \| \eta \| , \quad \| (\phi^{-n})^\bullet \eta \| \geq a e^{n\lambda} \| \eta \| , & \text{if } \eta \in Y_q . \end{array} \tag{10.11}$$

X_q is called the *expanding* linear subspace of TM_q, and Y_q is called the *contracting* linear subspace. Locally, the Anosov system looks like a linear automorphism of the torus.

If these conditions are applied to a flow ϕ_t, the tangent space TM_q decays into three linear subspaces, $TM_q = X_q \oplus Y_q \oplus Z_q$, where Z_q is along the direction of the flow. A vector $\zeta \in Z_q$ stays constant in length, whereas the vectors $\xi \in X_q$ grow exponentially with t, and the vectors $\eta \in Y_q$ decay exponentially with t. The dimensions of X_q and of Y_q are at least 1, whereas the dimension of Z_q is always 1. The prime example of these Anosov systems are the *geodesic flows on surfaces of negative curvature* which are discussed in a monograph by Anosov (1969), and will be the topic of Chapter 19.

The expanding and contracting subspaces in the tangent space of each point can be tied together into smooth submanifolds of the phase space M, called the *unstable (expanding)* and the *stable (contracting) submanifolds*. The phase space carries two families of, roughly speaking, parallel leaves where each leaf in one family is *transverse* to the leaves of the other family, i.e., intersects each leaf in a point (discrete automorphisms) or in a line (continuous automorphisms). Each family of submanifolds is said to *foliate* the manifold M, i.e., decompose it completely into a family of non-intersecting, smooth submanifolds. A trajectory can be viewed as the intersection of two leaves, one from each family; neighboring trajectories approach exponentially along the stable submanifold, and diverge exponentially along the unstable submanifold.

This structure of phase space is as clean and simple as the foliation into invariant tori for integrable systems. The dimension of the leaves equals the number of degrees of freedom in both situations; but the function of the foliations is entirely different. The trajectories in integrable systems stay on one leaf, and only one family of leaves is necessary, whereas in Anosov systems each trajectory belongs to one leaf from each family and is defined by the intersection of those two leaves. The latter construction allows for a lot of possibilities, whereas very few different designs are compatible with a single family of invariant tori. The double foliation of the Anosov systems is stable against small perturbations, whereas the invariant tori are easy victims of the small denominators as described in the KAM theorem.

Anosov systems and dynamical systems of a similar nature, such as the *Axiom-A systems* of Smale (1967), have received wide attention in the mathematical literature. They are in some vague sense the opposite extreme to the integrable systems, because they are almost equivalent to Bernoulli schemes, and represent, therefore, something that looks totally random, although it remains deterministic. It seems as important to understand what they are doing as it is to study the integrable systems; unfortunately, the physicists have grown up to believe in the virtues of being integrable. A main motive for writing this book is to wean them away from the misguided attachment to that rather exceptional set of circumstances.

The great bulk of dynamical systems looks like an intimate mixture of the two extremes; the left-over KAM tori separate regions where stable and unstable submanifolds govern the behavior of the trajectories. This simplified picture still needs to be substantiated by looking at many examples more closely than has been done so far. We called such a fusion of opposite behavior soft chaos in the last chapter, in contrast to the extreme situation which was described in this section, and which will be called *hard chaos* from now on.

CHAPTER 11

The Anisotropic Kepler Problem

The physically most appealing example of a conservative Hamiltonian system with hard chaos is the analog of the hydrogen atom inside a crystal of silicon or germanium. The significant difference with the ordinary Kepler problem is the anisotropy of the mass tensor, i.e., the electron moving in the crystal has a much larger inertia along one axis than along the two other axes. Although the trajectories cannot be written in terms of simple functions, the Poincaré surface of section has a simple structure very close to an Anosov system. The energy surface foliates into two families of smooth submanifolds, the stable and unstable ones. All the trajectories can be coded uniquely with the help of binary sequences, and in particular, all the periodic orbits can be effectively enumerated.

11.1 The Donor Impurity in a Semiconductor Crystal

Since solid-state physics is not a prerequisite for studying chaos, the experimental origin of the *Anisotropic Kepler Problem (AKP)* will be briefly explained in this section. Further details, especially concerning the band structure of solids, have to be gleaned from any of the standard textbooks (cf. Kittel 1966, p.316; Burns 1985, p.312).

The elements carbon, silicon, and germanium are chemically fourvalent; they arrange themselves naturally in the beautiful *diamond lattice* where each atom sits at the center of tetrahedron with exactly four neighbors at the vertices. These crystals are insulators, unless some of the atoms are replaced by their five-valent neighbors in Mendeleev's table of elements, such as phosphorus, arsenic, antimony, or bismuth. Each of these impurities carries effectively one more positive charge in its nucleus and brings one more electron with it so as to maintain electric neutrality.

The extra electron is weakly bound to its parent atom and is easily liberated from its bound state to roam freely through the crystal. The silicon or germanium then becomes a semiconductor, i.e., not as good an electric conductor as a metal, with the help of the electrons 'donated' by the five-valent impurities, thus their name of *donor impurities.* Our problem is to understand the bound states of the extra electron, but that requires a more precise idea of the way the isolated donor impurity and its additional electron can be accommodated in the crystal lattice.

The additional nuclear charge in the donor impurity generates an electric field which deforms (polarizes) the regular atoms in its neighborhood. The extra electron feels the combined effect of the nuclear charge and the polarization of the surrounding lattice. The resulting net force is still attractive, and decreases with the inverse square of the distance; but it is weakened by a factor κ, the *dielectric constant* of the lattice, whose value is 11.4 in Si and 15.4 in Ge. The potential energy of the electron is, therefore, given by $-e^2/\kappa\sqrt{q_1^2 + q_2^2 + q_3^2}$.

The unusual feature of the AKP comes from its kinetic energy; a few explanatory words have to suffice. Since the chemical bonds between the immediate neighbors are saturated, the pure crystal is an insulator, and no further electrons can be accommodated. Additional slots for electrons are available at a price of at least 1 electron-volt; but these states are extended over the whole crystal like plane waves with a well-defined wave-vector k. Their energy depends on k in a rather complicated manner that is not easy to understand a priori and differs drastically in Si and Ge; this dependence is expressed in the so-called *dispersion function* $E_c(k)$. All these states form the so-called *conduction band,* which is empty in the ideal crystal, in contrast to the *valence band,* which is normally occupied.

The extra electron makes up a linear combination of the plane-wave like conduction states so as to be localized near its donor impurity. In order to minimize its kinetic energy, wave vectors k near the minimum k_0 in the dispersion function $E_c(k)$ are selected. The kinetic energy of the electron is then assumed to be

$$E_0 + \hbar^2 \left(\frac{(k_1 - k_{01})^2}{2m_1} + \frac{(k_2 - k_{02})^2}{2m_2} + \frac{(k_3 - k_{03})^2}{2m_3} \right), \tag{11.1}$$

where the quadratic form in the vector $k - k_0$ has already been diagonalized by choosing the appropriate directions in position space. The coefficients are written so as to give them the physical dimension of a mass. A classical momentum p can be associated with the wave vector $k - k_0$ according to the *de Broglie relation* $p = \hbar(k - k_0)$.

The expression (11.1) and the de Broglie relation is all we need for the discussion of the AKP; but two further complications will be mentioned, although we will not consider them because they are well understood and are not involved in the chaotic features of the problem. First, there are six equivalent minima in Si, and four equivalent minima in Ge, due to the cubic symmetry of the crystal lattice. This degeneracy can be fully discussed and treated by the standard methods of group theory, and will be ignored. Second, the extra electron takes a glimpse at the exact chemical properties of the donor impurity to see whether it is P, As, Sb, or Bi. The resulting shifts affect all the energy levels in a global manner without influencing their differences; again these shifts will not be considered any further.

The Hamiltonian for the AKP now becomes

$$\frac{p_1^2}{2m_1} + \frac{p_2^2 + p_3^3}{2m_2} - \frac{e^2}{\kappa\sqrt{q_1^2 + q_2^2 + q_3^2}}, \tag{11.2}$$

where the crystal symmetry in Si and Ge is responsible for the equality of the second and the third mass, $m_2 = m_3$. The so-called cyclotron experiments in the semiconductors give precise values for these *effective masses*. In terms of the usual free electron mass m_f, one finds that

$$\begin{aligned} m_1 &= .916\, m_f, \quad m_2 = .1905\, m_f \text{ for Silicon}, \\ m_1 &= 1.588\, m_f, \quad m_2 = .0815\, m_f \text{ for Germanium}. \end{aligned} \tag{11.3}$$

Again, the large differences in the values of the effective masses are not easy to explain although they can be obtained from elaborate numerical band calculations. The large mass-anisotropy makes it impossible to treat the AKP as a perturbation of the usual Kepler problem.

When the energy levels of a donor impurity became of interest in the 1950s, the energies of the lowest states for the Schrödinger equation corresponding to the Hamiltonian (11.2) were first found by Kohn and Luettinger (1954), with the help of very simple variational wave functions. Their results were improved by Faulkner (1969) who used a basis of 9 such functions, thereby getting a number of excited states. Wintgen, Marxer, and Briggs (1987) have recently extended the basis to well over a thousand states, with the purpose of looking for chaotic features in the spectrum. Recent experiments by Navarro, Haller, and Keilmann (1988) give energy levels with three-figure accuracy.

The author (Gutzwiller 1971, 1973) was the first to examine the classical behavior of the AKP, and to discover its hard chaos. This chapter is devoted to the exclusive discussion of the classical trajectories, but the AKP will be taken up in Chapter 17 as the prime example where the transition to quantum mechanics can be examined in great detail. Nevertheless, many questions are still open even for this special example, as will be pointed out at the end of this chapter.

The anisotropy of the mass-tensor is a rather common feature in mechanics, although the reader may not have noticed it before in connection with electrostatic or gravitational forces. The Helium atom can have a configuration where the nucleus is the vertex of an isoceles triangle and the two electrons form the base. The relative motion of the two electrons with respect to the nucleus has a much larger mass than their relative motion between each other. Similarly, in the ammonia molecule NH_3, the three hydrogen atoms form the base in the shape of an equilateral triangle and the nitrogen is the vertex. The regular shape of this pyramid is preserved, but the two degrees of freedom have quite different masses. The first demonstration of maser action was achieved in this system with exactly this kind of vibration. Devaney (1980, 1981, 1982) went a long way in showing the close similarity of these systems with the AKP.

11.2 Normalized Coordinates in the Anisotropic Kepler Problem

The natural units for the AKP are the geometric mean of the masses $m_0 = \sqrt{m_1 m_2}$, the 'Rydberg' $E_0 = m_0 e^4/2\kappa^2\hbar^2$ for the energy, and Planck's quantum $\hbar$. Further, we use the ratios $\mu = \sqrt{m_1/m_2}$ and $\nu = \sqrt{m_2/m_1}$ for which the product $\mu\,\nu = 1$. The main parameter in the whole treatment is the mass ratio $m_1/m_2 = \mu/\nu$ whose value is $\simeq 4.8$ for Si and $\simeq 20$ for Ge.

The Hamiltonian (11.2) shares with the Hamiltonian of the ordinary Kepler problem its homogeneity: the kinetic energy is homogeneous of the second degree in the momenta, and the potential energy is homogeneous of degree -1 in the position corrdinates. If all momentum components are multiplied with some positive number λ, and all the position coordinates are divided by λ^2, then the value of the energy gets multiplied by λ^2. All the trajectories are preserved as solutions of the equations of motion, if the time variable is divided by λ^3. Thus, we can arbitrarily set the normalized energy $= -1/2$ in order to investigate the behavior of the trajectories.

The components of the normalized momentum will be called (u, v, w), and the normalized Cartesian coordinates are (x, y, z), with the x-axis along the "heavy" direction of mass m_1. The normalized Hamiltonian becomes

$$H = \frac{u^2}{2\mu} + \frac{v^2 + w^2}{2\nu} - \frac{1}{\sqrt{x^2 + y^2 + z^2}} = -\frac{1}{2} . \quad (11.4)$$

The all-important action integral S is given in ordinary units by

$$\int p\,dq = \sqrt{m_0 e^4 / -2\kappa^2 E}\ \Phi , \quad \Phi = \int (u dx + v dy + z dw) , \quad (11.5)$$

where the quantity Φ now has a purely geometric meaning, while all the physics is contained in the factor $\sqrt{m_0 e^4 / -2\kappa^2 E}$. Of course, Φ depends on the trajectory.

Since $m_2 = m_3$ in (11.2), the angular momentum $M = yw - zv$ is a constant of motion. In cylindrical coordinates (x, ρ, ϕ) around the x-axis, (11.4) becomes

$$H = \frac{u^2}{2\mu} + \frac{v^2}{2\nu} + \frac{M^2}{2\nu\rho^2} - \frac{1}{\sqrt{x^2 + \rho^2}} = -\frac{1}{2} , \quad (11.6)$$

where now (v, ρ) and (M, ϕ) form conjugate pairs. At fixed M the system has only two degrees of freedom.

If $M \neq 0$, then $\rho > 0$; the centrifugal potential $M^2/2\nu\rho^2$ stabilizes the trajectories. As M increases from 0 to the maximum $\sqrt{\nu}$, the Poincaré section shows an increasing number of islands; only the limit $M = 0$ shows hard chaos.

The soft chaos when $M \neq 0$ has not been studied as yet. We will concentrate exclusively on the case $M = 0$ with the hard chaos. Each trajectory is then confined to a plane through the x-axis; henceforth, we will write y for the radial coordinate ρ in order to emphasize the purely plane character of the trajectories. But we will also find at the very end that the three-dimensional nature of the problem shows up anyhow in the count of conjugate points and in the admissible symmetries of the periodic orbits.

11.3 The Surface of Section

The equations of motion of the AKP in two dimensions follow from the normalized Hamiltonian (11.4), and are deceptively simple,

$$\dot{u} = -\frac{x}{r^3}, \quad \dot{v} = -\frac{y}{r^3}, \quad \dot{x} = \frac{u}{\mu}, \quad \dot{y} = \frac{v}{\nu}. \tag{11.7}$$

Notice that the force is always directed toward the origin; but the acceleration $(\ddot{x}, \ddot{y})$ tends to have a larger component in the y-direction because the y-component of the force gets divided by ν, while in the x-direction the force gets divided by μ where $\mu > \nu$.

The trajectories, therefore, intersect the x-axis more often than the y-axis, contrary to the usual Kepler problem where any two axes through the origin are intersected the same number of times. The choice of the x-axis as the surface of section is then inevitable. If $y = 0$, the condition (11.4) for the kinetic energy to be positive becomes

$$|x| \leq 2/(1 + u^2/\mu), \tag{11.8}$$

where $-\infty < u < +\infty$. The corresponding region in the (x, u) plane has an awkward shape; its total area is $4\pi\sqrt{\mu}$.

This region can be transformed into a rectangle by stretching every slice $(u, u + du)$ to go from $x = -2$ to $x = +2$, and thinning it accordingly. The area-preserving transformation is given by the formulas

$$X = x(1 + u^2/\mu), \quad U = \sqrt{\mu}\ \mathrm{arctg}(u/\sqrt{\mu}), \tag{11.9}$$
$$u = \sqrt{\mu}\ \mathrm{tang}(U/\sqrt{\mu}), \; x = X \cos^2(U/\sqrt{\mu}),$$

where $|X| \leq 2$ and $|U| \leq \sqrt{\mu}\,\pi/2$. The surface of section is now a rectangle.

The most important property of the AKP is contained in the following description of the trajectories: Consider the sequence of consecutive intersections for a particular trajectory, $\ldots(X_{-2}, U_{-2})$, (X_{-1}, U_{-1}), (X_0, U_0), (X_1, U_1), (X_2, U_2),..., and associate with it a sequence of binary numbers $a = \{\ldots, a_{-2}, a_{-1}, a_0, a_1, a_2, \ldots\}$, where $a_j = \mathrm{sign}(X_j) = \mathrm{sign}(x_j)$. The binary sequence is mapped into two real numbers,

$$\xi = \sum_{j=0}^{\infty} a_{-j}\, 2^{-j-1}, \quad \eta = \sum_{j=1}^{\infty} a_j\, 2^{-j}, \tag{11.10}$$

which are obviously contained in the square $-1 \leq \xi, \eta \leq +1$.

At this point, the map from the *rectangle,* i.e., the surface of section in the (X, U) plane given by (11.9), to the *square,* i.e., the region in the (ξ, η) plane defined in (11.10), is no more than a very schematic description of any particular trajectory. If this map were sufficiently smooth in any generous sense, then we could use it to define another invariant measure in the surface of section. Notice that the sequence of consecutive intersections (ξ_{-2}, η_{-2}), (ξ_{-1}, η_{-1}), (ξ_0, η_0), (ξ_1, η_1), (ξ_2, η_2),... obeys the simple rule,

$$\xi_{j+1} = (\xi_j + \text{sign}(\eta_j))/2 \ , \eta_{j+1} = 2\,\eta_j - \text{sign}(\eta_j) \ . \quad (11.11)$$

This map is none other than the inverse of the so-called *baker's transformation,* i.e., stretching by a factor of 2 in the horizontal direction and thinning by 1/2 in the vertical direction; the area in the (ξ, η) plane is preserved.

Some further comments are obvious, but they quickly get us into the middle of the AKP dynamics. Two binary sequences which start in the past with a uniform binary, either $a_j = -1$ for $j < k < 0$ and $a_k = +1$, or $a_j = +1$ for $j < k < 0$ and $a_k = -1$, but with the same binaries for indices $j > k$, are mapped into the same point (ξ, η). The same happens for two binary sequences whose binaries are identical all the way to some index $k - 1 \geq 0$, but which differ from then on in the same manner with a uniform sequence going to $+\infty$.

If the parameters (ξ, η) are of any use, the two binary sequences of the preceding paragraph should correspond to a single trajectory. Such is indeed the case; the trajectory comes out of *a collision with the origin* if the binary sequence is uniform in the past, and it goes into a collision if the binary sequence is uniform in the future. In more detail, consider the initial conditions (X_0, U_0) as a function of a single parameter varying in some limited interval; the signs of all the intersections out to some index $k - 1 > 0$ stay the same, but sign(X_k) changes, say from $-1$to $+1$, at the critical value (X_{0c}, U_{0c}). As this critical initial condition is approached from 'below', with sign$(X_k) = -1$, the numerical integration of the trajectories shows an ever increasing string of sign $(X_j) = +1$ for $j > k$. Conversely as the critical initial condition is approached from above, i.e., with sign$(X_k) = +1$, there is an ever lengthening string of indices $j > k$ where sign$(X_j) = -1$. The critical initial condition (X_{0c}, U_{0c}) yields a trajectory that goes into the origin at its k-th intersection.

Collision trajectories are characterized by the values of ξ, or η, or both, rational with a power of 2 in the denominator. They form a dense grid in the square, which will be used in the next section to construct numerically the map from the rectangle into the square. At this point, such a map is still entirely based on numerical results, but some general propositions have been proved mathematically.

Devaney (1978 a, b, and c) and the author (Gutzwiller 1977) have shown independently, and by different arguments, the following theorem: *Each binary sequence a, and therefore every point*

$-1 \leq \xi, \eta \leq +1$, can be realized by at least one initial condition (X_0, U_0), provided the mass ratio μ/ν is larger than 9/8; the only exceptions are the totally uniform sequences, either $a_j = +1$ or $a_j = -1$ for all indices j.

The proof is largely based on a detailed study of the trajectories in the neighborhood of a collision with the origin, which also shows that such a collision is an isolated event, i.e., does not take place in an open interval of the initial conditions. The AKP is more smooth in this respect than two other Hamiltonian systems where binary sequences yield a good qualitative description: both, the *isoceles three-body problem* of Devaney (1980, 1981, 1982) and the *bouncing-ball model* of Hénon (1988) must assign open intervals to binary sequences with a uniform past or future. All our further applications of the binary sequences are not possible in these systems. Atela (1988) has recently studied the isoceles three-body problem with both gravitational and electrostatic interactions; the equations of motion become identical with the AKP if the gravitational attraction between the two identical masses is exactly compensated by their electrostatic repulsion.

The mathematical proof for the converse of the above theorem is still missing, however; in spite of continued efforts by a number of people, we are only able to state as a conjecture: *Each pair (ξ, η) is realized by no more than one initial condition (X_0, U_0), with the exception of $\xi = \eta = \pm 1$. The map from the rectangle into the square is a homeomorphism, i.e., a one-to-one transformation that is continuous in both directions.*

The evidence for this sweeping statement is a vast amount of numerical computation for mass ratios > 2, coupled with checking various special cases of the above conjecture. For example, the symmetry of the binary sequence is reflected in the symmetry of the corresponding trajectory, if indeed the relation between them is one-to-one; in particular, certain simple periodic orbits must be unique, as well as symmetric with respect to the *x*-axis, or the *y*-axis, or both. Nevertheless, the situation is more complicated for mass ratios < 2, as was shown for the first time by Broucke (1985), with more detailed information to be found in the author's paper (Gutzwiller 1989); these results are discussed in Chapter 20.

11.4 Construction of Stable and Unstable Manifolds

The (ξ , η) parameters tell the 'story' of the trajectory; ξ for the past, and η for the future. The Bernoulli sequences in Section 10.1 were interpreted in the same manner. Once the binary expansions (11.10) for ξ and η have been calculated, the exact order in which the trajectory crosses the positive or negative x-axis is known, and a good picture of the trajectory exists.

Instead of varying the initial conditions (X_0, U_0), we can vary ξ and η to see how the trajectory changes. In particular, we can compare neighboring trajectories by watching how the displacement $(\delta\xi_0 , \delta\eta_0)$ develops as we go from one intersection with the surface of section to the next. Since the coordinates (X_0, U_0) are in a one-to-one continuous relation with (ξ_0, η_0), at least the immediate neighborhood shows the same qualitative behavior.

The transformation of the 'square' is given by the explicit formulas (11.11). Therefore, $\delta\xi$ decreases with every intersection by a factor of 2 as one goes forward in time, i.e., with increasing index, while $\delta\eta$ increases by a factor of 2. A change in ξ becomes less and less significant, while a change in η becomes more and more important, as one goes forward in time. Just the opposite happens when going backward in time. The trajectory is stable with respect to changes in ξ, and unstable with respect to changes in η. Of course, two trajectories which approach each other in the future, tend to diverge in the past, and vice versa.

The lines ξ = *constant* and η = *constant* are eactly the unstable (expanding) and stable (contracting) submanifolds of Section 10.6 which are characteristic of hard chaos. The conditions (10.11) are explicitly given by (11.11), and are almost trivial when applied to the binary sequences a. The conditions (10.11) are no longer obvious when going back to the original surface of section with the coordinates (X, U); but the map into the (ξ, η) coordinates allows us to construct the stable and unstable submanifolds also for (X, U).

If a trajectory originated in a collision at the intersection labeled $-k$, it has $\xi = \text{integer}/2^k$; if it ends in a collision at the intersection labeled $k + 1$, then $\eta = \text{integer}/2^k$. The *collision trajectories* form a one-parameter family that can be represented explicitly by solving the equations of motion (11.7) in the neighborhood of the origin,

$$\begin{aligned} 2u &\simeq \mu^{3/2} A\beta\zeta^{\beta - 3/2} , & 2v &\simeq \nu^{1/2} B(2\beta - 1)\zeta^{2\beta - 5/2} , \\ x &\simeq A\zeta^{\beta} , & y &\simeq 2\zeta + B\zeta^{2\beta - 1} , \end{aligned} \qquad (11.12)$$

in terms of the parameter $\zeta > 0$, and with the abbreviations

$$\beta = 3(1 + (1 - 8/9\mu^2)^{1/2})/4\ , \ B = 3A^2/4(2\beta - 3)(4\beta - 1). \quad (11.13)$$

The right-hand sides in (11.12) are the lowest terms of an expansion in powers of ζ; the higher-order terms become rapidly more complicated, and these formulas serve only to get the trajectory out of the collision far enough from the origin so that the ordinary numerical integration can take over. The value of β is slightly below 3/2 for Si and Ge; the collision trajectories are characterized by the single parameter A. For more details, cf. Gutzwiller (1977).

Given the mass-ratio μ/ν and the parameter A, the values (11.12) of the momenta and the positions for, say, $\zeta = .02$ serve as initial conditions in the numerical integration of the equations of motion (11.7). The $k-$ th intersection of the resulting trajectory yields the coordinates (X_0, U_0) of the surface of section. The corresponding binary representation $\{0, a_{-k+1}, a_{-k+2}, ..., a_{-1}, a_0, ...\}$ is known from the numerical integration for the particular value of A; this binary sequence has been truncated on the left by setting $a_{-k} = 0$ to indicate the collision. This simplification in the notation is consistent with the formulas (11.10), and the discussion of the collisions following (11.11); the value of ξ_0 corresponding to (X_0, U_0) is, therefore, known.

As the parameter A is varied very carefully, the coordinates (X_0, U_0) trace out a smooth curve in the rectangle. The value of $\xi_0 = \text{integer}/2^k$ does not change as long as the k-th intersection in our trajectory does not end up in a collision. The smooth curve in the rectangle runs from a well-defined limit point on the lower boundary to an equally well-defined limit on the upper boundary as A increases. In fact, U_0 increases monotonically with A, while X_0 may vary either way although it stays away from the values $X_0 = -2, 0, 2$.

The last description is again the result of the computational experience; some of these features can be shown analytically. If they could be proven in full generality, the basic conjecture concerning the one-to-one continuous map between the square and the rectangle would follow immediately. The above construction gives a series of smooth curves in the rectangle as A is varied from $-\infty$ to $+\infty$, each with a label ξ_0. These curves do not intersect; since they run from the bottom to the top of the rectangle, they can be ordered from left to right. Their label ξ_0 is then found to increase monotonically from -1 to $+1$.

The k-th intersections for the full range of A generate 2^k such curves of constant value ξ in the (X, U) plane. As k is increased the rectangle gets covered with an ever finer set of smooth non-intersecting lines. These are unstable (expanding) submanifolds; if one chooses two neighboring points (X, U) and $(X + \delta X, U + \delta U)$ in the same smooth line, the corresponding trajectories share the same value of ξ, but have different values of η. As the trajectories with these initial conditions are integrated forward in time, their consecutive intersections with the surface of section will move further apart just as described in (10.11).

The stable (contracting) submanifolds are constructed in the same manner, by integrating the equations of motion (11.7) backward in time assuming a collision to occur at the $(k+1)$-th intersection. When the parameter A is varied smoothly, the intersection (X_0, U_0) lies on a smooth curve with a constant value for η_0. Everything works out in complete analogy to the unstable submanifolds. Actually, it is not necessary to repeat the calculation because the equations of motion (11.7) are symmetric with respect to time reversal. The stable submanifolds in the rectangle are obtained by taking the mirror image of the unstable ones with respect to the X-axis.

The advantage of the (X, U) coordinates over the original (x, u) is quite clear when looking at Figure 26. The ends of the stable and unstable submanifolds are spread out along the upper and the lower boundaries of the rectangle; but they would all be squeezed into the infinitely far portion of the original domain (11.8) in the (x, u) plane. Also, the stable and unstable submanifolds are clearly seen as transverse to one another in Figure 26.

The binary parameters (ξ, η) define new coordinates in the rectangle; but while the ordering of the ξ and η labels shows that these new coordinates are continuous with respect to the old ones, a close look reveals some rather irregular spacings. A detailed study in the neighborhood of the origin in position space gives more information on the local continuity. If a displacement in the (ξ, η) plane has the absolute value ε, and the corresponding displacement in the (X, U) plane has the absolute value δ, one can define the *Hölder exponent* α by setting

$$\varepsilon \simeq \text{constant}\ \delta^{\alpha} \ , \text{or} \ , \ \alpha \simeq \log\varepsilon / \log\delta \ , \qquad (11.14)$$

in the limit of δ going to zero.

The values of α have a rather complicated *multifractal* distribution, a term that will be discussed in Section 20.7. At this point, it is enough to realize that α may be locally anywhere between 1/2 and 2. If the initial conditions of a trajectory are sought for a given associated binary sequence, i.e., given (ξ, η), the usual, linear interpolation schemes break down. Even if a trajectory has already been found such that only a small correction ε is needed, the corresponding correction δ in the initial conditions may be quite large.

The construction of the stable and unstable manifolds in phase space is the main avenue to a better understanding of the hard chaos in a Hamiltonian system. The crucial ingredient in the AKP is the relation with the binary sequences; they constitute a sort of *code*, which

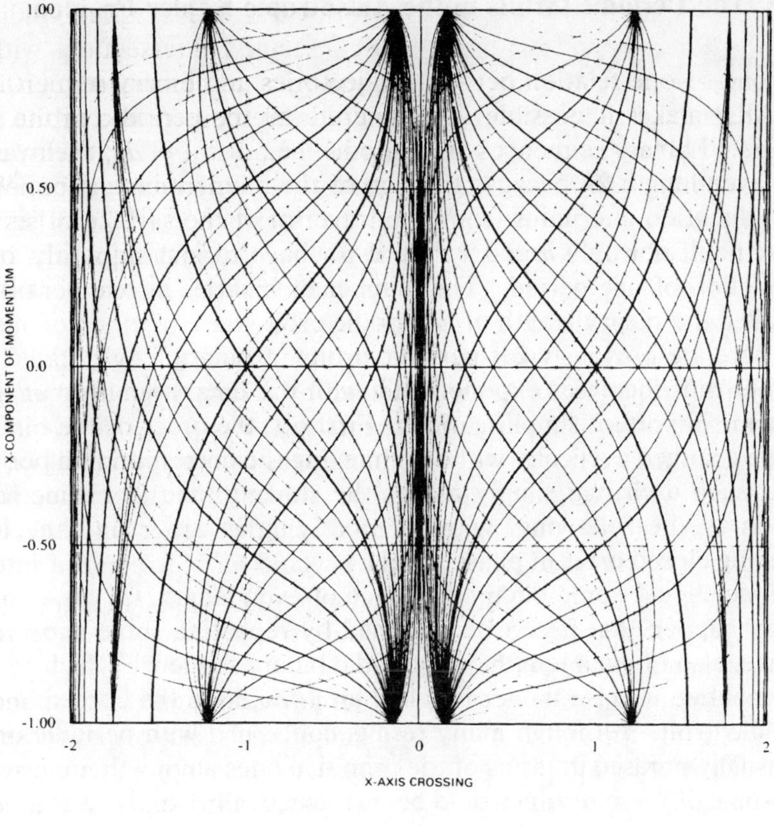

Figure 26 The stable and unstable lines in the surface of section ('rectangle') of the AKP; they are uniquely labeled with parameters ξ (past) and η (future) whose binary expansions give the qualitative description of the corresponding trajectory.

gives the most important physical features of every trajectory. Finding the appropriate code seems the most important task when facing a dynamical system with hard chaos. The AKP demonstrates that such a code can be unexpectedly simple, although the mathematical properties of the code may still be difficult to establish. Numerical computations are essential in finding a particular code and testing whether it is useful.

11.5 The Periodic Orbits in the Anisotropic Kepler Problem

The one-to-one relation between trajectories and binary sequences in the AKP makes it possible to enumerate all the periodic orbits. The associated binary sequences are periodic, i.e., $a_{j+2k} = a_j$; their period is of even length $2k$ because the sign of the momentum component in the y-direction alternates from one traversal of the surface of section to the next; not only u and x, but also v have to join smoothly at the completion of one period. The number of binaries in one period will be called the binary length of a periodic orbit.

There are obviously 2^{2k} periodic orbits of binary length $2k$. Their number thus increases exponentially with the binary length, but there are some important simplifications. First off, the order of the binaries can be changed cyclically without changing the corresponding periodic orbit; such a change simply moves the starting time from one intersection to the following, without modifying in any way the corresponding closed orbit in phase space.

If we choose some binary sequence of even length, $(a_1, a_2, ..., a_{2k})$, it may happen that it can be obtained by repeating some shorter sequence, also of even length. A periodic binary sequence which cannot be simplified in this manner is called *primitive,* as is the corresponding periodic orbit. Although many results concerned with periodic orbits are usually phrased in terms of the primitive ones among them, it seems that some of these results would be mathematically simpler if they were phrased so as to cover both primitive and nonprimitive periodic orbits.

Further reductions in the number of different periodic orbits come from the symmetries in the equations of motion (11.7). As an example, changing the signs of all the binaries gives the same trajectory reflected with respect to the y-axis, thus a new periodic orbit. But if a trajectory intersects the y-axis at a right angle immediately after its start, then it is symmetric with respect to the y-axis. Moreover $a_j = - a_{-j+1}$ for $j > 0$; its binary sequence is antisymmetric. From what was said in the preceding paragraph, any place in the binary sequence can be taken as the beginning of a periodic orbit. Thus, if there is some place where the sequence of binaries is antisymmetric, then the corresponding periodic orbit is *symmetric with respect to the y-axis.* Similar criteria apply to the symmetries with respect to the x-axis, and to time reversal. Therefore, different periodic binary sequences may have different multiplicities attached to them.

The search for the appropriate initial conditions is largely simplified by these reductions due to symmetries. The number of new and different periodic orbits for the binary lengths $2k$ =2, 4, 6, 8, 10, 12 is 1, 2, 6, 14, 42, 112. As a rule, it is easy to locate the symmetric ones because they require the variation of a single parameter; e.g., for symmetry with respect to the y-axis, it is best to start the trajectory on the y-axis with the y-component of the momentum $v = 0$, so that only the initial position y has to be found. As $2k$ increases, however, most of the periodic sequences are asymmetric, and their initial conditions must be searched in the whole rectangle of (11.9).

The construction of stable and unstable submanifolds in the preceding section is crucial for this purpose. Periodic orbits have no collisions so that each one can be localized with increasing precision inside a small parallelogram whose sides describe trajectories that come out of or go into a collision. Since the collision trajectories are found from varying a single parameter, such a procedure is at least systematic. Nevertheless, the underlying instability of all trajectories continues to make all numerical work very time-consuming. For example, with the mass-ratio 5, the periodic orbit (+ + − + + − − − + +) in Figure 27 has an stability exponent of 6.875, i.e., an error in the initial conditions gets blown up by a factor of $\exp(6.875) \simeq 1000$ after one period.

The periodic orbits are used in the present context mainly to calculate the classical approximation for the energy levels with the help of the trace formula. The most important ingredient there is the normalized action integral Φ in (11.5) to be calculated for each periodic orbit. This task is feasible only if one can find some general effective formula that yields Φ directly as a function of the binary code for the periodic orbit. The author has established such an expression which works unexpectedly well, although it is not exact.

The action integral Φ for the periodic orbits of a given binary length $2k$ varies over a large range. The maximum value belongs to the repetition k times of the shortest periodic orbit (+ −); the minimum value is taken by the somewhat arbitrary, but consistent assignment of 0 to the only non-realizable periodic sequence (+++ ... ++). The maximum is, therefore, k times the value $\tau = \Phi(+ -) = 5.74272$ for the mass ratio 5, instead of 2π for the ordinary Kepler problem. The minimum value of Φ is realized by (−++ ... ++), and goes to the limit 8.95 in the limit of $k \to \infty$. The distribution is narrowly peaked and has a well-defined mean.

Finding an effective expression for $\Phi(a_1, ..., a_{2k})$ requires a sophisticated fit of the computational results. In the first try (Gutzwiller 1981a), the numbers for all the periodic orbits up to binary length 10 were used. In spite of the many parameters available, such as 42 of them for $2k = 10$ in addition to the known maximum and assumed minimum, a single parameter γ was used to get the correct value for the mean. The second try (Gutzwiller 1988a) is based directly on the construction of the stable and unstable submanifolds in the preceding

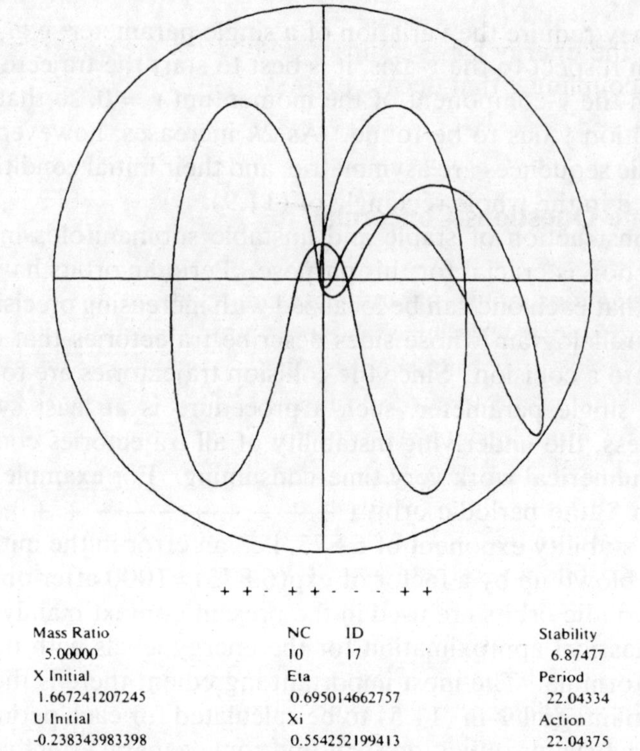

Figure 27 Periodic orbit of the AKP for the mass ratio 5, and the binary code (+ + − + + − − − + +) corresponding to the sequence of signs of x at the intersection with the (heavy) x-axis; the circle is the boundary for the classical motion in a bound state; the initial values are X = 1.66724121 and U = -0.23834398 with the period T = action S = 22.04334.

section and has the advantage of using only the data from the collision trajectories. Since they are dense just as the periodic orbits, it is indeed reasonable to get the same information from either of them.

The numerical data for the action Φ of the periodic orbit $(a_1, ..., a_{2k})$ is fitted by the formula

$$\Phi \simeq 2k\tau \cosh^2(\gamma/2) - \frac{\tau}{2} \sinh \gamma \sum_{i=1}^{2k} \sum_{j=-\infty}^{+\infty} a_i a_j \exp(-\gamma |j-i|), \quad (11.15)$$

whose maximum is indeed $2k\tau$ for $(+-+-\cdots+-)$, the minimum is 0 for $(+++\ldots+)$, and the mean is $k\tau(1+\exp(-\gamma))$ yielding $\gamma = .610$ for the mass ratio 5. The quality of the fit is seen from the root of the mean square deviation, which is less than .21, while the mean is 22.16 for the orbits of binary length 10. This formula corresponds to the energy for a chain of *classical spins* with ferromagnetic exchange coupling τ that decays exponentially at the rate γ.

11.6 Some Questions Concerning the AKP

Several questions come up quite naturally about the nature of the Anisotropic Kepler Problem, whose structure in phase space is really quite simple, and yet totally unfamiliar. We will try to answer some of them in this section; but the reader still may not feel comfortable, unless many more results are presented requiring a lot of additional work.

A first question concerns the connection with the ordinary Kepler problem: how can one understand the transition from the great variety of periodic orbits when the mass ratio differs from 1, to the Kepler ellipses, which make just one simple loop around the origin ? The periodic orbit (+ −) has the binary representation $\xi = 1/3,\ \eta = -1/3$, which is close to the point $X = .5,\ U = 0$ when $\mu/\nu = 5$. This point lies inside a parallelogram that is bounded by lines $\xi = 5/16$, $\xi = 3/8$, $\eta = -1/4$, $\eta = -3/8$ in the Figure 26.

These curves of constant ξ or η can be drawn in the (X, U) rectangle for decreasing values of the mass ratio μ/ν. The parallelogram surrounding the point $\xi = 1/3$, $\eta = -1/3$, is found to take an ever bigger portion of the whole rectangle. When the mass ratio is 1.2, almost one half of the rectangle is contained inside the parallelogram. Therefore, almost half of the initial conditions lead to a trajectory that has the binary sequence (... − + − + − + − ···), where the dots indicate an arbitrary sequence; the corresponding trajectory looks like a Kepler orbit for at least three intersections preceding and following the start.

The next question concerns the behavior of the trajectories for mass ratios close to 1. The hard chaos which was described in this chapter so far is limited to the mass ratio $\mu/\nu > 9/8$. No work has been done to my knowledge to investigate the region of mass ratio from 1 to 9/8. The peculiar limiting value 9/8 arises in the detailed study of the trajectories close to the x-axis; they can be represented by an expansion similar to (11.12), where the exponent β is given by a formula like (11.13) with ν replacing μ; there is obviously a singularity when $\mu/\nu = 9/8$. Such a trajectory has a long sequence of identical binaries, indicating a near-collision. The collisions yield the clue to the chaotic features, a situation that has been known for some time in the three-body problem.

The region of mass ratios below 2 will be discussed in more detail in the last chapter. There is at least one isolated island (cf. Gutzwiller 1989), to be called Broucke's island after its discoverer (Broucke 1985), in the surface of section for a mass ratio 1.5; it surrounds the periodic orbit with the binary sequence $(+ \; + \; - \; + \; + \; -)$. The existence of this island is hard to understand, because it does not interfere with the existence of all the trajectories belonging to the other possible binary sequences, so that the theorem of Devaney and Gutzwiller in Section 11.3 is still correct. Also the basic (Keplerian) periodic orbit with the code $(+ \; -)$ remains unstable all the way down to the mass ratio 1, although its stability exponent goes to 0.

Some mathematical methods have been developed recently to show that there exists no integral of motion in addition to the Hamiltonian. Yoshida (1987a and b) has applied them to the AKP; but the proofs are fraught with rather forbidding technical details. The existence of Broucke's island for mass ratios below 2, and presumably other islands that could escape even a high-precision numerical search, seems to exclude any clear-cut and simple, mathematical result. Nevertheless, as a physicist, one is tempted to accept that the AKP behaves effectively like a system with hard chaos, in particular in its transition to quantum mechanics.

The special trajectories along the two main axes were studied more closely by Yoshida (1987c); the collision orbits are exponentially unstable. It does not seem feasible to continue such a trajectory through a collision in a natural and unambiguous manner, as it is in the ordinary Kepler problem. The various qualitative aspects of the AKP have been reviewed by Casasayas and Llibre (1984).

A final remark concerns the two kinds of entropies. Formula (11.15) gives good approximate values for the lengths Φ of the periodic orbits. Unfortunately, there is no simple relation between the length of the binary sequence and the value of Φ; in particular, the periodic orbits with only one crossing of the negative x-axis have a finite upper bound for Φ which is independent of the number of binaries in the symbolic sequence. The count of periodic orbits as a function of Φ becomes tricky, and the topological entropy is not well defined; on the other hand, these orbits can be incorporated properly into the transition from classical to quantum mechanics, for instance, in the trace formula of Chapter 17, because the instability of these special orbits makes their contribution negligible.

CHAPTER 12

The Transition from Classical to Quantum Mechanics

Since our physical intuition is so firmly grounded in classical mechanics, we have little choice but to advance as far as we can into quantum mechanics along the trails that can be laid out with the help of classical mechanics. To be of help in our context, they have to be usable for regular as well as chaotic dynamical systems, and, therefore, they differ from the ones in most textbooks. The two main guideposts are the classical approximation for the quantum-mechanical propagator in position space and time, as first proposed by Van Vleck in 1928, and the consistent use of the stationary phase method whenever an integral has to be evaluated.

The topics in this chapter to be discussed in this manner include the change of variables from position space and time to other coordinates, and the composition of propagators for consecutive times. Special attention is given to Green's function whose controlling parameter is the energy rather than time. The hydrogen atom is treated in momentum space to illustrate this approach to quantum mechanics, starting with Rutherford scattering, but then using the same formulas to get the approximate Green's function for bound states.

The usual name for this type of approach to quantum mechanics is *semiclassical,* suggesting a mixture of classical and quantal ideas. We will not follow this usage; a quantum-mechanical object like Green's function will be called classical if it is calculated purely with the help of classical mechanics. Although it could not have been conceived without understanding quantum mechanics, its explicit computation becomes a technical problem in classical mechanics and requires a much better grasp of what is going on there than most physicists have right now.

12.1 Are Classical Mechanics and Quantum Mechanics Compatible?

Classical mechanics has served humanity well for three centuries. It has been confirmed to very high precision particularly in celestial mechanics, from the naked eye observations of Tycho Brahe all the way to the intricate orbital maneuvers of space probes. The relevance to very small systems like atoms and molecules, however, has been severely questioned for about 100 years. The most incisive criticism is contained in Heisenberg's *uncertainty relations,* which seemed to resolve the issue once for all times by setting up unbeatable limitations to the usual classical interpretation of nature.

The boundary between classical and quantal behavior lost some of its interest after Heisenberg's decisive results. Quantum mechanics itself has become the object of intensive studies, because it seems to accommodate all kinds of paradoxes, i.e., situations which offend our most sincerely held beliefs about what nature is or is not able to do, such as *Schrödinger's cat* and the *Einstein-Podolsky-Rosen experiment.* Again, a clearcut quantitative answer to these problems exists in the form of *Bell's inequality,* which has been confirmed experimentally at least for photons; a comprehensive review was provided by Wheeler and Zureck (1983). Nevertheless, quantum mechanics remains somewhat of a mystery, although it is well confirmed by all the experimental evidence available.

This book does not intend to discuss these issues; the reader will undoubtedly see many connections between, on the one hand, the mainly technical problems of understanding chaotic behavior in dynamical systems, and, on the other hand, the more philosophically oriented efforts to find the conceptual basis of quantum mechanics. Since the author does have fairly clearcut opinions on some of these questions, the reader is entitled to know in what spirit they are getting short shrift in the present context. For ease of further reference, the relevant points are numbered 1 through 7.

1) In all cases of a real experiment, in contrast to an abstractly made-up situation (sometimes called a Gedanken or "thought" experiment), two things seem to hold without exception: (a) the practical

rules of quantum mechanics give unambiguous quantitative answers, and (b) these answers have always been found to be correct, i.e., in agreement with the measurements.

2) The discussion of the so-called 'thought experiments' in most cases is singularly crude, i.e., removed from any awareness of the practical considerations in a real experiment. Time and effort is spent on purely logical discussion, along with entirely formal manipulations of mathematical relations. With few exceptions, the hard work of writing down, and then solving the relevant equations for a specific laboratory set-up has not even begun; in particular, the inevitable presence of noise is ignored most of the time. The work of Leggett (1978, 1980), Caldeira and Leggett (1983), as well as Grabert and Weiss (1984) and their collaborators are notable exceptions.

3) Experimental techniques have advanced well beyond what the early masters in the field of quantum mechanics could have imagined. Some of the standard answers are now open to question, in spite of the continued uncritical acceptance in most textbooks. For example, the wave function of a small quantum system could conceivably be measured experimentally, as long as relativistic effects such as pair production can be ignored.

4) The exact place in a specfic experiment where quantum mechanics interferes with our often dogmatic and simplistic philosophical prejudices (everyone has some) is very hard to pinpoint. For example, the breakdown of classical mechanics is probably more subtle and remote than most of the theoretical discussions so far. A similar situation prevails in trying to understand the second law of thermodynamics, which started well over a century ago with *Maxwell's demon,* and where the crucial failure of time reversibility is no longer ascribed to the immediate physical processes, but to the handling and, in particular, the erasure of the relevant information (cf. Bennett 1987).

5) Classical mechanics, with some simple, but important modifications in its interpretation, can get us a long way in the treatment of particular problems. It is essential to understand as much of quantum mechanics as feasible, and as explicitly as possible on that basis. The hard questions of classical mechanics can, therefore, not be dismissed as irrelevant because they are presumably superseded by modern physics. For example, the study, and in particular the effective enumeration, of periodic orbits, as suggested first by Poincaré, has to be met head-on.

6) The chaotic features of classical mechanics seem to destroy much of its practical usefulness in the more difficult problems of physics and chemistry; e.g., it would be close to impossible to define a sensible cross-section for the scattering of an electron from a molecule as if it

were a purely classical phenomenon. The main point to recognize, however, is that even what we now call chaos has a well-defined, if unfamiliar structure which is perhaps no more difficult to handle than the familiar, but exceptional, invariant tori in phase space. If this structure is interpreted in the light of quantum mechanics it gives useful, if approximate results.

7) Quantum mechanics mitigates the destructive influence of classical chaos on simple physical processes. Indeed, quantum mechanics is sorely needed to save us from the bizarre aspects of classical mechanics; but most paradoxically, this process of softening the many rough spots is entirely in our grasp as soon as the nature of the roughness is well understood.

The required tools for translating classical chaos into quantum mechanics, and thereby evading the bad classical features, are assembled in this chapter. They will be used later; I believe that there is a great opportunity for more work on specific examples to be treated in this manner.

12.2 Changing Coordinates in the Action

The course of a dynamical system may be most easily described in one coordinate system, say position and time; but a measurement on the system may test for the value of some other variable, like the momentum. Such a situation in quantum mechanics requires the discussion of the relevant operators and their expectation values, whereas in classical mechanics no more than a transformation of coordinates is involved. Nevertheless, this transformation has all the earmarks of the quantal situation as soon as it is carried out on the action function $R(q''q't)$ of (1.4) and the associated density $C(q''q't)$ of (1.19).

The transformation from the canonically conjugate pair (p, q) to the new pair (μ, ρ) will be performed on the final position q''; the double primes will be left out in this section so as to simplify the formulas. The final position coordinate q will be replaced by the (so far not specified) coordinate ρ, while the initial position coordinate q' will not be changed. Thus, the old action $R(q\,q'\,t)$ becomes the new action $\overline{R}(\rho\,q'\,t)$. Notice that only one half of each conjugate pair appears at time t, so that there is no conflict with Heisenberg's uncertainty relation.

As long as we do classical mechanics, there is no impediment to the use of the canonically conjugate variable whose construction was discussed in Chapter 7. The change of coordinates from q to ρ can, therefore, proceed with the help of the generating function W whose properties are established as follows. The first variation of each action is written in the form (1.9), $\delta R = p\,\delta q - p'\delta q' - E\,\delta t$ and $\delta \overline{R} = \mu\,\delta\rho - p'\delta q' - E\,\delta t$, so that their difference becomes $\delta\overline{R} - \delta R = \mu\,\delta\rho - p\,\delta q = \delta W$. The natural variables in W are, therefore, ρ and q, with the canonical transformation

$$\mu = \partial W/\partial\rho\,, \quad p = -\,\partial W/\partial q\,, \tag{12.1}$$

where the generating function W has nothing to do with the dynamical system to which the transformation is applied.

In order to make the appropriate replacement in the action, the old variable q has to be given as function of the new one ρ; the second equation in (12.1) is combined with first equation (1.8) to yield the condition

$$p = \frac{\partial R(q, q', t)}{\partial q} = -\frac{\partial W(\rho, q)}{\partial q}\,. \tag{12.2}$$

Notice that the solution of this equation, $q(\rho, q', t)$, depends on the initial coordinate q' and the time t.

The new action is obtained from the old by writing

$$\overline{R}(\rho, q', t) = R(q, q', t) + W(\rho, q)\,, \tag{12.3}$$

and inserting $q = q(\rho, q', t)$. One checks that this definition yields indeed the first derivatives

$$\mu = \partial\overline{R}/\partial\rho\,, \quad p' = -\,\partial\overline{R}/\partial q'\,, \quad E = -\,\partial\overline{R}/\partial t\,, \tag{12.4}$$

as required in complete analogy to (1.8).

The determinant of the mixed second derivatives, $C(q\,q'\,t)$ as in (1.19), gives the spread after time t of the trajectories that started in q' and end near q. The corresponding spread in the space with the coordinates ρ is given the determinant $\overline{C}$ of the second-order mixed derivatives of $\overline{R}$ with respect to ρ and q'. Using the first equation (12.4), and taking the derivative of μ with respect q' demands a little juggling act involving both (12.1) and (12.2); at the end, the determinant has to be calculated.

Besides $\overline{C}$ and C, two new determinants appear:

$$CW = \left|\frac{\partial^2 W}{\partial\rho_i\partial q_j}\right|, \quad CR = \left|\frac{\partial^2 R}{\partial q_i\partial q_j} + \frac{\partial^2 W}{\partial q_i\partial q_j}\right|; \tag{12.5}$$

the complete relation between the two densities now becomes

$$\overline{C}(\rho\,q'\,t) = CW(\rho, q)\,[CR(\rho, q, q', t)]^{-1}\,C(q\,q'\,t)\,, \tag{12.6}$$

where q on the right-hand side has to be expressed in terms of ρ, q', t. This formula, though computationally straightforward, has two factors whose presence is expected, namely CW to make the transition from ρ

to q, and C to express the density in the old coordinates; but the middle factor is strange. Its significance will appear only after the classical approximation to quantum mechanics has been discussed in Section 5.

As an example of the general formulas in this section, let us make the transition from position coordinates q to momentum coordinates $\rho = p$. The generating function is $W = -\rho q$ from which follows $\mu = -q$ and $p = \rho$ according to (12.1). Leaving aside the names (μ, ρ) for the new coordinates, one can write directly the new action as

$$\overline{R}(p\, q'\, t) = R(q\, q'\, t) - pq \ , \quad p = \partial R/\partial q \ , \qquad (12.7)$$

where the second equation is the condition to validate the first equation. This is no more than an ordinary Legendre transform; the factor CW in (12.6) becomes 1; and CR simplifies to the determinant of the second derivatives $\partial^2 R/\partial q_i \partial q_j$, which is the Jacobian $\partial(p)/\partial(q)$ as would be expected.

The same process of classical transformations can also be used to change the initial coordinates q', while leaving the final coordinates $q = q''$. For example, the energy E and the angular momentum L can take over the function of q'. As we shall see shortly, the resulting action function and its density yield directly the classical approximation to the wave- function in q-space for the energy E and the angular momentum L.

12.3 Adding Actions and Multiplying Probabilities

The development of a dynamical system takes place in consecutive time-steps, at least in what we called the Lagrangian view of nature at the end of Chapter 1. It is then natural to ask for the Lagrange action $R_{12} = R(q''\, q'\, t'' - t')$ over a total time from t' to t'', if it is already known as $R_1 = R(q\, q'\ \ t - t')$ from the beginning t' to the intermediate time t, and as $R_2 = R(q''\ q\ t'' - t)$ from t to the final time t''. The intermediate point q has not been specified; but it is clear that it has to be chosen so as to allow for one continuous trajectory in phase space to go all the way from q' to q'' in the alotted time $t'' - t'$.

Another manner of presenting this requirement is to say that the value of the total action, $R_2 + R_1$, is stationary with respect to the intermediate point q. Thus, we find the condition

$$\frac{\partial R(q''\ q\ t'' - t)}{\partial q} + \frac{\partial R(q\, q'\ \ t - t')}{\partial q} = 0 \ , \qquad (12.8)$$

where the two endpoints, q'' and q', and the time intervals are given, and there is assumed to be a solution $q = q_0$. According to (1.8), this equation stipulates the equality of the momentum at arrival in q_0 with the momentum at the departure from q_0.

The total action becomes the sum of the partial actions,

$$R(q''\, q'\; t'' - t') \;=\; R(q''\, q_0\; t'' - t) \;+\; R(q_0\, q'\; t - t') \;, \qquad (12.9)$$

provided q_0 satisfies (12.8). Since we have already invoked the variational principle, we might just as well calculate the second variation in terms of the displacements $\delta q = q - q_0$,

$$R_2 + R_1 = R(q''q'\; t'' - t') + \frac{1}{2}\,\delta q_j\Big(\frac{\partial^2 R_2}{\partial q_j \partial q_k} + \frac{\partial^2 R_1}{\partial q_j \partial q_k}\Big)\delta q_k \;. \qquad (12.10)$$

The deviation from the extremum of the action is given by a quadratic form in δq whose matrix is given by the last term in (12.10), and whose elements will be called c_{jk} henceforth.

Again the density of trajectories can be calculated with the help of the expression (1.19) in terms of the mixed second derivatives of the action. Call these determinants C_{12} for the whole trajectory, with C_1 and C_2 for the two partial trajectories. The same juggling act as in the preceding section then leads to the relation

$$C_{12} = \; C_2\, [\det\, c_{jk}]^{-1}\, C_1 \;, \quad c_{jk} \;= \frac{\partial^2 R_2}{\partial q_j \partial q_k} + \frac{\partial^2 R_1}{\partial q_j \partial q_k} \;, \qquad (12.11)$$

which has the same structure as (12.6).

The interpretation of (12.11), however, is more transparent: The density of trajectories $C(q''\, q'\; t'' - t')$ is viewed as the probability to get to the final point q'' when starting from the initial point q'. If this process were to take place in two consecutive, independent time-steps, then the total transfer probability C_{12} would be the product of the partial ones, C_2 and C_1; but in classical mechanics these two partial processes are not independent, because the intermediate point q_0 is fixed by the condition (12.8). The necessary correction depends on the second variation according to (12.11); if the second variation is small, the matrix elements c_{ij} are small and so will be the value of their determinant. The product $C_2 C_1$ will be enhanced thereby, as if, indeed, there were more freedom in the choice of the intermediate point q.

The essence of Feyman's path integral in quantum mechanics is to release the intermediate point q from the restriction (12.8). Nevertheless, the formula (12.11) still holds, at least approximately; the enhancement or reduction by the factor in the middle, according to the second variation of the action integral, stems from the constructive or destructive interference of waves, as will be explained in Section 5.

12.4 Rutherford Scattering

The existence of a positively charged, practically point-like nucleus in each atom was demonstrated by Geiger and Marsden in a famous experiment which was successfully interpreted by Rutherford in 1911: α-particles were sent into a gold-foil; some of them came out at large angles with respect to the incoming direction. This can happen only if they make an almost head-on collision with a very concentrated, heavy, electrically charged object, namely the nucleus of the gold atom.

Rutherford derived a formula for the angular distribution of the outgoing α-particles on the basis of classical mechanics. He assumed an ordinary repulsive Coulomb potential acting between the α-particle and the point-like heavy nucleus of the gold atom. The trajectories are hyperbolas with the nucleus at one of the foci.

Rutherford's argument can be couched in terms of the probabilities in the preceding section, but this time applied to the momentum rather than the position coordinates. The resulting formula is identical with the corresponding result in quantum mechanics; it seems that Rutherford sometimes expressed particular pleasure at having discovered about the only formula that holds both in classical and in quantum mechanics. In the present context, it shows that classical mechanics can be phrased in correct quantum-mechanical terms.

The Kepler problem in momentum space is geometrically simpler than in position space. *The trajectory in momentum space, the so-called hodograph, i.e., the plot of the momentum as a function of time, is a circle.* This startling fact is unknown to most physicists although it must have been well understood by Kepler, Huygens, and Newton who were supreme geometers; it is mentioned in Sommerfeld's lectures on mechanics (1942). Milnor (1983) discusses the relevant geometry for a mathematical audience, and even physicists could profit from his presentation.

Lambert's formula (1.14) with its corollaries (1.15) and (2.10) could be used together with (12.7) to calculate the relevant action integrals and probabilities as function of the momenta p' and p''. Since the trajectories are so simple, however, their action integral can be obtained directly. The most important formulas will now be listed, with the task of checking left to the reader.

The trajectory of the α-particle in polar coordinates around the gold nucleus is given by (1.10), which is now written as

$$r = \frac{M^2}{2mZe^2(1 + \varepsilon \cos(\phi - \phi_0))}, \tag{12.12}$$

where the interacting electric charges are $2e$ and Ze, and the angular momentum is M. The eccentricity ε is

$$\varepsilon^2 = 1 + M^2E/2mZ^2e^4, \tag{12.13}$$

where the energy $E > 0$ for a scattering experiment, and, therefore, $\varepsilon > 1$. The momentum in the plane of the trajectory has the components

$$p_1 = -\frac{2mZe^2}{M} \sin(\phi - \phi_0), p_2 = \frac{2mZe^2}{M} (\varepsilon + \cos(\phi - \phi_0)), \tag{12.14}$$

which is indeed a circle of radius $2mZe^2/M$ centered on $p_1 = 0$, $p_2 = ((2mZe^2/M)^2 + 2mE)^{1/2}$, as shown in Figure 28.

The geometric construction of the hodograph is straightforward: a circle of radius $\sqrt{2mE}$ is drawn in the momentum plane around the origin; any point outside this base circle can be the center of a particular hodograph, which is then found by drawing the tangents to the base circle. The hodograph intersects the base circle at right angles. The base circle divides the hodograph into two separate arcs: the arc inside the base circle is used for a repulsive potential, while the outside arc is the hodograph for an attractive potential. The endpoints of these two arcs on the base circle correpond to the positions of the scattering particle infinitely far away from the scattering center.

Since the experiment is done for a fixed energy E, all the further calculations will be based on the Hamilton-Jacobi action (2.3). In analogy with (12.7) we have to set

$$\begin{aligned} \overline{S}(p''\, p'\, E) &= S(q''\, q'\, E) - p''q'' + p'q' = \\ &= -\int_0^t q\,\dot{p}\, d\tau = -\int_{p'}^{p''} q\, dp\, . \end{aligned} \tag{12.15}$$

The integral is positive when q and $\dot{p}$ have opposite signs, as in an attractive potential; but it is negative in Rutherford scattering. If one sets $\dot{p} = -\operatorname{grad} V$ where V is a homogeneous potential, i.e., $V(\lambda q) = \lambda^\kappa V(q)$ so that Euler's relation $q \operatorname{grad} V = \kappa V$ holds, then $\overline{S}(p''\, p'\, E) = \kappa \int V(q)\, dt$, which is called the *virial* in statistical mechanics.

The trajectories in momentum space, i.e., circles intersecting a fixed based circle at a right angle, will reappear in Chapter 19 as geodesics on a surface of constant negative curvature. The relation between the two constructions is easily recognized if we write down the virial for the Coulomb problem with positive energy. Since the coordinate q and the increase in momentum dp are parallel, the integrand in (12.15) re-

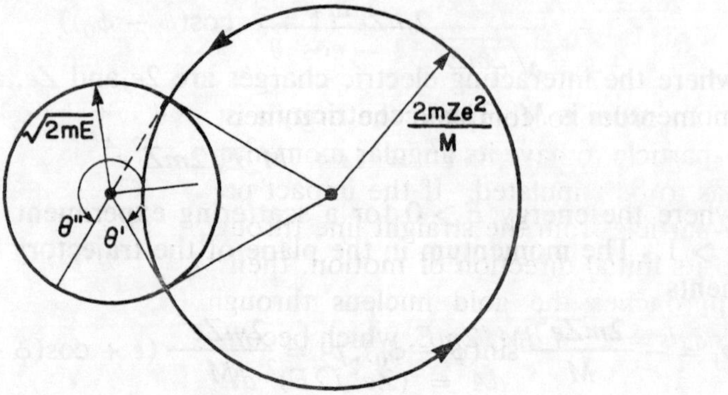

Figure 28 The hodograph (plot of the trajectory in momentum space) for the scattering from θ' to θ'' in an attractive (repulsive) Coulomb field is the outer (inner) part of a circle of radius $2mZe^2/M$, which intersects the reference circle of radius $\sqrt{2mE}$ at a right angle.

duces to $2me^2|dp|/(2mE-p^2)$, which is exactly the Riemannian metric on a surface of Gaussian curvature $-1/2mE$; notice that $p^2<2mE$. The variational principle of Euler and Maupertuis (cf. Section 2.3) applies to momentum space as well as to the usual position space.

The explicit calculation proceeds in two steps. First, the angular momentum M has to be expressed directly in terms of p'' and p'; if η is the counterclockwise angle from p' to p'', then

$$\frac{M}{2mZe^2} = \sin\eta\,(P''^2+P'^2-2P''P'\cos\eta-2mE\sin^2\eta)^{-1/2}, \quad (12.16)$$

where $2P=|p|+2mE/|p|$. Secondly, the integral (12.15) is calculated; the manipulations are elementary, but tricky, and yield

$$\bar{S}(p''p'E) = -\int_{\phi'}^{\phi''}\frac{d\phi}{1+\varepsilon\cos(\phi-\phi_0)} = -\frac{2mZe^2}{\sqrt{2mE}}\log\frac{1+\zeta}{1-\zeta},$$

$$\zeta^2 = \frac{2mE|p''-p'|^2}{(p''^2-2mE)(p'^2-2mE)+2mE|p''-p'|^2}. \quad (12.17)$$

These two formulas have been written explicitly because they will be used in Section 6.

The Rutherford scattering formula could be obtained systematically by calculating the density $D(p''p'E)$ from (2.7); but the result follows more directly from (12.16). At large distance from the gold foil, the α-particle has only kinetic energy so that $p^2=P^2=2mE$; the angle η is then the total deflection suffered by the α-particle, i.e., η is the scattering angle θ. Formula (12.16) simplifies to

$$M = \frac{2mZe^2}{\sqrt{2mE}} \frac{\sin\theta}{1 - \cos\theta} = \frac{2mZe^2}{\sqrt{2mE}} \frac{\cos\theta/2}{\sin\theta/2} . \quad (12.18)$$

In order to complete the argument, the initial probability for the α-particle to have its angular momentum in the interval $(M, M + dM)$ has to be stipulated. If the impact parameter s is the distance of the α-particle from the straight line through the gold nucleus and parallel to its initial direction of motion, then $M = s\sqrt{2mE}$. The α-particle approaches the gold nucleus through the annulus of area $dA = 2\pi s\, ds = 2\pi M\, dM/2mE$, which becomes

$$dA = (Ze^2/2E)^2\, d\Omega/(\sin\theta/2)^4 , \quad (12.19)$$

where $d\Omega = 2\pi \sin\theta\, d\theta$ is the solid angle into which the α-particle scatters. All the physics of the Rutherford scattering formula (12.19) is contained in (12.16); the rest is kinematics.

Although this approach to the scattering problem seems unnecessarily lengthy, or even artificial, it is not so much the fact that the result happens to be quantum-mechanically correct, as the possibility to solve quantum-mechanical problems classically, which makes this whole exercise worthwhile. As a consequence of the remark at the end of Section 12.2, the initial probability distribution in terms of the impact parameter s and the angular momentum M can be given a more consistent formulation; since one measures the initial coordinates s and the direction of motion, it is natural to use them in the action S rather than the initial momentum p'. In order to carry out the classical analysis, they have to be considered as one half of a set of canonically conjugate variables in phase space. But as we emphasized earlier, these extra variables do not enter into the expression for the action or the density of trajectories, so that the Heisenberg uncertainty relations are not violated. Rowe (1987) has recently given a detailed picture of the classical limit of quantum-mechanical Coulomb scattering, where the surfaces of constant action are constructed to show the progressive waves.

12.5 The Classical Version of Quantum Mechanics

The fundamental tenets of quantum mechanics cannot be derived from classical mechanics. Once they are known, however, it is natural to look for the opening that leads from the narrow classical confines to the wide open quantum fields. The usual passageway is built on the assumption that the classical system is integrable, or has no more than one degree of freedom, whereas the trail to be used here will also allow chaotic systems to pass. *The crucial formula was first written down by Van Vleck in 1928, shortly after the discovery of Schrödinger's equation.* The transformations of the action and the density of trajectories in Sections 2 and 3 can now be interpreted very convincingly.

The important step is to replace the density $C(q''\, q'\, t)$ of (1.19) by its square root. More precisely, *a complex-valued function, $K_c(q''\, q'\, t)$, called the (quasi-)classical propagator, is defined by Van Vleck's formula,*

$$K_c(q''q'\, t) = (2\pi i\hbar)^{-n/2}\sqrt{C(q''q'\, t)}\quad \exp[(i/\hbar)R(q''q'\, t) - i\phi], \tag{12.20}$$

where $(2\pi i\hbar)^{1/2}$ is always an abbreviation for $(2\pi\hbar)^{1/2}\exp(i\pi/4)$. The *phase* ϕ will be specified later as a multiple of $\pi/2$; it was not part of Van Vleck's original work and was first introduced by the author (Gutzwiller 1967) in his derivation of (12.20) from Feynman's path integral. More detailed studies along these lines are due to Möhring, Levit, and Smilansky (1980); cf. other references there.

In the special case of a *free particle* in Euclidean space, the expression (1.5) for the action integral can be used in (12.20), which then yields

$$K(q''\, q'\, t) = (m/2\pi i\hbar t)^{n/2}\ \exp[im(q'' - q')^2/2\hbar t]\,. \tag{12.21}$$

We have written K instead of K_c, because it will turn out in the next chapter that the expression (12.21) is already correct in quantum mechanics; it is not an approximation like the more general (12.20).

The absolute square of K_c differs from C in (1.19) by $(2\pi\hbar)^n$ in the denominator. As we pointed out at the end of Section 2.4, the double differential $d^nq''\, d^nq'\, C(q''\, q'\, t)/(2\pi\hbar)^n$ can be understood as the probability of finding the system after the time t in the volume element d^nq'' of position space, if it was in d^nq' at the beginning $t = 0$.

The propagator for K_c describes a wave which originates in q' and spreads to q''; its name comes from its main property which is contained in the integral formula

$$K_c(q''\, q'\, t'' - t') \simeq \int dq^n\, K_c(q''\, q\, t'' - t)\, K_c(q\, q'\, t - t')\ , \tag{12.22}$$

where $t'' > t > t'$, and the $\simeq$ sign has been inserted because the integral has been evaluated by the stationary phase method. The proof of (12.22) will now be sketched.

The Van Vleck expressions are inserted on the right-hand side of (12.22). The integrand consists in the product of the two roots, $\sqrt{C_1 C_2}$, and an exponential whose exponent is the sum $(R_2 + R_1)/\hbar$ of the actions that were defined at the beginning of Section 3. If the variation of this sum covers many multiples of 2π as the intermediate position q ranges over its domain, then the integral depends only on the neighborhood of a stationary point q+q_0, in the nomenclature of Section 3. The two roots are assumed to vary slowly in that neighborhood because they are not gauged against Planck's quantum $\hbar$. This way of approximating the integral is called the *stationary phase method.*

The value of the integral is now obtained by inserting the expansion (12.10) in the neighborhood of the stationary point q_0. The computation has thereby been reduced to the integral

$$(2\pi i\hbar)^{-n/2} \int d^n q \; \exp[i \; \delta q_j \; c_{jk} \; \delta q_k/2\hbar] \; , \qquad (12.23)$$

which is a slight generalization of the *Fresnel integral.* The matrix c_{jk}, given by (12.11), is real and symmetric; it can be diagonalized by an orthogonal transformation of the variables of integration δq. The integral becomes then a product of n Fresnel integrals, each of which looks like

$$\int d\xi \; \exp(i\lambda\xi^2/2\hbar) = \frac{(2\pi i\hbar)^{1/2}}{(|\lambda|)^{1/2}} \exp[i\pi(sign(\lambda) - 1)/4] \; , \qquad (12.24)$$

where λ is one of the n eigenvalues of the matrix c_{jk}. The product of the eigenvalues is the determinant which appears in the second factor of (12.11).

The right-hand side of (12.22) is now compared with the left-hand side where we insert again the Van Vleck formula (12.20): The various roots match exactly because of (12.11), and the exponents match because of (12.9). The only possible discrepancy arises because of the phase factor in (12.24), which yields a factor $\exp(-i\pi/2)$ for every negative eigenvalue in the second variation (12.10). Accordingly, we now adopt the following rule for the definition of the phase ϕ in (12.20):

The root in the Van Vleck formula (12.20) is always taken on the absolute value of the determinant $C(q'' \; q' \; t)$, and the phase ϕ is defined as $\pi/2$ times the number κ of conjugate points along the trajectory from q' to q''.

In view of Morse's proposition in Section 1.5, relating the signs of the eigenvalues in the second variation to the number of conjugate points, the integral formula (12.22) has, therefore, been proven under the assumptions of the stationary phase method.

In most situations, there is more than just one trajectory going from q' to q'' in the fixed time t. The classical propagator is then simply assumed to be the sum of terms like (12.20), one for each trajectory,

$$K_c(q''q'\,t) = \sum_{class.traj.} (2\pi i\hbar)^{-n/2}\sqrt{C}\,\exp[iR/\hbar - i\kappa\pi/2]\,. \qquad (12.25)$$

At this point, *the superposition principle of quantum mechanics* has been used; but the computation of each term in (12.25) is still entirely done within classical mechanics. The trigger for the system in the position q' has produced different waves, each following its own mainly classical trajectory to the position q''; the result is simply the sum of the individual waves.

If the whole sum (12.25) is inserted for each occurrence of K_c in (12.22), there will be all kinds of mixed terms on the right-hand side, coming from partial trajectories which do not form a complete continuous trajectory in phase space; call them broken trajectories, because their direction of motion changes abruptly. The stationary phase method eliminates them all because they do not satisfy the condition (12.8). The associated exponents never settle to a condition where the expansion (12.10) can be used; in other words, the integral vanishes because of the destructive interference between these broken trajectories even over small domains of the intermediate position q.

12.6 The Propagator in Momentum Space

The expression (12.25) is usually called *semiclassical,* because it is a mixture of classical and quantal ideas; as mentioned earlier, we will call it simply *classical,* because the burden of computing it explicitly for any special example lies completely within classical mechanics. With the help of the stationary phase method, (12.25) can be transformed into the classical approximations for all kinds of useful quantum-mechanical objects. Although we will not give the details of the calculations, the reader is encouraged to do the hard work of deriving the general formulas before using any of them in some special example.

Bound states in quantum mechanics often have a better classical approximation in momentum space than in position space. The hydrogen atom is a particularly glaring example, as will be seen in Section 12.8, where the reasons for this peculiar situation will be explained.

The experience from the preceding section together with the discussion in Sections 2 and 3 makes the required formulas for the transformation from position to momentum coordinates almost obvious. We will, therefore, only write down the results, and point out the relations with the earlier work which could have motivated the formulas in the first place.

If the correct quantum-mechanical propagator in position space is called $K(q''\,q'\,t)$, without the index c in (12.20), which indicates the classical approximation, then the Fourier transform $K_F(p''p'\,t)$ is given by

$$\frac{1}{(2\pi\hbar)^n}\int d^n q'' \int d^n q'\, K(q''\,q'\,t)\,\exp[i(p'q' - p''q'')/\hbar]\,, \qquad (12.26)$$

and represents exactly what is needed in momentum space. This rather abstract expression becomes more understandable if it is applied to the classical propagator (12.20), or more generally to (12.25).

If these expressions are inserted into (12.26), the integrand again consists of an amplitude, the root of a determinant, and an exponential; the exponent is $[R(q''\,q'\,t) + p'q' - p''q'']/\hbar$. Notice that this exponent corresponds exactly to the action function (12.7), except that both space coordinates, q'' and q', are now transformed, whereas only $q'' = q$ was transformed at the end of Section 2.

The stationary phase method for doing the integral (12.26) requires that the exponent be stationary with respect to the variation of both q'' and q', not only q as in (12.7). The second variation around the stationary point is, therefore, a quadratic function in $2n$ variables, and the Fresnel-type integral corresponding to (12.23) leads to the root of a $2n$ by $2n$ determinant. This determinant has to be manipulated somewhat, and then yields the expected result, namely the Van Vleck formula (12.20) with the variables p'' and p', instead of q'' and q', and in terms of the action function $\overline{R}(p''\,p'\,t)$ rather than $R(q''\,q'\,t)$.

There are now two forms for the classical propagator, $K_c(q''\,q'\,t)$ and $K_{Fc}(p''\,p'\,t)$, which apply to position space and to momentum space respectively. They can both be composed as in (12.22) because they depend on time t rather than energy E. Both of them will now be converted so as to have the energy E as their parameter rather than the time t.

12.7 The Classical Green's Function

The conversion from time to energy is accomplished quite generally by the transformation that defines *Green's function,*

$$G(q'' q' E) = (i\hbar)^{-1} \int_0^\infty dt\, K(q'' q' t) \exp(iEt/\hbar) , \quad (12.27)$$

with an entirely analogous formula to relate the propagator $K_F(p'' p' t)$ of (12.26) to the Green's function $G_F(p'' p' E)$. Notice that the integral is extended only over positive times; it is a Laplace integral, since one can add a small positive imaginary part $i\varepsilon$ to the energy, $E \rightarrow E + i\varepsilon$, and insure the convergence of the integral. The Green's functions are, therefore, defined in the whole upper half of the complex energy plane.

The Van Vleck formula (12.20) is inserted on the right-hand side, and the stationary point of the exponent, $R(q'' q' t) + Et$, as a function of time is determined; the condition to be satisfied is exactly the last equation (1.8), $\partial R/\partial t = -E$. The stationary value of the exponent is the Hamilton-Jacobi action $S(q'' q' E)$; the second variation is simply $(\partial^2 R/\partial t^2)\delta t^2/2$.

The Fresnel integral (12.24) in one dimension can be used. The determinant C has to be expressed in terms of the action $S(q'' q' E)$, a task, that was already carried out in Section 2.4. In combination with $\partial^2 R/\partial t^2 = -[\partial^2 S/\partial E^2]^{-1}$ from the Fresnel integral, the amplitude becomes the square root of the density $D(q'' q' E)$, which was defined by (2.7).

The *classical Green's function* $G_c(q'' q' E)$ becomes

$$\frac{2\pi}{(2\pi i\hbar)^{(n+1)/2}} \sum_{\text{cl.tr.}} \sqrt{(-1)^{n+1} D}\, \exp[\frac{i}{\hbar} S(q'' q' E) - i\mu\pi/2] , \quad (12.28)$$

where the factor $(-1)^{n+1}$ in front of D insures that the expression under the square root is positive for short trajectories. The phase ψ is again defined as $\pi/2$ times the number μ of conjugate points, but the relevant conjugate points in (12.28) are obtained from varying the trajectory at constant energy E, rather than at constant transit time t as in (12.25).

In order to understand the difference, we have to return to (2.7), which relates the amplitude $C(q'' q' t)$ for (12.25) to the amplitude $D(q'' q' E)$ for (12.28). The singularities in C determine the number κ of conjugate points in (12.25); these are at fixed q' the exceptional points q'' and times t that the system can reach with different energies E, i.e., $\delta E \neq 0$ while $\delta q' = \delta q'' = \delta t = 0$. In going over to D with the help of (2.7), however, the amplitude C is divided by $\partial^2 R/\partial t^2$

which is just $-\,\delta E/\delta t$ according to (1.8). The singularities along the trajectory from q' to q'' where $\delta t = 0$ are thereby canceled, and new singularities are introduced every time $\delta E = 0$. These are the places where the trajectory can be displaced infinitesimally while keeping both endpoints as well as the energy fixed; they are the conjugate points when the trajectory is restricted to vary on the surface of constant energy.

The expression (2.10) for D shows that there are two varieties of conjugate points entering into the total count μ in (12.28): ordinary caustics on the energy surface come from the singularities in the determinant of (2.10); but the points of classical return where $\dot{q} = 0$ also contribute, although they are comparatively rare in a system with two or more degrees of freedom. On the other hand, these classical return points, where the potential energy $V(q) = E$, yield exactly the phase loss of $\pi/2$ that is known in one-dimensional systems from Kramers' connection formulas. All these considerations apply in momentum space, provided the Hamilton-Jacobi action S and its second derivatives in D are replaced everywhere by the virial $\overline{S}(p''\,p'\,E)$, given by (12.15), and its second derivatives.

Again, one can work out the classical Green's function for a *free particle* in Euclidean space by inserting the expression (2.5) of the classical action $S(q''\,q'\,E)$. The resulting classical Green's function $G_c(q''\,q'\,E)$ is

$$\frac{\pi}{E}\,\frac{(2mE)^{(n+1)/4}}{(2\pi i\hbar)^{(n+1)/2}\,|q''-q'|^{(n-1)/2}}\exp[i\sqrt{2mE}\;|q''-q'|/\hbar] \tag{12.29}$$

In contrast to the propagator (12.21), this expression is only an approximation to the correct quantum-mechanical Green's function for the free particle, if the Euclidean space has an even number n of dimensions; it is correct, however, when n is odd.

The formula (12.28) for G_c is the basis for the so-called trace formula, which relates the spectrum of energy levels to the collection of periodic orbits of the corresponding classical system and which will be discussed in some detail in Chapter 17. Before calculating G_{Fc}, the classical approximation of Green's function in momentum space, for the Kepler problem in the next section, a major failure of all Green's functions as compared with the propagators has to be mentioned.

There is no analog for Green's functions to the composition (12.22) that holds for the propagators. The role of the propagator, as its name indicates, is to describe the evolution of the dynamical system at the time t'', by taking the appropriate average over all the relevant features at the intermediate time t. When performing this averaging operation over the relevant features of the system at the time t, one can fall back to the description for an even earlier time t'. This simple idea is expressed in (12.22) for the special case of K_c; but it is true for all the other propagators. Quite in contrast, there is no sensible modification of this principle that holds for any of the Green's functions.

The calculation that comes to mind would be something like

$$\int d^n q \, G_c(q'' \, q \, E) \, G_c(q \, q' \, E) \;,$$

in the hope that it can be shown to yield $G_c(q'' \, q' \, E)$. The stationary phase method should work in the present context; but the reader will find that the various determinants that arise will not combine appropriately to give the desired result. The main consequence of this failure is that *the Green's functions cannot be calculated by anything resembling Feynman's path integral for the propagators.* This crucial difference will be discussed later in more detail; it is mentioned here because it is already visible in the classical approximations (cf. Gutzwiller 1988c).

12.8 The Hydrogen Atom in Momentum Space

All the questions about a particular dynamical system in quantum mechanics can be answered if either the propagator or Green's function is understood well enough. The reader is asked to accept this statement on faith for the time being, because there will be ample arguments in favor of this view in the remaining chapters. Of course, one could dismiss such a claim as devoid of any significance, since there is no use trying the impossible, i.e., to get either the propagator or Green's function for a complicated system if indeed that would answer all the questions.

In practice, we will settle for something less, namely the trace of Green's function; but in the case of the hydrogen atom, one can do more, and thereby demonstrate the usefulness of this approach, which works for chaotic systems as well as for integrable ones. In particular, the sum over the classical trajectories in (12.28) can be carried out explicitly for the Kepler problem. The calculations for Green's function could be worked out in position space with the help of Lambert's formula (2.11); but the detailed expressions would be complicated just like the series (1.14). There is a good reason for these complications in position space, which does not apply to the treatment of the Kepler problem in momentum space.

If the total energy $E < 0$ in an attractive potential, the classical trajectories are restricted to the volume in position space where the potential energy $V(q) < E$. In general, the situation is actually worse as was shown in Section 2.5 for the Kepler problem; the Kepler orbits are restricted to the inside of a critical ellipse, which depends on the given energy E and the point of departure q'. The classical Green's function G_c is expected to be singular on the boundary of the accessible volume, whereas the correct quantum-mechanical Green's function G is well behaved there, and continues beyond. Thus, aside from being mathematically more complicated, G_c is not even a good approximation to G.

A similar difficulty arose when we discussed Rutherford scattering in momentum space in Section 12.4; the trajectories had to stay inside a circle of radius $\sqrt{2mE}$. Such a restriction also arises in position space when a particle scatters from a repulsive potential, since the trajectories are then restricted to the volume $V(q) < E$, which excludes a neighborhood of the scattering center itself. Most remarkably, however, no limitation of this kind comes into play in momentum space for the bound states of an attractive potential.

The equations (12.12) through (12.14) are still valid, except that the product of the charges $2Z = -1$ and $E < 0$. The hodograph remains a circle, which, however, is now characterized by intersecting the base circle of radius $\sqrt{-2mE}$ in diametrically opposite points, as shown in Figure 29. The hodograph goes around the origin of the momentum plane; the trajectory does not stop on the base circle as it did in the repulsive potential.

The formula (12.16) for the angular momentum M can be rederived, but turns up unchanged. The integral for the virial is no more difficult, and the final result as written in (12.17) still holds; but the quantity ζ is now purely imaginary. Notice that ζ does not depend on the electric charge, whereas the sign of the interaction appears in the factor in front of the logarithm. As ζ becomes imaginary the logarithm changes into an arctang, and the factor i gets canceled by the $\sqrt{2mE}$ in the denominator. Altogether, the virial becomes

$$\overline{S}(p''\, p'\, E) = 2\,\sigma \operatorname{arctang}\sqrt{-\zeta^2}\ , \sigma = \frac{me^2}{\sqrt{-2mE}}\ , \tag{12.30}$$

where ζ^2 is defined in (12.17); the negative sign in the square root is welcome because now $E < 0$.

For a brief discussion of this formula, let us start the trajectory on the base circle. As p'' moves away from p' the arctang increases, until its argument becomes ∞ when p'' reaches the point opposite p' on the base circle. The virial $\overline{S}$ then becomes $\pi\,\sigma$; and twice that amount upon completing the whole hodograph, i.e., Kepler ellipse.

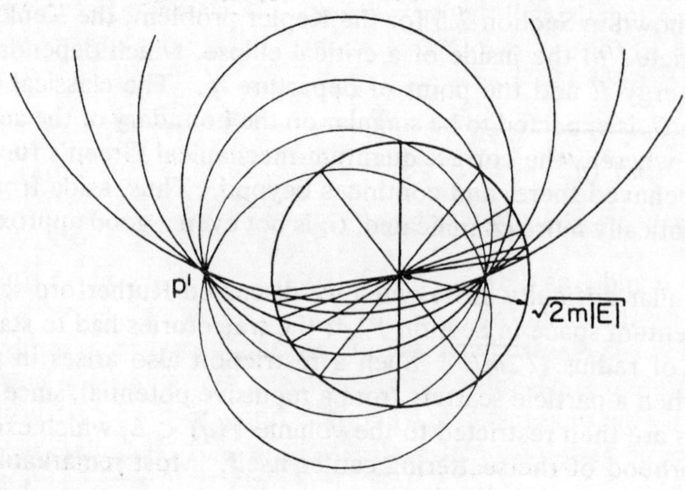

Figure 29 The hodographs in a Coulomb field for fixed $E < 0$ through a given initial momentum p' are circles which intersect the reference circle of radius $\sqrt{-2mE}$ in diametrically opposite points.

An arbitrary endpoint p'' can be reached from the equally arbitrary starting point p' in two distinct trajectories: directly, i.e., with a virial less than $\pi\,\sigma$ and given by (12.30); or indirectly, i.e. with a virial greater than $\pi\,\sigma$ and given by $2\pi\,\sigma$ minus (12.30). In the latter case, the trajectory also passes a double counting conjugate point (single counting in two dimensions). The whole situation is exactly the same as for the geodesics on a sphere at the end of Section 1.5; indeed, Fock (1935) treated the hydrogen atom in momentum space as the free motion of a particle on a sphere.

In addition to these two simple trajectories, the final stop at p'' can always be delayed until the particle has gone through an arbitrary number of complete orbits. Each time the virial is increased by $2\pi\,\sigma$, an amount that does not depend on either p' or on p''. Therefore, when the amplitude factor $\sqrt{(-1)^{n+1}\overline{D}}$ is calculated from (2.7), it will not depend on the number of times the orbit has been traversed. With each completed orbit the trajectory picks up four conjugate points when moving in three dimensions (only two in two dimensions), as was discussed at length at the end of Chapter 2.

Finally, the determinant $\overline{D}(p''\,p'\,E)$ has to be worked out by inserting (12.30) into (2.7), replacing everywhere q by p. That is obviously a fairly messy affair, which has to be carried out with some judgement; the main ingredient is to use polar coordinates in momentum space and to notice that $\overline{S}$ depends only on the absolute values of p' and p'', as well as their scalar product.

Putting together all the pieces yields the classical Green's function $G_{Fc}(p''\ p'\ E)$ for the hydrogen atom in three dimensions,

$$-\frac{4\,m\,\sigma^2\sqrt{2mE\,\zeta^2}}{\pi\hbar^2(p''^2-2mE)(p'^2-2mE)(-2mE)\,|p''-p'|^2}$$

$$\sin\,(2\,\sigma\,\mathrm{arctang}\sqrt{-\zeta^2}\,)/\sin(\pi\,\sigma)\,, \tag{12.31}$$

with the σ and ζ as defined in (12.30) and (12.17). This formula was derived by the author (Gutzwiller 1967) in his first paper on the relation between classical and quantum mechanics. It yields not only the correct bound state energies, but also the correct bound eigenstates, apparently the only instance where this happens on the basis of a classical approximation, except in the trivial case of plane waves. It illustrates the basic idea of using Green's function to find classical approximations for the spectrum, an idea that works also in classically chaotic systems. Norcliffe and Percival (1968), and then with Roberts (1969), obtained essentially the same results independently of the author; they coined the name 'correspondence identities' to convey the idea that in the case of the hydrogen atom the classical approach can yield exact quantum-mechanical answers.

The poles in the complex σ-plane are found where $\sin\pi\sigma = 0$, i.e., σ is a non-zero integer. (When $\sigma = 0$, the numerator in (12.31) vanishes, thus canceling the zero in the denominator.) Therefore, the poles of the classical Green's function are found at the energies,

$$E\ =\ -\ me^4/2n^2\hbar^2\ . \tag{12.32}$$

as Bohr was the first to find out.

More remarkable, however, is the expression for the residue at this pole; in terms of the Bohr momentum $\gamma_n = me^2/n\hbar$, and the angular variable $\beta = 2\ \mathrm{arctang}\sqrt{-\zeta^2}$ where E has been replaced by (12.32), the residue can be written as

$$(-1)^{n+1}(8n/\pi^2)\gamma_n^5(p'^2+\gamma_n^2)^{-2}(p''^2+\gamma_n^2)^{-2}\ (\sin n\beta_n/\sin\beta_n) \tag{12.33}$$

The quotient $\sin n\beta/\sin\beta$ is a rational function of p' and p'', as can be seen by first expanding $\sin n\beta$, and then expressing the sines and cosines rationally in terms of $\mathrm{arctang}\sqrt{-\zeta^2}$. Finally, the residue is written as the sum over n^2 products $\phi(p'')\ \phi(p')$; each function $\phi(p)$ is exactly one of the normalized eigenfunctions of the hydrogen atom in momentum space. (See Bethe and Salpeter (1957) for the details of these wave functions.)

CHAPTER 13

The New World of Quantum Mechanics

The study of possible chaos in quantum mechanics is less than two decades old, and the results so far are not nearly as clearcut and systematic as in classical mechanics. The interesting examples have all been quite simple, however, and the essence of quantum mechanics is not in question. The reader is assumed to be familiar with most of the general precepts as well as the standard methods of wave mechanics. They will be reviewed briefly in this chapter with special emphasis on the later applications.

13.1 The Liberation from Classical Chaos

The need for a new kind of mechanics arose at the beginning of this century for reasons that had nothing to do with the discovery of chaos in classical mechanics. Quantum mechanics developed for the next 75 years and succeeded in solving many important problems in physics, while blissfully unaware of anything more complicated than perturbation theory. A 1977 conference in Como, Italy, was probably the first to bring together the two strands (cf. Casati and Ford 1979). Therefore, why not leave a good thing alone since it has served us so well without any lapse in the comparison between experiment and theory ?

The main reason for trying to bring about some kind of reconciliation between classical chaos and quantum mechanics is strange, indeed. While classical mechanics gives good results on the atomic scale in certain exceptional situations as in Rutherford scattering, its answers in most cases are clearly absurd. For example, the classical scattering of an electron from a simple molecule consisting of more than three atoms is a case of hard chaos as we shall see in Section 20.1, and no sensible cross-section can be obtained. If nothing else, quantum mechanics is needed to provide at least reasonable answers where classical mechanics fails to satisfy our most elementary expectations for an acceptable, let alone correct, result.

Quantum mechanics has liberated us from the scourge of classical chaos, and we will find that the symptoms of chaos are hard to pin down in this new environment. So again, why should we even bother to track them down if they are so elusive? First, the problems in quantum mechanics are hard to solve even though they are well defined in terms of mathematical relations and conditions to be satisfied. Second, our intuitions and instincts have as yet to adjust to the new mechanics; it almost looks as if a real understanding of a complicated situation on the atomic scale requires a close tie with classical mechanics.

The principles of quantum mechanics enter, of course, into this mixed interpretation in terms of classical trajectories; the final answers, while acceptable or even good, cannot be expected to be correct to the last decimal; but they are at least understandable intuitively, rather than being the result of a monstrous numerical calculation. The basic ingredients to this uneasy compromise are the formulas of the last chapter; nobody could have guessed them without first getting acquainted with the new wave mechanics.

Aside from these philosophical considerations, there are good reasons to study quantum-mechanical systems whose classical behavior is chaotic. Not only are they more typical than the integrable systems; but they reveal significant differences in the character of their wave functions, the distribution of their energy levels, the response to outside perturbations, the dependence of the scattering phase-shift on the momentum, and so on.

Many applications of quantum mechanics have had the primary purpose of solving special problems, and comparing the results with the outcome of the corresponding measurements in the laboratory. In contrast, the concern with chaos is based on the hope of gaining some systematic awareness for all the possibilities in quantum mechanics. Many special cases are investigated and compared with one another, although very often they are not directly related to a particular experimental set-up; they are chosen because their classical behavior is chaotic, and they differ from one another in some important qualitative feature.

This work is now in full swing approximately for a decade, and many partial results have been obtained; but no systematic treatment is in sight, even compared with the rather moderate generalizations concerning soft chaos in classical mechanics. The remaining chapters of this book constitute, therefore, an effort to organize what is known so far under some general headings, rather than to present a coherent overall view. The field is very productive right now, and a painful selection among many worthwhile results is unavoidable.

The importance of these recent researches lies in the questions they ask and the kind of answers they are trying to establish. Just as the examples of chaos in the earlier chapters are bound to find their way eventually into the textbooks on classical mechanics, the physicists in general will become aware of the issues which are discussed in the next chapters. Quantum mechanics may then become better appreciated in all its tremendous but subtle variety, and some of its mysterious ambiguities may become better understood.

13.2 The Time-Dependent Schrödinger Equation

Schrödinger's equation is the simplest statement of the principles of quantum mechanics. It requires the complex-valued wave-function $\psi(x, y, z, t)$ to satisfy the partial differential equation

$$i\hbar \frac{\partial \psi}{\partial t} = -\frac{\hbar^2}{2m} \Delta \psi + V(x, y, z, t)\psi , \qquad (13.1)$$

where Δ=div(grad) is the *Laplace operator,* $\partial^2/\partial x^2+\partial^2/\partial y^2+\partial^2/\partial z^2$ in Cartesian coordinates. The solution is supposed to be square integrable, with a finite integral over all space which can be normalized to 1, provided the right-hand side of (13.1) is self-adjoint. This condition presents no difficulties for the potential energies $V(x, y, z, t)$ that will be discussed in the coming chapters, including the scattering problems. In the presence of a magnetic field, the Laplace operator is modified in a rather obvious manner, which will be discussed in Chapter 18.

A case of particular interest arises when $V = 0$ inside some given domain D, and $V = +\infty$ outside D. Then, the wave function $\psi = 0$ on the boundary of D; the boundary of D acts like an ideal mirror in optics. Some popular examples of chaos are exactly of this type, like the Sinai billiard and the Bunimovitch stadium (cf. the end of Section 10.3).

The Laplace operator on a manifold with a *Riemannian metric* $ds^2 = g_{jk}\, dx_j\, dx_k$ has the form

$$\Delta = (1/\sqrt{g}\,)(\partial/\partial x_j)(\sqrt{g}\; g^{jk}\partial/\partial x_k) + \kappa/4\ , \tag{13.2}$$

where $g = \det(g_{jk})$, and g^{jk} is the inverse of the matrix g_{jk}. There is a quasipermanent discussion among the specialists whether or not to add a term proportional to the *Gaussian curvature* κ that can be calculated from the metric tensor g_{jk} (cf Section 19.1). In two dimensions, there is little doubt in my mind that it should be $\kappa/4$ as indicated in (13.2), while in three dimensions it is simply κ, at least provided the 3 by 3 curvature tensor is a multiple of the unit-tensor, as in the spaces of constant curvature such as the sphere. Some details on this controversy will be given as the opportunity arises in Section 19.5.

Schrödinger's equation plays the role of the equations of motion in classical mechanics, either in the Lagrangian form (1.3) or in the Hamilton-Jacobi form (2.2). The physical information, however, is contained in the solutions of these equations, particularly in the action integrals (1.4) and (2.3). Similarly, the physics of quantum mechanics is directly obtained from the *propagator* $K(q''\, q'\, t)$ that is the value of the wave function in q'' at the time $t > 0$ if it was concentrated in q' at the time $t = 0$. This formulation is only valid if the potential V in (13.1) does not depend on time; otherwise, it is necessary to use it two times to define the propagator, t' for the starting concentration in q', and $t'' > t'$ for the observation in q''.

In mathematical terms the propagator is defined by the conditions

$$\lim_{t\to 0} K(q''\, q'\, t) = \delta(q'' - q')\ , \tag{13.3}$$

$$i\hbar\frac{\partial K}{\partial t} - \left(-\frac{\hbar^2}{2m}\Delta'' + V(q'')\right) K(q''\, q'\, t) = 0\ , \tag{13.4}$$

where the double prime on the Laplacian indicates that the differentiation is done with respect to q''. Notice that the propagator is defined for times $t \geq 0$, and that for definiteness one may set $K(q''\, q'\, t) = 0$ for $t < 0$. For a *free particle,* i.e., if $V = 0$, the propagator takes the simple form (12.21), which results directly from Van Vleck's formula (12.20).

The name propagator has been chosen for this function because it tells us how the effect of the potential V spreads the wave functions from q' to q'' in the given time t. This spreading phenomenon is a stepwise process that can take the system first from the starting positions q' to the intermediate position q in the time $\tau > 0$, and then from q to the final position q'' in the time $t - \tau > 0$. The final result of this two-step process is obtained by compounding the partial results from all the possible intermediate positions. Mathematically, this crucial property of the propagator is expressed as

$$K(q'' \, q' \, t) \; = \; \int dq \, K(q'' \, q \; t - \tau) \, K(q \, q' \, \tau) \; , \qquad (13.5)$$

where the integral extends over all position space, and the fixed intermediate time τ can be chosen anywhere in the interval $0 < \tau < t$. This relation can be checked for the free particle propagator (12.21) with the help of the Fresnel integral (12.24).

The ability to compound the propagator as indicated in (13.5) is similar to the idea that was first proposed by *Huygens* as an explanation for optical phenomena (cf. Baker and Copson 1950; Born and Wolf 1959, chapter VIII; Gutzwiller 1988c). Light behaves like a sequence of spreading pulses; every point that has been reached by a pulse is the source of a new pulse whose amplitude is proportional to the amplitude of the arriving pulse.

The propagator K also represents a spreading pulse; but in contrast to the optical pulses, which spread at the speed of light, the quantum-mechanical pulses spread at all speeds. There is no limiting relation between the distance $|q'' - q'|$ and the time t for Schrödinger's equation; of course, its relativistic generalization, *Dirac's equation,* is subject to the finite speed of propagation.

13.3 The Stationary Schrödinger Equation

If the potential energy V does not vary with time, Schrödinger's equation (13.1) can be simplified by assuming that the wave function has an exponential time-dependence, $\psi(x, y, z, t) = \phi(x, y, z) \exp(-i\omega t)$. The stationary wave-function $\phi(x, y, z)$ satisfies the time-independent Schrödinger equation,

$$-(\hbar^2/2m) \, \Delta\phi + V(x, y, z) \, \phi \; = \; E \, \phi \; , \quad E \; = \; \hbar\omega \; . \qquad (13.6)$$

The condition on ϕ to be square integrable, with a finite value for the integral, limits the acceptable solutions of this partial differential equation.

Without going into the spectral theory of linear operators, we will assume henceforth that there exists a *spectrum,* i.e., an ordered set of *energy levels* $E_0 \leq E_1 \leq ... E_j \leq ...$ to each of which belongs a well-defined normalized *eigenfunction* $\phi_j(x, y, z)$. There may be *degeneracies* in the spectrum, i.e., consecutive energy levels may actually have the same value; but they affect only a finite number of energy levels at any given energy.

The spectrum is *denumerable* as indicated, provided the space available to the wave functions is compact. This condition is not satisfied in many realistic situations, e.g., the Kepler problem; above the denumerable spectrum, there is then a *continuous spectrum* where the energy E varies continuously, and the corresponding eigenfunctions are not easily normalizable. This continuous spectrum is of special importance for scattering problems, and the reader is asked to brush up on his or her knowledge of wave mechanics in this respect. The demands are quite elementary in the present context.

The principal property of the spectrum is its *completeness,* i.e., its ability to approximate to arbitrary accuracy any reasonable function in the form

$$\phi(x, y, z) = \sum_{j=0}^{\infty} c_j \, \phi_j(x, y, z) , \tag{13.7}$$

where the sum may have to include an integral over the continuous spectrum, if the space is not compact. We will, in general, not bother to write down this integral over the continuous spectrum, although we will not imply that it can be neglected. The completeness can be expressed in the formula for the *Dirac δ-function*

$$\delta(q'' - q') = \sum_{j=0}^{\infty} \phi_j(q'') \, \phi_j^+(q') . \tag{13.8}$$

The propagator can, therefore, be expanded as

$$K(q'' \, q' \, t) = \sum_{j=0}^{\infty} \phi_j(q'') \, \phi_j^+(q') \, \exp(-iE_j t/\hbar) . \tag{13.9}$$

This formula looks like a Fourier expansion of K with respect to time t; but it should be remembered that the time is restricted to $t \geq 0$. Indeed, if one inserts (13.9) into (12.27), it is better to think of the integral over time as a Laplace transform, where the energy E may have a positive imaginary part ε. When each term is integrated separately over t, one finds *Green's function* in the explicit form

$$G(q'' \, q' \, E) = \sum_{j=0}^{\infty} \frac{\phi_j(q'') \, \phi_j^+(q')}{E - E_j} , \tag{13.10}$$

where E can be complex with a positive imaginary part.

Green's function satisfies an inhomogeneous stationary Schrödinger equation,

$$(E - H_{op}(p'', q'')) \, G(q'' \, q' \, E) = \delta(q'' - q') . \tag{13.11}$$

We have taken the liberty of using the *Hamilton operator* $H_{op}(p , q)$, which is obtained from the classical Hamiltonian $H(p , q) =$

$p^2/2m + V(q)$ by replacing each component of the momentum according to the rule $p_k \rightarrow (\hbar/i)\partial/\partial q_k$.

In the Euclidean space of odd-numbered dimension, Green's function of a *free particle* is given by the formula (12.29); again, the classical approximation is exact. In the even-numbered dimensions, however, Green's function is a Bessel function in the distance $|q'' - q'|$ with a logarithmic singularity at the origin; formula (12.29) gives the asymptotic expression for large distances.

In contrast to the propagator K, Green's function G does not satisfy any simple relation like (13.5). Formula (13.11) is an elliptic partial differential equation; therefore, its solution in function of the coordinates q'' is analytic, i.e., it can be expanded everywhere in a convergent power series. The effect of even remote obstacles or boundaries is always felt; it does not have to wait for a wave to get there.

Many physicists live under the erroneous impression that Green's function can somehow be compounded, and that it can be interpreted as a step in the spread of the wave function, at the given frequency $\omega = E/\hbar$, from the starting point q' to the endpoint q''. To see the fallacy of this view, it suffices to replace everywhere the propagator K in (13.5) by Green's function G from (13.10) with the same value of the energy E throughout (cf. also the end of Section 12.7). The two sides of the presumed equality do not match (cf. Gutzwiller 1988c); Huygens idea does not apply at constant frequency !

13.4 Feynman's Path Integral

The propagator K can be constructed quite explicitly by making repeated use of the fundamental relation (13.5). This idea comes from Dirac, and was incorporated into the second edition of *The Principles of Quantum Mechanics* (Dirac 1935, p.125). A time interval from 0 to t is broken up into N pieces, of equal size for simplicity's sake, $0 < t_1 < t_2 < .... < t_{N-1} < t_N = t$. The propagator from $q' = q_0$ to $q'' = q_N$ now appears as an $(N-1)$-fold integral over all the intermediate positions $q_1, q_2, ..., q_{N-1}$.

It is hard to understand why Dirac stopped at this point of the argument, and left to Feynman (1948; for an updated version see also the monograph of Feynman and Hibbs 1965, as well as Schulman 1981 and 1988) the task of completing the picture, more than a decade later. Since each time step is very short, the propagator can be effectively approximated, in the hope that the cumulative effect of the errors is reduced by increasing the number N of intermediate time intervals.

The free-particle propagator (12.21) takes care of the Laplacian in (13.4). What is the effect of the potential energy V over a short time interval?

While the particle remains in the neighborhood of, say (x, y, z), for the time δt such that $\delta t\ V(x, y, z) << \hbar$, the propagator maintains its absolute value, and changes its phase by the factor $\exp(-iV(x, y, z)\ \delta t/\hbar)$. We can assume that a good value for the position (x, y, z) is the starting point q'. The propagator for short times δt, therefore, becomes

$$(m/2\pi i\hbar\delta t)^{n/2} \exp\{\frac{i}{\hbar}[\frac{m(q''-q')^2}{2\ \delta t} - V(q')\delta t]\}\ . \quad (13.12)$$

The $-$sign in front of the potential energy in (13.12) is crucial.

Feynman recognized that the expression inside the brackets is the *Lagrangian (not the Hamiltonian!)* of a classical particle which moves from q' to q'' in the short time δt. The factor in front is a somewhat bothersome normalization which guarantees the correct reduction in the amplitude as the particle moves; it is part of the free-particle propagator (12.21), and appears automatically, if the expression (1.5) for the free-particle action integral $R(q''\ q'\ t)$ is inserted into Van Vleck's formula (12.20).

Let us now define a *polygonal path in position space* by the sequence of points $(q', t') = (q_0, t_0), (q_1, t_1), \ldots, (q_{N-1}, t_{N-1}), (q_N, t_N) = (q'', t'')$. The Lagrangian action for this path is well approximated by the sum

$$R_N = \sum_{k=1}^{N} (t_k - t_{k-1})\ L(\frac{q_k - q_{k-1}}{t_k - t_{k-1}}, q_{k-1}, t_{k-1})\ ,(13.13)$$

where the formula is written for the general case of a time-dependent Lagrangian.

Feynman's Path Integral for the propagator $K(q''q't)$ is the expression

$$\lim_{N\to\infty} \prod_{1}^{N} (m/2\pi i\hbar(t_k - t_{k-1}))^{n/2} \int dq_1 \ldots \int dq_{N-1}\ \exp(iR_N/\hbar)\ .(13.14)$$

The $(N-1)$-fold integral can be interpreted as adding up indifferently the contributions from all the possible paths that lead from q' to q'' in the given time interval from t' to t''. Hamilton's principal function R_N was defined in (13.13) as a Riemann integral where the position q in the Lagrangian can be chosen arbitrarily in the interval (q_{k-1}, q_k). In the presence of a magnetic field, however, it is crucial to adopt the *midpoint rule,* which requires that $q = (q_{k-1} + q_k)/2$ (cf. Schulman's treatise on Path Integrals, 1981, Chapter 4).

Integrals like (13.14) were studied first by Norbert Wiener in the early 1920s, with the crucial difference that the combination $i\hbar$ in (13.14) is replaced by a positive number, basically a diffusion constant. The integral then becomes absolutely convergent with the proper precautions on the potential V, and can be shown to satisfy the conditions corresponding to (13.3) and (13.4), also known as the Fokker-Planck equation (cf. Wang and Uhlenbeck 1945, Kac 1959).

Feynman's integral (13.14) is very difficult to handle mathematically in a rigorous manner comparable to the *Wiener integral.* (For an introduction into this topic cf. Koval'chik 1963, and Kac 1966a.) The many efforts in this direction, while technically successful, seem physically wrongheaded. They do not account in a straightforward way for the constructive and destructive interference between the individual paths, which is due to their phase factors $\exp(iR_N/\hbar)$. Even some of the large-scale numerical computations which are based on the path integral circumvent this difficulty by falling back on the Wiener integral. The price for doing so is that only the lowest state of the quantum system can be obtained, as evidenced in the Feynman-Kac formula (Cf. Schulman 1981, chapter 7).

13.5 Changing Coordinates in the Path Integral

The path integral (13.14) will be used as it was originally conceived by Feynman, as a way of getting correct answers, even though the mathematical procedures are not easily justified. As an example of this philosophy, we propose a recipe for transforming the coordinates in the description of the paths from the Cartesian as in (13.14) to some other, such as polar coordinates. The subtleties in such a transformation have to be approached with great care; a unique source for the required procedures is the monograph by Schulman (1981).

The conditions to be satisfied in such a change concern the moments in the expression (13.12) for the short-time propagator. They are

$$\lim_{\delta t \to 0} (\delta t)^{-1} \left(\int dq''\, K(q''q'\,\delta t) - 1 \right) = V(q')/i\hbar \; ; \qquad (13.15a)$$

$$\lim_{\delta t \to 0} (\delta t)^{-1} \int dq''\, K(q''q'\,\delta t)\,(q''_j - q'_j) = 0 \; ; \qquad (13.15b)$$

$$\lim_{\delta t \to 0} (\delta t)^{-1} \int dq''K(q''q'\,\delta t)(q''_j - q'_j)(q''_\ell - q'_\ell) = (i\hbar/m)\delta_{j\ell} \,. \qquad (13.15c)$$

When a magnetic field is present, the right-hand side of (13.15b) differs from 0, and becomes the electromagnetic vector-potential ; the right-hand side of (13.15c) is the inverse-mass tensor which is non-trivial in the Anisotropic Kepler Problem.

Let us now try to express the short-time propagator (13.12) in terms of *spherical polar coordinates* (r, θ, χ) with the usual relations $x = r \sin\theta \cos\chi$, $y = r \sin\theta \sin\chi$, $z = r\cos\theta$. The conditions (13.15) on the first three moments with respect to q'' and q' can only be satisfied by the expression

$$(m/2\delta t)\,\{(r'' - r')^2 + r''r'(\theta'' - \theta')^2 + r''r' \sin\theta'' \sin\theta' (\chi'' - \chi')^2\}$$
$$- \delta t\,\{V(q') + (\hbar^2/8mr''r')(1 + 1/\sin\theta'' \sin\theta')\}, \quad (13.16)$$

to be placed inside the brackets in the exponent of (13.12). It is important to write the coordinates in the kinetic energy as symmetrically as possible.

The expression (13.16) is the integrand for Hamilton's principal function as resulting from Feynman's path integral in spherical coordinates; the classical approximation can be expected to require the new terms proportional to $\hbar^2$. In terms of the total classical angular momentum L, the Hamiltonian that comes out of (13.16) has the kinetic energy consisting of the radial part, $p_r^2/2m$, and the angular part, $(|\mathbf{L}|^2 + \hbar^2/4)/2mr^2$. If one proceeds from this modified Hamiltonian to apply the rules of classical quantization (cf. next chapter), one finds without any additional tricks that $|\mathbf{L}|^2 + \hbar^2/4 = (\ell + 1/2)^2\hbar^2$, and, therefore, the correct quantum-mechanical result, $|\mathbf{L}|^2 = \ell(\ell + 1)\hbar^2$ where ℓ is an integer ≥ 0.

The additional terms in (13.16) look like the potential energy due to an extra angular momentum of strength $\hbar/2$. That is exactly the term $\kappa/4$ to be added to the Laplacian on the two-dimensional sphere which was already included in (13.2). It arises here as the direct consequence of maintaining the first three moments of the short-time propagator as given in (13.15). This curvature correction arises also in spaces of negative curvature as was shown by Gutzwiller (1985b) as well as Grosche and Steiner (1988); the motion in such a space will be the subject of Chapter 19.

There does not exist at present a simple rule which would permit one to perform a general canonical transformation of the coordinates in the path integral. Some special cases have come up recently, such as the transformation of the Kepler problem with the help of the Kustanheimo-Stiefel coordinates (cf. Duru and Kleinert 1979: Ho and Inomata 1982). The new Lagrangian is quadratic in the coodinates, so that the path integral can be worked out analytically; but there is also a change in the time variable which has some experts worried.

13.6 The Classical Limit

The path integral is the ideal tool to find out what happens when Planck's quantum $\hbar$ becomes small compared with the prevailing values of the Lagrangian action integral R_N in (13.14). If the path is allowed to vary freely, the phase angle of the integrand in (13.14) goes many times through 2π, and the individual contributions destroy one another.

This destructive interference does not happen in the neighborhood of those paths whose Lagrangian action does not change under small variations of the path. Of course, these are exactly the classical trajectories because the first variation of their Lagrangian actions vanishes as explained in Section 1.2. The value of R_N for the neighboring paths is then well approximated by the second variation which was discussed in Section 1.5. The integral (13.14) becomes a Fresnel integral like (12.24) whose quadratic form is the second variation of the Lagrangian action R_N

Although the full calculations along this line of reasoning are tricky, and will not be carried out, their end-result is not in question. Morette (1951) was the first to obtain *Van Vleck's formula* (12.20) from Feynman's path integral; but the method of Papadopoulos (1975) is closer to our way of looking at this essential task. We shall mention only the main ideas without going into the full derivation; Albeverio and Höegh-Kron (1977) have given a full mathematical treatment, while Levit and Smilansky (1977) approach the problem more closely in the spirit of this book.

The second variation in the exponent of the path integral (13.14) is a quadratic form in the function space that describes all the possible displacements from the classical path; this quadratic form has to be diagonalized exactly as in the transition from (12.23) to (12.24). The integral then has an amplitude factor, $(\Pi|\lambda|)^{-1/2}$, and a phase factor that is simply $\exp[-i(\#\text{ of }\lambda < 0)\ \pi/2)$. The amplitude factor can be shown to become $(|C(q''\,q'\,t)|)^{1/2}$. Because of *Morse's theorem* (cf. Section 1.5), the phase factor subtracts $\pi/2$ from the phase angle for every conjugate point along the classical trajectory from q' to q''. A more traditional way to derive the phase loss near a conjugate point consists in expanding the solution of the wave equation near the caustic; the parallel motion is a plane wave, whereas the perpendicular motion is treated through the WKB approximation (cf. Ludwig 1975 and references therein).

In this straightforward manner, we end up with the *Van Vleck formula* (12.25) for the classical approximation $K_c(q'' q' t)$ to the quantum-mechanical propagator $K(q'' q' t)$. This derivation has been carried out in many forms by different people during the last 40 years (cf. Morette 1951; Choquard 1955); but they got stuck at the first conjugate point where the amplitude in Van Vleck's formula becomes singular, and it is then difficult to judge the quality of the approximation. The author was the first, however, to extend its validity beyond the first conjugate point, and to apply Morse's theorem to the phase loss of $\pi/2$ at each conjugate point (Gutzwiller 1967).

It was shown in the last chapter how the Laplace transformation of the classical propagator K_c with respect to time leads to the classical Green's function G_c, and to a formula (12.28) very similar to Van Vleck's. It is this last form of the classical approach to quantum mechanics which seems most useful: in particular, (12.28) will be interpreted as far as feasible as an aproximation to (13.10). The full-blown comparison between the exact (13.10) and the approximate (12.28) will be investigated in this book only for exceptional cases such as the hydrogen atom. We will, however, concentrate on a reduced version which is obtained by taking the trace in both (13.10) and (12.28). The resulting *trace formula* provides the most direct method of connecting classical chaos with quantum mechanics.

Although there is no simple analog for Feynman's path integral to yield the quantum-mechanical Green's function $G(q'' q' E)$, there is a direct and obvious analogy between Van Vleck's expression (12.25) for the quasi-classical propagator K_c and the quasiclassical Green's function G_c of (12.28). Both are sums over all the classical paths from q' to q''; both look equally like the reduction of an original sum over all paths, although that is true only for the K_c, and not for G_c.

Green's function $G(q'', q', E)$ can formally be written as an *integral over 'histories' in phase space* (cf. Garrod 1966), i.e., an alternating series of positions q and momenta p, such as $q' = q_0, p_{1/2}, q_1, p_{3/2}, \cdots, q_{N-1}, p_{N-1/2}, q_N = q''$ (for more details cf. Gutzwiller 1967). The trouble with this approach is, as far the author can tell, that the continuous 'histories' are no longer prevalent, as they are in Feynman's and Wiener's integral.

Daubechies and Klauder (1982, 1989) as well as Klauder (1988 and 1989) have proposed a path integral in phase space that works with complex values for both the position and the momenta. It is not clear whether this formal generalization simplifies the problem of justifying the path-integral mathematically, and of finding the classical limit. Phase space is a problematic concept in quantum mechanics because of Heisenberg's uncertainty relations; it seems preferable, at least in the present context, not to face all the issues that are raised by, and the many efforts that are concerned with, the phase-space version of the path integral.

While the summation over all paths requires some kind of integration in a space of functions, the sum over all classical trajectories remains a sum over a denumerable set of terms. That set is entirely determined by the classical mechanics of the system and represents the best as well as the worst of chaos depending on the problem. Its application to (12.25) or (12.28) constitutes an entirely novel view of classical chaos, which subjects its structure to a stringent test, of the kind which could never have been proposed by a classical mechanician before the arrival of quantum mechanics.

If the classical expression (12.28) is indeed an approximation to $G(q'', q', E)$, then its functional dependence of q'', q' and E has to mimic (13.10). In particular, there have to be poles on the real E-axis in the sum (12.28), which can then be interpreted as energy levels, with the residues yielding the corresponding wave functions, exactly as we found in the Kepler problem at the end of the preceding chapter. There is no guarantee known at this time that anything like simple poles will occur as the result of some arbitrary Hamiltonian, but we will see cases later on where this turns out to be true.

The author hopes that the direct evaluation of the sum (12.28), where each term keeps its complex value, consisting of an amplitude and a phase factor, may eventually become feasible for most systems of interest. If the structure of classical chaos is indeed as rigid as one has every reason to suspect, the necessary mathematical manipulations, and perhaps even their rigorous justification, may be easier than today's approaches to the full-fledged Feynman path integral.

CHAPTER 14

The Quantization of Integrable Systems

Quantum mechanics was first developed and tried out on a few integrable systems, such as the harmonic oscillator, the hydrogen atom, and the electron in a rectangular box. In spite of the conceptual differences between the classical and the quantal regime, the mathematical connection is quite close in these special cases. Some of the relevant methods can be carried over into systems with soft chaos, and seem to yield reasonable results, as if quantum mechanics were capable of riding smoothly over the rough spots in classical mechanics.

This chapter presents first the construction of wave functions in the integrable cases with the help of classical mechanics, as proposed by Keller on the basis of Einstein's critical insights. Different ways of applying these ideas to softly chaotic systems are then discussed, along with their presumed justification. As in most of the topics in this book from here on, the final verdict is not in as yet; the reader is expected to disagree with some of the loose speculation, and is encouraged to think of new ways to understand the examples.

Many theoreticians have tried to find a systematic expansion in ascending powers of $\hbar$; for instance, Voros (1977 and 1986) discusses a mathematically sophisticated method for obtaining asymptotic $\hbar$-expansions of stationary quantum states. Reluctantly, I have concluded, however, that no method has succeeded as yet in yielding higher-order corrections to the lowest approximation that was proposed in Chapter 12, except for a simple quantity like the density of states (cf. the corrections to Weyl's formula in Section 16.2). But it seems that, even there, it is only possible to improve on some average over many quantum states, whereas the discrete nature of the spectrum depends entirely on our primitive first order for which no useful corrections are available at this time.

14.1 Einstein's Picture of Bohr's Quantization Rules

Niels Bohr proposed a number of ad hoc rules in 1913 to modify the classical mechanics of the Kepler problem, in order to obtain the energy levels of the hydrogen atom. The spectacular success of Bohr's ideas, in all their apparent contradiction, was explained 12 years later with the advent of wave mechanics. Meanwhile, Bohr's rules of quantization were generalized and tried out on other systems. Among the best-known contributors were Sommerfeld and Epstein, but they as well as Kramers got stuck when it came to considering more complex atoms such as helium.

Einstein (1917) turned the whole question around and asked which mechanical systems could be subject to the Bohr-Sommerfeld-Epstein rules. In his customary low-key, intuitive manner he came up with a much more direct view of the basic ingredients in terms of the invariant tori which so far no physicist had ever talked about. He went on to point out that the absence of these tori in phase space prevents any use of the quantization rules and removes the system from being 'quantized' in this way. Such systems form the majority in Einstein's view, and he quotes Poincaré to back up this conclusion.

There seems to be only one reference, Lanczos (1949), to this paper in the ensuing 40 years. Such total neglect of an incisive comment on a 'hot subject' by the world's best-known physicist is almost beyond comprehension, in particular, since many close colleagues wrote large and learned reviews of the whole topic in the early 1920s. Most outstanding among these is the great treatise by Sommerfeld (1919, 1922, 1931) which went through several editions; but there is also a book by Born (1925, 1927), and two lengthy survey articles by Pauli (1926b, 1929) for two separate Handbooks; even Van Vleck (1926) wrote a report of more than 300 pages for the (American) National Academy of Sciences.

The existence of invariant tori is the hallmark of integrable systems; their construction was discussed in the first three sections of Chapter 3. Einstein did not need these general considerations, since he got his insight from studying a special case. Einstein's example, a particle in a cylindrically symmetric potential, was presented in Section 6.1; the reader should review this section in order to appreciate the next steps.

The value of the action integral I_j in (3.2) does not depend on the exact choice of the contour C_j. Indeed, if there are two topologically equivalent loops, C and C', the difference of the corresponding action integrals $I - I'$ is the integral $\int p\, dq$ taken over the combined path $C - C'$. Since the loops are equivalent, this last contour integral is the same as a contour integral over a loop Γ, which can be contracted to a point. According to Stokes's theorem, the line integral over Γ is equal to the surface integral of the curl, extended over the inside of Γ. But the integrand for the line integral is $p\, dq = \Sigma\, I_j dw_j$ in action-angle variables and the curl for the latter vanishes on the invariant tori because the value of each I_j is a constant.

The last step in this argument can be phrased in more general terms. *Stokes's theorem,* converts any line integral $\int p\, dq$ over a closed contour into a surface integral $\int \Omega(\delta, \Delta)$, where Ω is the canonical two-form (7.1). The surface is bounded by the closed contour, and the contour can be contracted to a point inside the surface, exactly as $C - C'$ is contracted on the invariant torus. The two-form Ω represents the curl of the *one-form* $\omega(d) = \Sigma\, p_j\, dq_j$. The two-form becomes $\Omega(\delta, \Delta) = \Sigma(\delta I_j\, \Delta w_j - \delta w_j\, \Delta I_j)$ in action-angle variables. The invariant torus has, therefore, the special property that $\Omega(\delta, \Delta) = 0$ for any two tangent vectors δ and Δ. A manifold of phase space with this property is called a *Lagrangian manifold;* its dimension can be at most equal to the number of degrees of freedom. The Lagrangian character of a manifold is maintained as time develops because of (7.3).

As soon as the invariant tori are established, e.g., with the help of a surface of section, any closed loop C_j on a given torus can be used to compute I_j. There are as many topologically independent loops as there are degrees of freedom. In contrast to the earlier versions of the quantization rules, it is not necessary to perform explicitly the separation of variables; indeed, in some cases like the Toda lattice the separation is difficult to carry out. More importantly, the whole idea of separating variables loses its rigorous foundation as soon as a small perturbation has destroyed even a small portion of the invariant tori in conformity with the KAM theorem.

One of the action variables in the cylindrically symmetric potential of Section 6.1 was the angular momentum M ; Bohr's original rule was to make the conserved angular momentum equal to a multiple of $\hbar$ = Planck's quantum divided by 2π. *Einstein's quantization condition* assigns a multiple of $2\pi\hbar$ to every conserved action integral, e.g., the radial action N in Section 6.1. Therefore,

$$\int_{C_j} p\,dq \;=\; 2\pi\,\hbar\,n_j\;, \qquad (14.1)$$

where the contour C_j can be any closed loop on an invariant torus.

The Hamiltonian of the system can be written as a function $H(I_1, ..., I_n)$ of the actions I, as was explained in Section 3.2 and 3.3; e.g., $H_0(M, N)$ was obtained by solving the equation (6.2) for E as a function of M and N. In the space of the actions (M, N), the quantization conditions (14.1) determine a lattice of mesh-size $\hbar$. Since the function $H_0(M, N)$ is smooth, the resulting *spectrum of energy levels,* i.e., the set of energies $H_0(n_1\hbar, n_2\hbar)$ obtained in this way, has certain characteristic properties which will be discussed in Chapter 16.

Chapter 6 was devoted to the classical periodic orbits, and their construction was described quite generally for an integrable system in Section 6.2. Each *periodic orbit* was characterized by as many integers as degrees of freedom. Notice that each energy E has a full complement of periodic orbits; the action integral $S = \int p\,dq$ for each is a well-defined function of the energy. There is no direct relation, however, between the periodic orbits and the spectrum of energy levels, although an indirect connection will be established in Chapter 16. The quantization condition (14.1) does not imply that the trajectory on this particular torus is a periodic orbit. Figure 30 shows the typical situation on an invariant torus.

Einstein's derivation of the quantization conditions (14.1) makes it clear that only integrable systems can be treated in this manner. The *Helium atom* is not integrable as the theorem of Bruns and Poincaré shows (cf. Section 4.1); the formulas (14.1) cannot even be written down. Since it is possible to write down Schrödinger's equation, however, and its energy levels can be found with the help of some large-scale computations, we are left to ask whether the classical mechanics of the He atom can tell us anything about its quantum-mechanical properties. While this special problem is still unresolved, important progress has been made in the last two decades on certain modifications of the hydrogen atom (presence of a magnetic field as explained in Chapter 18, anisotropic mass tensor as in Chapter 11, perturbation by a periodic electric field, cf. the experiments of Bayfield and Koch 1974, as well as the theoretical work mentioned in the review of Bayfield 1987) to connect classical trajectories with the quantal energy levels in spite of chaotic behavior.

This fundamental issue concerning classically chaotic systems was raised for the first time in Einstein's 1917 paper, in the more narrow context of Bohr's quantization rules. Even when Keller (1958) finally discovered this paper, and interpreted it in the light of Schrödinger's

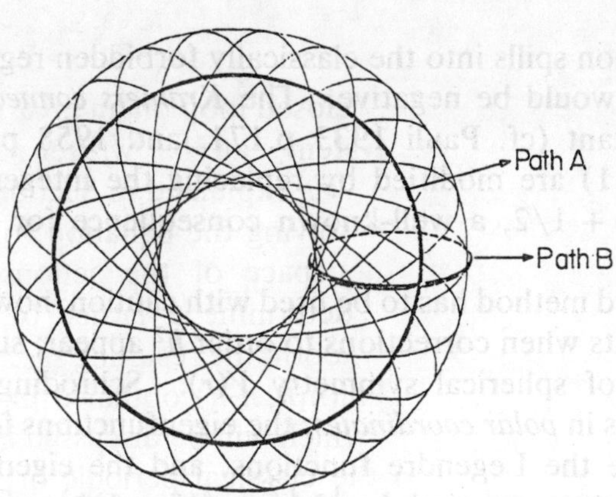

Figure 30 Two topologically independent contours (closed paths A and B) on an invariant torus for Einstein's quantization condition [from Noid and Marcus (1977)].

equation, he was only concerned with integrable (and possibly non-separable) systems of which he and Rubinow (1960) gave many novel and interesting examples. Most textbooks in quantum mechanics are still stuck at an even earlier stage, where the explicit separation of variables seems the necessary condition for the classical approximation.

14.2 Keller's Construction of Wave Functions and Maslov Indices

A *separable dynamical* system can be integrated simply by choosing the appropriate coordinates in position space; the hydrogen-ion molecule (Section 3.4) and the motion of a particle on a triaxial ellipsoid (Section 3.5) are of this type, but not the Toda lattice as explained in the last two sections of Chapter 3. The separation of variables carries over into Schrödinger's equation without difficulty; the classical kinetic energy remains quadratic in the momenta so that the new kinetic energy operator contains no more than second-order derivatives of the wave function.

The degrees of freedom are properly separated by writing the wave function as a product of functions, each of which depends only on one coordinate. Each factor then satisfies a one-dimensional Schrödinger equation whose solution can be approximated by the *Wentzel-Kramers-Brillouin (WKB)* method (cf. Dicke and Wittke 1960 p. 245). The quantization conditions (14.1) are now modified in some circumstances, because the particle may be confined to a potential well, and the wave function spills into the classically forbidden region where the kinetic energy would be negative. The *Kramers connection formulas* become important (cf. Pauli 1933 p.171, and 1958 p.92), and the conditions (14.1) are modified by replacing the integers n_j with the half-integers $n_j + 1/2$, a well-known consequence for the harmonic oscillator.

This standard method has to be used with caution, however, and can yield poor results when corrections to order $\hbar^2$ appear; such is the case in a potential of spherical symmetry $V(r)$. Schrödinger's equation (13.6) separates in *polar coordinates;* the eigenfunctions for the angular dependence are the Legendre functions, and the eigenvalue for the angular momentum squared is $|\mathbf{L}|^2 = \ell(\ell + 1)\hbar^2$ with an integer $\ell \geq 0$. Schrödinger's equation for the radial motion, therefore, has the centrifugal potential $\ell(\ell + 1)\hbar^2/2mr^2$; its eigenvalues for the Coulomb potential are given by $m\, e^4/n^2\hbar^2$ where n is a positive integer, as found originally by Balmer. Everything is fine!

A rather common procedure, however, is to discuss the radial (three-dimensional) Schrödinger equation in terms of the WKB approximation even though the the angular dependence of the wave function was taken into account exactly, with $|\mathbf{L}|^2 = \ell(\ell + 1)\hbar^2$. The relevant classical Hamiltonian is then (6.1), and the quantization condition, including the phase loss at the classical turning points due to the Kramers connection formulas, yields $N = (\nu + 1/2)\hbar$ for the radial action (6.2) with an integer $\nu \geq 0$. The energy levels are obtained from the (two-dimensional) formula (6.3) where $|M|$ in the denominator is replaced by $|L| = \sqrt{\ell(\ell + 1)}\ \hbar$; the denominator is no longer the square of an integer as in the Balmer formula. The mixture of exact solution for the angular motion and WKB approximation for the radial motion causes a mistake of order $\hbar^2$.

Various tricks have been invented to avoid this pitfall, the best known is attributed to Langer (1937); but as far as the philosophy in this book is concerned, there are only two possible ways out. The first has been mentioned in Section 13.5: Feynman's path integral has to be written in the new (polar) coordinate system in order to find out what the equivalent classical Lagrangian has to be. An additional term equivalent to $\hbar^2/8mr^2$ appears which has to be added to the Hamiltonian (6.1) from the very start in order to treat the three-dimensional problem. The angular momentum M in the centrifugal term of (6.1) is, therefore, replaced by $M^2 \rightarrow |\mathbf{L}|^2 + \hbar^2/4$. The second way out is to avoid the separation of the coordinates altogether, except for the computational work, and stay with the Van Vleck formula (12.25) and its constant-energy version (12.28).

The construction of the wave function based on classical mechanics, independent of the explicit separation of the variables, starts from the following remark: Every invariant torus provides a double covering of the classically allowed region in position space, unless the torus has a complicated shape in phase space and its projection into position space covers certain parts more than twice, as shown in Figure 31. Each piece of this projection is characterized by a unique assignment of a momentum p to a position q. In Einstein's example (6.1), the energy E and the angular momentum M fix the value of the azimuthal component of the linear momentum at a given distance r; but the radial component (called p in Section 6.1) has two possible values, differing in sign, and corresponding to the two pieces of the invariant torus which cover the annulus between r_1 and r_2.

The wave function on each piece of the projection is obtained by writing

$$\phi(q) = A(q) \exp[iS(q)/\hbar] , \tag{14.2}$$

where both $A(q)$ and $S(q)$ are real, and $A \geq 0$. To lowest order in $\hbar$, they satisfy the first-order partial differential equations,

$$H(\partial S/\partial q, q) = E , \quad \operatorname{div}(A^2 \partial S/\partial q) = 0 , \tag{14.3}$$

where the first is the Hamilton-Jacobi equation (2.8), and the second insures the continuity of the flow in position space for a fluid of local density A^2 and local momentum $\partial S/\partial q$.

This flow is continuous on the invariant torus, although the particular representation (14.2) in the position-space projection has singularities in the higher derivatives of $S(q)$ along the boundaries. A closer look (cf. Figures 30 and 31) shows that the boundaries are the caustics for the trajectories of the flow; the momentum p is tangential to the boundary; the component at right angles to the boundary vanishes. The wave function in the neighborhood of the caustics can be separated locally into a motion parallel and perpendicular to the caustic. Whereas the parallel motion proceeds as a plane wave, the perpendicular motion is approximated in the same manner as in the Kramers connection formulas, with the help of Airy functions, and yields a phase-loss of $\pi/2$ just as in Section 12.5.

The phase of the wave function consists, therefore, of two (or more) parts corresponding to the pieces of the invariant torus that project into position space: inside each piece, the phase is approximated by the classical action $S(q)/\hbar$ which is continuous and finite at the boundary. There is a loss of $\pi/2$ at each transition from one piece of the projection into another with the flow on the torus.

Instead of thinking of the approximate wave function as defined in position space, Keller puts it on the torus and defines it by (14.2) on

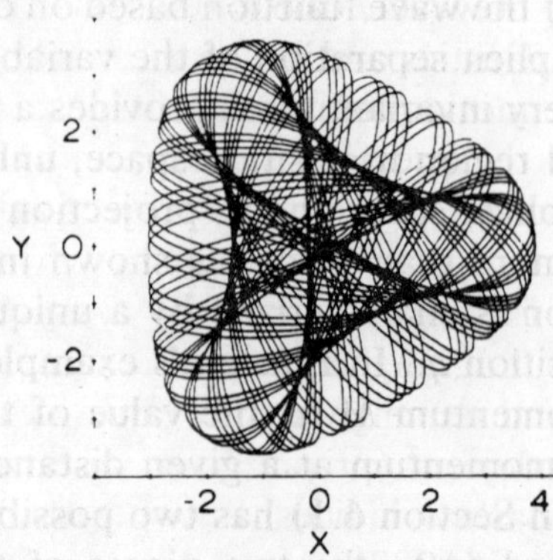

Figure 31 Trajectory in position space from a Hénon-Heiles-like Hamiltonian; the projection of the invariant torus in phase space leads to multiple coverings, and the corresponding number of component wave-functions in Keller's construction [from Noid and Marcus (1977)].

each piece of the projection. If $\phi(q)$ has to be determined for the position q, then the sum over all the projections into q is formed. The phase factor of the wave function on the torus, with all the phase losses on the caustics, has to be well defined; and this in turn requires that the cumulative phase changes for any closed contour on the torus be a multiple of 2π. In this way the slightly more general conditions arise,

$$\int_{C_j} p\, dq = 2\pi\hbar(n_j + \beta_j/4) \,, \qquad (14.4)$$

where $n_j \geq 0$ and $\beta_j \geq 0$ are integers. Notice that the quantum number n_j can indeed vanish as in the ground state of the harmonic oscillator; but then the number β_j of boundary crossings cannot be 0 at the same time.

That is exactly the detail that Pauli could not know in 1919 when he treated the hydrogen molecule ion in his Ph.D. thesis for Sommerfeld, as was mentioned in Section 3.4. Instead of setting $n = 0$ and $\beta = 2$ for the radial motion with respect to the axis of the molecule, he chose $n = 1$ and, of course, $\beta = 0$. The resulting energy came out to be positive, and he concluded that the hydrogen molecule ion is only metastable.

The integer β_j is called a *Maslov index,* in honor of V.P. Maslov who did basic work on classical approximations to quantum mechanics during the 1960s (Maslov 1972, 1981). It can also be understood as the number of conjugate points, or *Morse Index* for trajectory, as discussed in Section 1.5. The latter concept is more general and fundamental, since it applies to any Hamiltonian system.

The Maslov index, however, can be defined independently of the trajectories with their caustics and conjugate points, provided the system is integrable. The motion then takes place on a Lagrangian manifold, as mentioned in the previous section, and the Maslov index is determined by the topology of the Lagrangian manifold in phase space with respect to position space. From this more general perspective, even the Morse index for non-integrable systems can be defined, provided one studies the Lagrangian manifold which is generated by a particular trajectory and its neighbors in phase space. This approach is investigated by Littlejohn (private communication); the practical calculation of the Maslov indices has been studied recently by Littlejohn and Robbins (1987), as well as Robbins and Littlejohn (1987).

The quantization condition (14.4) for integrable systems, and the construction of the wave function that goes with it, has been given the name of *EBK quantization,* for Einstein, Brillouin, and Keller; similar ideas were proposed independently by Laudauer (1951 and 1952). Keller's work is particularly valuable because it uses Einstein's invariant tori and offers many new examples. Related ideas have been promoted mainly by mathematicians under the heading of *geometric quantization;* a readable first account of this development can be found in the book by Souriau (1970).

14.3 Transformation to Normal Forms

The surface of section (cf. Chapter 7) shows quite clearly whether a particular dynamical system is integrable or not. If there are only few invariant tori left on the surface of constant energy, the construction of wave functions in the preceding section cannot be carried out. How can one still get some useful information about the energy spectrum from the classical system in such a situation of soft chaos? An appeal to Van Vleck's formula (12.25) or (12.28) may be very difficult because it requires a complete enumeration of the classical trajectories. This enumeration can be performed in hard chaos, as we shall see later, but it has defied most efforts in soft chaos.

The KAM theorem assures us that the transition from integrable to chaotic is not discontinuous, in general. The break-up of the invariant tori takes place near trajectories with rational frequency ratios, and leaves the tori with algebraic frequency ratios intact. The transition regime presents, therefore, an intimate mixture of invariant tori and chaotic regions.

Neighboring, concentric-looking, invariant tori in the surface of section can be used to estimate the degree of dissolution due to chaos. Let us take the case of two degrees of freedom where the surface of section has two dimensions. If the area between two invariant tori is smaller than Planck's constant h, we can speculate that quantum mechanics is able to ignore the complications of the chaotic behavior, and continue its relation with classical mechanics as for integrable systems. This idea has been used with rather unexpected success in different ways; the most convincing is the method of transforming the Hamiltonian to a normal form, which will be discussed in this section.

As usual, the guiding principle is not diffficult to understand, although the technical details have to be worked out carefully, and may lead to some significant differences depending on the system at hand. We will follow the work by Delos and Swimm (1977 and 1979), which goes to back to the paper by Gustavson (1966), who in turn relied on the seminal work by Birkhoff (1927). The results of the last two authors was briefly reported in Section 8.4; canonical transformations are used to eliminate systematically all unwanted perturbations (cf. Section 5.3).

The starting Hamiltonian describes the motion of a system in the neighborhood of a point of stable equilibrium. The kinetic energy is quadratic in the momenta, and the potential energy is expanded in powers of the position coordinates with the point of equilibrium at the origin. The lowest term, is therefore a positive definite, quadratic form of the position coordinates. It can be diagonalized by an appropriate linear orthogonal transformation, so that the Hamiltonian takes the form $p^2/2m + m\omega^2 q^2$ in each degree of freedom. The additional canonical transformation $p \rightarrow p/\sqrt{m\omega}$ with $q \rightarrow \sqrt{m\omega}\, q$ yields the starting Hamiltonian

$$\sum_{k=1}^{n} \frac{\omega_k}{2}(p_k^2 + q_k^2) + H^{(3)}(p, q) + H^{(4)}(p, q) + \cdots \quad , (14.5)$$

where $H^{(s)}$ is a homogeneous polynomial of s-th degree in the coordinates p and q.

At the beginning, the polynomials $H^{(s)}$ are functions of the positions q only; after the first transformation, however, and all the later ones, the form (14.5) remains the same, but the polynomials $H^{(s)}$ contain both, momenta and positions. Our aim is to give these polynomials the simplest form which can be achieved. Birkhoff's idea is to make $H^{(s)}$ a function of the variables $\rho_k = (p_k^2 + q_k^2)/2$. This requirement can be expressed in the form of a Poisson bracket,

$$[H^{(2)}, H^{(s)}] = \sum_{k=1}^{n} \omega_k \, (q_k \partial/\partial p_k - p_k \partial/\partial q_k) \, H^{(s)} = 0 \, . \quad (14.6)$$

If this condition is fulfilled, the partial Hamiltonian $H^{(s)}$ is said to be in its *normal form.*

The first transformation is designed to do the job for $H^{(3)}$; the second transformation does it for $H^{(4)}$ while leaving $H^{(3)}$ unchanged, and so on. Just as in Section 5.3, the generating function that takes care of $H^{(s)}$ depends on the new momenta P_k and the old positions q_k. In analogy to (5.6), it is written as

$$F^{(s)} = \sum_{k=1}^{n} P_k q_k + W^{(s)}(P, q) \, , \quad (14.7)$$

where $W^{(s)}$ is a homogeneous polynomial of degree s in P and q.

The resulting transformation,

$$Q_k = q_k + \partial W^{(s)}/\partial P_k \, , \; p_k = P_k + \partial W^{(s)}/\partial q_k \, , \quad (14.8)$$

does not affect the terms in the Hamiltonian of degree lower than s. The explicit calculation of the coefficients in $W^{(s)}$, however, on the basis of condition (14.6) is not a trivial matter. We will not go into the computational detail, and point out only that the number of parameters available in $W^{(s)}$ matches the number of coefficients in $H^{(s)}$ that are supposed to vanish after the transformation.

As explained in Section 8.4, the above normalization procedure is equivalent to finding the n new constants of motion $\rho_k = (p_k^2 + q_k^2)/2$ in the new coordinates. After transforming these expressions back into the original coordinates, as many constants of motion for the original coordinates are established as there are degrees of freedom. By assigning each some fixed value, a corresponding invariant torus in the original phase space is found.

This method yields good approximations to the invariant tori in phase space wherever they happen to lie; in addition, it produces some fictitious invariant tori where the surface of section clearly indicates chaotic behavior, as shown in Figures 12, 13, and 14. The normalization does not converge in general and has to be truncated. The algebraic manipulations are done on a computer; truncation at the eighth order is apparently a good stopping point for Hamiltonians of the Hénon-Heiles type. At every order, the transformed Hamiltonian is a polynomial $\Gamma(\rho_1, \rho_2, \ldots)$.

Problems arise when some of the frequencies ω_k are *commensurate,* or nearly so; the procedure has to be modified in a manner that was first employed by Gustavson in the discussion of the Hénon-Heiles model (cf. Section 8.4). Swimm and Delos (1979) also consider a system with two degrees of freedom and a third-order coupling term, but they avoid the difficulty of commensurate frequencies by choosing the ratio $\omega_1/\omega_2 = 13/7$. This ratio cannot be changed with the help of canonical transformations. The frequencies are set equal to 0.7 and 1.3, rather than 1.0 and 1.0 as in the Hénon-Heiles model (8.4).

In order to appreciate the effect of the third-order term $H^{(3)}$, the reader has to recall that classical mechanics still allows the choice of a length scale, or the range of the coupling term, even after the quadratic terms have been brought into the form (14.5). Since the coupling range has been set at 1 in the Hénon-Heiles model, Planck's constant has a numerical value that can no longer be scaled to 1. Swimm and Delos choose the opposite route by setting $\hbar = 1$, and then writing the coupling term as

$$H^{(3)} = \lambda\, q_2(q_1^2 + \eta\, q_2^2) \ . \tag{14.9}$$

If λ and η are small enough, the lowest energy levels are very close to independent harmonic oscillators. The effect of soft chaos in this Hamiltonian is then felt only in the higher levels; it is the same as making Planck's constant small, and the coupling parameters λ and η of order 1.

The quantization of this system with the help of the EBK scheme of the preceding section requires that each

$$\rho_k = (p_k^2 + q_k^2)/2 = (n_k + 1/2)\hbar \ , \tag{14.10}$$

where the integer $n_k \geq 0$. The 1/2 in addition to the integer n_k corresponds to a Maslov index of 2 for each degree of freedom; or equivalently, 2 conjugate points on the trajectory for each turn around the invariant torus along the (p_k, q_k) direction; or finally, the $\pi/2$ loss of phase at each classical turning point of the k-th oscillator.

The spectrum of energy levels is, therefore, given by the formula

$$E(n_1, n_2, ...) = \Gamma((n_1 + 1/2)\hbar, (n_2 + 1/2)\hbar, ...) \ , \tag{14.11}$$

where Γ is the Hamiltonian after the normalization and after truncating at some high power. This simple result is not valid for the commensurable case, and has to be replaced by a more complicated method where one degree of freedom after another has to be quantized according to (14.4). The quantization of the subsequent degree of freedom depends on the quantum number in the preceding one, and the number of conjugate points has to be obtained by going back to the trajectory in the original coordinates.

The final results are remarkable because they show that formula (14.11) represents almost the whole spectrum for the (practically) incommensurate frequencies $\omega_1 = 1.3$ and $\omega_2 = 0.7$ with $\lambda = -0.1$ and $\eta = 0.1$. The error in the comparison with the exact quantum-mechanical levels does not go beyond 0.5% for 82 bound states out of the 83 levels found by solving Schrödinger's equation. These energy levels have actually come down from their values in the absence of the third-order term by some 15%, and yet the errors compared with the exact quantum-mechanical values amount to less than 0.5%.

The comparison of the exact quantum-mechanical energies with the ones from the normalized Hamiltonian is not as favorable in the commensurate case. Some levels are not described by the above method; but the normalized Hamiltonian still yields a large number of levels whose energies are in good agreement with some quantum-mechanical niveau. There is a much higher fraction of such levels than could be suspected from the volume in classical phase space which is covered by invariant tori. The conclusion seems inevitable that the eigenfunctions accommodate themselves to a limited amount of chaos between the remaining invariant tori. Quantum mechanics is able to ride over some of the rough spots in classical mechanics, and effectively to make up its own invariant tori between the real ones which the KAM theorem provides.

The pseudoinvariant tori in phase space were constructed by algebraic means, under the assumption that the procedure converges. The original method by Birkhoff, its refinements by Gustavson, and its application by Swimm and Delos used polynomials of increasing order; but the same idea can be carried out if the perturbation H' of the integrable Hamiltonian H_0 is a trigonometric function of the angular variables as (3.5).

The first canonical transformation is then exactly given by the method in Section 5.3, except that the explicit time-dependence is dropped, i.e., $m_0 = 0$. The generating function W in (5.6) now has a sum over all the relevant multiples (m_1, m_2, m_3), each given by the formulas (5.7) and (5.8) with the resonance denominators. The calculation of the new Hamiltonian is the difficult part: its perturbing part is smaller by one order than before the transformation. The whole procedure can be repeated again, until the truncation seems reasonable.

This alternative route was first suggested by Chapman, Garret, and Miller (1976) who tried it on the combination of the independent oscillators (14.5) with the perturbation (14.9). Jaffé and Reinhardt (1979) succeeded in accelerating the convergence, using Newton's idea for solving equations numerically, which was mentioned in Section 9.6 to explain the KAM theorem. Again most of the quantum-mechanical energy levels are well approximated, as if the system were integrable.

14.4 The Frequency Analysis of a Classical Trajectory

The projection of the trajectories into position space is sometimes found to yield complicated patterns, with all kinds of caustics that may form sharp tips, as in Figure 31. It is then not easy to find contours C of a simple shape so as to apply the EKB conditions (14.4). Nevertheless, the structure of the tori in phase space stays the same for a certain range of energy, and appropriate search procedures can be worked out. A survey of these endeavors was given by Noid, Koszykowski, and Marcus (1981).

The tori satisfying the quantization conditions do not generally carry closed trajectories. A 'trajectory closure method' is used to complete the contour C. Most of C consists of the numerical trajectory, which is eventually closed by a short straight link when it passes through the neighborhood of its starting point. An ordinary, but carefully chosen, surface of section gives the second contour needed to quantize a system with two degrees of freedom. This method gives good eigenvalues wherever there are invariant tori in phase space; but no energy levels corresponding to the chaotic area are obtained in this way.

An alternative criterion for integrability is found in the multiperiodic nature of the trajectories. Since the coordinates of a trajectory can be expanded in a Fourier series like (3.4), with as many frequencies as degrees of freedom, it suffices to find these frequencies directly from a numerical integration. The classical quantization rules (14.4) can then be applied directly, without pinning down the corresponding invariant tori. Such an analysis does not require a surface of section, and is, therefore, applicable to systems with more than two degrees of freedom.

This idea was explored by Marcus and coworkers (cf. Noid et al 1977 and 1980), and some interesting examples were treated numerically (cf. Dumont and Brumer 1988). Since these investigations are closely tied in with more general applications of classical trajectories, we shall discuss their basis in some detail. Rice (1944, 1945) wrote an exceptionally clear and simple exposition of these ideas in the *Bell System Technical Journal* under the title "Mathematical Analysis of Random Noise"; we will only give the bare outline of the relevant results.

In this section, we are interested in making the Fourier analysis of a particular trajectory; but more generally, we will get the *power spectrum* $I(\omega)$ of the *autocorrelation function* $C(\tau) = \langle x(t)x(t+\tau) \rangle$. The function $x(t)$ stands for some dynamical variable, such as a position coordinate $q(t)$, or a momentum coordinate $p(t)$, or a dipole moment $\mu(t)$ in response to an external electric field, as a function of time t. The averaging operation $< ... >$ refers to either one or possibly both of the following features: (i) the existence of a whole ensemble of equivalent dynamical variables, all participating equally in the measurement to be carried out and characterized by some external parameters like different initial conditions; (ii) the same particular system, but measured at different times. In the latter case we define

$$C(\tau) = \frac{1}{2\pi} \lim_{T\to\infty} \frac{1}{2T} \int_0^{2T} x(t)\, x(t+\tau)\, dt\,.$$

The power spectrum is obtained from the two complementary formulas,

$$I(\omega) = \frac{1}{2\pi} \int_{-\infty}^{+\infty} dt\, C(\tau)\, \exp(-i\omega\tau)\,, \tag{14.12}$$

$$C(\tau) = \int_{-\infty}^{+\infty} d\omega\, I(\omega)\, \exp(i\omega\tau)\,.$$

When applying these definitions to a trajectory calculation, it helps to make some additional assumptions: the correlation function $C(\tau)$ is stationary, i.e., $< x(0)x(\tau) > \; = \; < x(t)x(t+\tau) >$ for any value of t; the dynamical variable $x(t)$ is real, so that $C(\tau)$ is an even function of t, i.e., $C(-\tau) = C(\tau)$. The power spectrum $I(\omega)$ is then a real and non-negative function of ω. It is calculated in practice as the limit,

$$I(\omega) = \frac{1}{2\pi} \lim_{T\to\infty} \frac{1}{2T} \left| \int_0^{2T} dt\, x(t)\, \exp(-i\omega t) \right|^2. \tag{14.13}$$

This expression justifies the name 'power spectrum': if a signal is processed in a frequency-sensitive apparatus such as an audio-amplifier, $I(\omega)$ gives us an idea of what is going to happen.

As a special case, we can insert for $x(t)$ the expansion (3.4) of the momentum coordinate $p(t)$ in a multiple Fourier series; the power spectrum becomes

$$I(\omega) = \sum_{k_1,\ldots,k_n} |P_{k_1,\ldots,k_n}|^2\, \delta(\omega - k_1\omega_1 - \cdots - k_n\omega_n)\,. \tag{14.14}$$

There are sharp lines at the fundamental frequencies $\omega_1, ..., \omega_n$, and all their linear combinations with integer coefficients $k_1, ..., k_n$. Each line corresponds to a transition from one energy level to another, if there is some external stimulation. The strength of each line is related to a physical process and involves the coordinate $x(t)$, such as $p(t)$ in (14.14). It is given by the absolute square of its coefficient in the Fourier expansion (3.4) or (3.5). Figure 32 shows such a spectrum for an integrable system with *three degrees of freedom,* one of the few in-

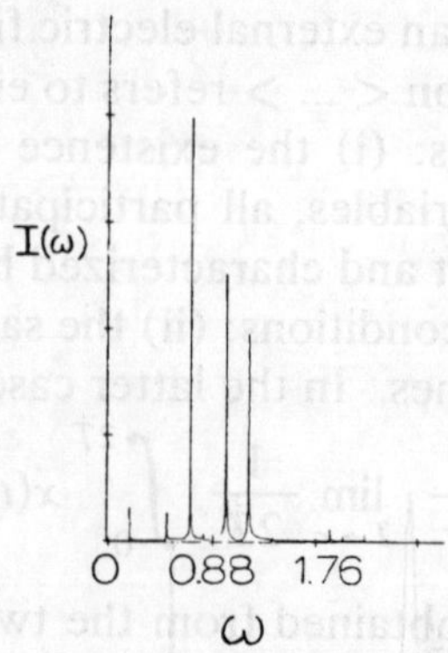

Figure 32 Frequency spectrum for a particular trajectory in a system with three degrees of freedom [from Noid, Koszykowski, and Marcus 1977)].

stances so far where a system with more than two degrees of freedom has been studied in this context.

The contrast between the power spectra for integrable and for ergodic trajectories is striking. Instead of the sharp peaks, there appear broadened lines which cannot be narrowed by extending the time of integration T in (14.13). Nevertheless, the spectrum has a complicated structure, as shown in the example of Figure 33. Different trajectories of the same energy, but belonging to separate ergodic regions in phase space, lead to very different-looking spectra. Their relation to the corresponding parts in the surface of section has not been investigated in detail as yet.

How can we extract classical approximations to the energy levels of a conservative system from the frequency spectrum of the integrable trajectories in a conservative system? If we try many initial conditions, we will eventually get a function $E(\omega_1, ..., \omega_n)$ which gives the energy of the trajectory whose frequencies are found to be $\omega_1, ..., \omega_n$. This function is the dual of $H(I_1, ..., I_n)$ in Sections 3.3 and 6.2, which gives the energy in function of the action integrals $I_1, ..., I_n$, and whose derivatives are the frequencies according to (3.3), i.e., $\omega_k = \partial H/\partial I_k$. As discussed in Section 14.2, the energy levels are given by the discrete values $H((n_1 + 1/2)\hbar, (n_2 + 1/2)\hbar, ...)$ in terms of the quantum numbers $(n_1, n_2, ...)$.

The procedure for finding $H(I_1, ..., I_n)$ from $E(\omega_1, ..., \omega_n)$ goes through the following steps: consider the I's as functions of the ω's and show that the mixed derivatives are equal, i.e., $\partial I_1/\partial\omega_2 = \partial I_2/\partial\omega_1$, and so on; therefore, a function $F(\omega_1, ..., \omega_n)$ exists, such that $I_k = \partial F/\partial\omega_k$. Now the usual argument for Legendre transforms shows that the condition

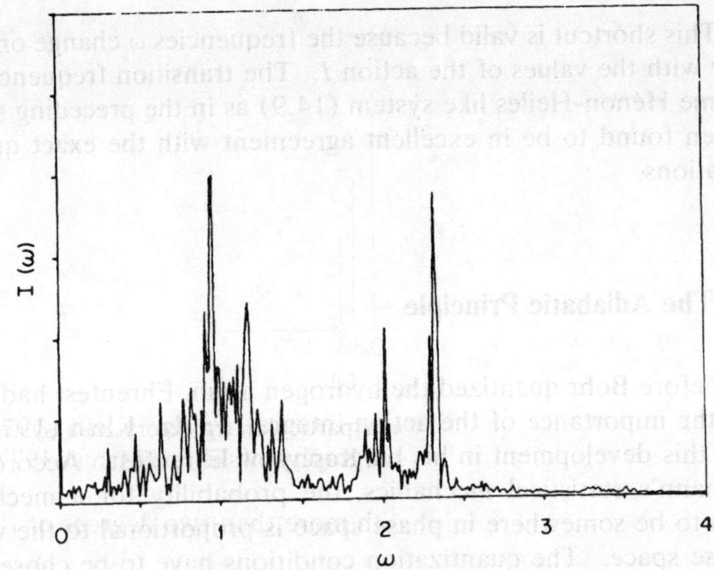

Figure 33 A chaotic trajectory in the Hénon-Heiles model and its frequency spectrum [from Noid, Koszykowski, and Marcus (1981)].

$$\omega_1 \frac{\partial F}{\partial \omega_1} + \omega_2 \frac{\partial F}{\partial \omega_2} + \ldots - F = E(\omega_1, \omega_2, \ldots) = H(I_1, I_2, \ldots) \quad (14.15)$$

holds; the first equality determines $F(\omega)$ when $E(\omega)$ is known, while the second equality tells us how to calculate $H(I)$.

The transition from frequencies to actions requires, therefore, the solution $F(\omega)$ of the first-order linear partial differential equation (14.15). The standard methods show that (14.15) determines F in frequency space only along straight lines through the origin. An initial value is needed at some point along each straight line in order to fix the value of F uniquely. Equivalently, any two solutions of (14.15) differ by a solution $f(\omega_1, \ldots, \omega_n)$ of the same equation, but with 0 on the right-hand side. This last condition says simply that f is any homogeneous function of first degree in the variables $(\omega_1, \ldots, \omega_n)$, such as $[\omega_1^2 + \ldots + \omega_n^2]^{1/2}$.

The function $E(\omega)$ alone, therefore, does not determine the function $H(I)$ uniquely. For each relevant direction in frequency space, the values of the corresponding actions I have to be obtained independently. A limited knowledge of the invariant tori is required after all, although only on a submanifold of codimension 1.

Marcus and coworkers circumvent this problem. They choose initial conditions for the perturbed system which satisfy the conditions (14.4) for the action integrals with $\beta_i = 2$ in the unperturbed system. The action integrals I are thereby fixed already for the perturbed system. This shortcut is valid because the frequencies ω change only very slowly with the values of the action I. The transition frequencies, for the same Hénon-Heiles like system (14.9) as in the preceding section, are then found to be in excellent agreement with the exact quantum calculations.

14.5 The Adiabatic Principle

Even before Bohr quantized the hydrogen atom, Ehrenfest had recognized the importance of the action integral, $\int p\,dq$. Klein (1970) discusses this development in his biography of Ehrenfest. According to Boltzmann's statistical mechanics, the probability for a mechanical system to be somewhere in phase space is proportional to the volume in phase space. The quantization conditions have to be chosen such that reversible, adiabatic changes in the dynamical system bring the quantized energies back to their original values upon the completion of a closed cycle.

The action integral for a periodic orbit satisfies this requirement as can be seen from the harmonic oscillator. If the spring constant is varied very slowly, i.e., it takes many oscillations for the frequency to change substantially, the ratio E/ω is found to remain constant. Since the energy E is twice the kinetic energy, and $\omega = 2\pi/T$ in terms of the period T, one finds indeed that $E/\omega = \int p\,dq/2\pi$ = multiple of $\hbar$ as claimed by Planck.

This argument is valid provided the classical trajectories are periodic orbits, as in the harmonic oscillator and the Kepler problem. Moreover, they have to preserve this periodicity as the parameters in the problem, like the spring constant, are changed. These conditions can be generalized to include integrable systems that keep their invariant tori in phase space when the parameters are modified. Certain natural precautions have to be taken to avoid any of the frequencies going through zero, as will be explained below. In spite of these severe restrictions, however, the adiabatic principle has some limited validity, and can be put to good use, even when the dynamical system becomes chaotic while changing its parameters (cf. Reinhardt and Dana 1987).

In order to appreciate the adiabatic invariance of the action integral, a short return to Chapter 7 is necessary. If any closed loop C in phase space is chosen, and a trajectory is started at each point of C by integrating the equations of motion for some (possibly time-dependent) Hamiltonian H, a whole sequence of loops $C(t)$ is obtained. The integral $\int p dq$ taken over $C(t)$ was shown in Section 7.1 to keep its initial value. The proof of this result is straightforward and very general. By contrast, the adiabatic principle, although somewhat reminiscent, makes a different claim, and its proof depends on many subtle assumptions (for a careful discussion, cf. Jaffé and Watanabe 1988).

Consider a time-independent Hamiltonian $H(p, q, \lambda)$ which depends on a parameter λ, and which remains integrable for all values of λ in a certain fixed interval, say from 0 to 1. Let us now replace λ in the $H(p, q, \lambda)$ by a monotonically increasing function of time, called $\lambda(t)$ for simplicity's sake. The Hamiltonian has now become time-dependent, and one can apply the construction of the preceding paragraph; whatever the function $\lambda(t)$, e.g., in (14.9), the value of $\int pdq$ over $C(t)$ remains the same.

Since $H(p, q, 0)$ is integrable, it is natural to choose the initial loop $C(0)$ to lie on one of its invariant tori. Let us now follow the fate of the contour $C(t)$ as the trajectories in phase space are calculated on the basis of the Hamiltonian $H(p, q, \lambda(t))$. Does the loop $C(1)$ lie on an invariant torus of the Hamiltonian $H(p, q, 1)$? No, unless the function $\lambda(t)$ increases infinitely slowly from 0 to 1, or equivalently, the steady increase from 0 to 1 is stretched over a very long time τ. This time τ has to be great with respect to any natural time scale of the time-independent Hamiltonian $H(p, q, \lambda)$ at a fixed value of $0 \leq \lambda \leq 1$.

In the further discussion, we follow Reinhardt and Gillilan (1986) who were among the first to exploit this idea in order to obtain the energy levels of a dynamical system with the help of the adiabatic principle. Their starting Hamiltonian $H(p, q, 0)$ has its degrees of freedom fully separated, e.g., by setting the third and higher-order terms in (14.5) equal to 0, or equivalently, $\lambda = 0$ in (14.9). The choice of $C(0)$ is then obvious, corresponding to a periodic orbit in one of the degrees of freedom, with the value of $\int pdq = (\nu + 1/2)\hbar$. As λ increases to the end of its interval, the quantum number ν remains the same; but the energy can change significantly and yields the classical aproximation to a quantal energy.

The energy for an individual trajectory can vary quite drastically, and non-monotonically, if the time τ for the adiabatic transition is not extremely long. The simultaneous integration over many trajectories, all starting on the same contour $C(0)$, leads to an average energy $E(t)$ with a more bounded variation. The whole process is sped up if the trajectories are sampled on the whole initial torus, which is then characterized by two quantum numbers. The resulting average $E(t)$ varies monotonically; the accuracy of the resulting approximate energy level can be judged by calculating the root-mean-square deviation from the mean of the energies for the individual trajectories. If it is substantially smaller than the energy difference between neighboring levels, the comparison with the exact quantum-mechanical computations is very good.

The validity of the adiabatic principle has been extensively studied in classical mechanics. Recent work, which includes some detailed numerical investigations, was done by Dana and Reinhardt (1987) on the standard map, and by Brown, Ott, and Grebogi (1987) on billiards with slowly moving walls as well as anharmonic oscillators. The reader will find there many references to earlier work.

The application to quasiclassical quantization was first made by Skodje, Borondo, and Reinhardt (1985); we can only hint at some of their results, and encourage the reader to study their fascinating paper. A critical aspect is again related to the occurrence of resonances as the parameters of the Hamiltonian are changed. As we saw in Chapter 9, the typical resonance generates a chain of islands in the surface of section with a *separatrix* between them, exactly as in the phase space of the pendulum. The motion near the separatrix is arbitrarily slow; if the loop $C(t)$ and the corresponding torus try to cross the separatrix, the variation of the parameter $\lambda(t)$ will always be too fast, and the adiabatic principle breaks down. Such a disaster can be avoided by choosing an initial torus, or an initial Hamiltonian, so that no crossing of resonances occurs. Obviously, such a strategy requires a good understanding of the dynamical system ahead of time; otherwise the theory of Cary and Skodje (1989) has to be used.

Although all these arguments are limited to integrable systems, the numerical calculations were carried out for the Hénon-Heiles type model with a slowly increasing $\lambda(t)$ in the coupling term (14.9), starting at 0 and going to the value chosen by Marcus and coworkers in their Fourier analysis of trajectories (cf. the preceding section). About half of phase space becomes chaotic eventually, but many energy levels which lie clearly in the ergodic region are still obtained with good accuracy. The agreement with the exact quantum-mechanical eigenvalues seems directly related to the root-mean-square deviation of the energy $E(t)$ for the individual trajectories.

Lest the reader think that complete agreement can be achieved by taking a large sample or a sufficiently slow variation, the numerical experience shows otherwise. The deviation can be reduced to some extent, but as the time for the adiabatic transition is lengthened beyond some critical interval, the deviation starts increasing again.

Reinhardt (1985) proposes an interpretation for this behavior in terms of *vague tori,* or equivalently, regions of phase space which are not quite separated any longer by a KAM torus. Indeed, as we saw in Chapter 9, when a last surviving torus with an irrational frequency ratio vanishes, the trajectories cross the ancient boundary in phase space only rather reluctantly. It now appears that such a temporary reprieve in the classical motion is sufficient for a quantum-mechanical wave function to settle down and define an energy level.

14.6 Tunneling Between Tori

The phase space of the ordinary pendulum foliates itself into a set of nested loops up to a separatrix which divides the oscillatory and the rotational motions; the latter form a layered structure, and the energy increases monotonically as one goes further away from the stable equilibrium. If the potential energy has two distinct minima, however, as in a double well, the phase space foliates itself into two sets of nested loops. The separatrix has the energy of the local maximum between the two minima. As the energy increases, one finds again the layered structure of a single kind of motion, namely an oscillation which runs over both minima, between the outside walls of the potential.

The same pictures arise in the surface of section for a system with two degrees of freedom, e.g., in the Hénon-Heiles model as shown in Figures 12 and 13. The various constructions in the preceding sections always applied to a single torus and its immediate surroundings. In particular, Keller's wave function is tied to one torus that satisfies the quantization conditions (14.4). Therefore, it may happen that two different tori end up with the same energy, or at least with an energy difference small compared to the difference with other levels.

Quantum mechanics proceeds to 'split' this degeneracy by making up eigenstates with significant amplitude in both wells, or more generally, on both tori. Nothing of the sort can happen in classical mechanics because there is no trajectory connecting the two troughs in the double well if the energy is insufficient to overcome the local maximum. Similarly, there is no classical passage from one torus to another of the same energy, although the situation is now less transparent since the obstacle is not simply a wall too high to climb with the available energy. As we saw already in the Kepler problem (cf. Section 2.5), when the particle starts in some given position q' with a fixed total energy E , it is still unable to reach all the other positions q whose potential energy $V(q) \leq E$; only the positions inside the critical ellipse are dynamically accessible.

A particle is said to *tunnel quantum-mechanically* from one well to the other in one dimension; this effect is discussed in most textbooks, although not in its full generality. If the potential has a simple rectangular shape, the wave functions can be written down in terms of trigonometric and exponential functions. When the shape is arbitrary, however, although there is still only one degree of freedom, the way to proceed is not obvious because Kramers' connection formulas between the classically allowed and the classically forbidden region do not give a unique prescription.

A systematic start on this general problem could be made as in Chapter 13 with the help of Feynman's path integral; but the tunneling problem is defined only if a particular value of the energy is given, whereas the path integral exists only as a function of time. Indeed, the transition across the (static or dynamical) barrier is in general possible classically for sufficiently short times, at a price in energy. The integral (12.27) of the propagator K over time, in order to get Green's function G, can not be performed by stationary phase for a sufficiently low energy to require tunneling. It is then not clear how the exponential decay of the probability amplitude comes about (cf. Freed 1972), unless the time t is allowed to have a negative imaginary part (cf. among others, Weiss and Haeffner 1983).

Conventional wisdom now takes the expression (12.28) for the classical Green's function, and inserts the integral (2.6) for the action integral, with the endpoints q' and q'' at the boundary of the classically allowed region. Since $V(q) \geq E$, however, the integrand $\sqrt{E - V(q)}$ becomes imaginary; its sign is chosen in agreement with E having a small positive imaginary part $i\,\varepsilon$, so that the phase factor $\exp[iS(q''q'E)/\hbar]$ in (12.28) becomes exponentially small. The wave function in the forbidden region is also assumed to decay in this exponential fashion, and the splitting of the degeneracy is, therefore, proportional to the 'overlap',

$$\Delta E \simeq \exp\left[- \int_{q''}^{q'} dq \sqrt{2m(V(q) - E)} \;/\hbar\right] . \qquad (14.16)$$

Whereas there is little doubt about this expression for tunneling in one dimension, the analogous problem in two or more dimensions requires a more detailed investigation. Wilkinson (1986), as well as Wilkinson and Hannay (1987) have presented a particularly convincing way to proceed, which seems applicable in many different problems. They start from an identity due to Herring (1962): consider a typical tunneling situation which results in a near-degeneracy between two states with the wave functions ϕ_a and ϕ_b. Let the surface Σ divide the two regions in position space where the two wells are located, so that both wave functions are very small near Σ. Since both wave functions are eigenfunctions of the Schrödinger operator, Green's identity can be applied to yield the exact equation,

$$\frac{2m(E_a - E_b)}{\hbar^2} \int_V dV \, \phi_a \phi_b = \int_\Sigma d\Sigma (\phi_a \operatorname{grad}\phi_b - \phi_b \operatorname{grad}\phi_a) \, , \quad (14.17)$$

where the volume V is either one of the half spaces which are created by Σ, and the gradient points toward the outside.

The tunneling situation now gives the two nearly degenerate wave functions the following special features: both are very small near the dividing surface Σ. Their main amplitudes are equally distributed between the two halves of space. They differ because they have different relative signs in the two halves. For definiteness sake, ϕ_a has the same sign, while the prevailing signs of ϕ_b are opposite in the two halves. Except for the region near Σ, and the obvious differences in overall sign, ϕ_a and ϕ_b are very closely the same in each half. Therefore, the volume-integral in (14.17) is very close to 1/2, and

$$E_a - E_b = \pm \frac{\hbar^2}{m} \int_\Sigma d\Sigma \, (\phi_a \operatorname{grad}\phi_b - \phi_b \operatorname{grad}\phi_a) \, , \quad (14.18)$$

which is already Herring's formula. In order to use (14.18), it is enough to know the wave functions ϕ in only one-half of the position space, i.e., comprising only one of the two wells, or one of the two tori. This one-half of a wave function depends, of course, on the whole potential, as we shall see immediately.

The basic idea is to solve the equations (14.3) in the region where $E < V(q)$ by assuming the action function $S(q)$ to be purely imaginary. The wave function $\phi(q)$ in (14.2) has then a real exponent which can be chosen so that ϕ becomes exponentially small near the dividing surface Σ, in complete analogy to (14.16). Wilkinson and Hannay phrase the whole argument more carefully by falling back on Green's function in the form (12.28) where the action integral can be written as in (2.6). If the energy E under the square root is interpreted as a complex number with a small positive imaginary part ε, so that $E \rightarrow E + i\varepsilon$ as we had already done in (12.27) and (13.10), then the integral (2.6) also acquires a positive imaginary part, and (14.16) is quite consistent.

Since the wave function on the dividing surface Σ is now known, at least by a classical approximation, the integral over Σ can be calculated by the method of steepest descent. Thus, one has to find a trajectory which goes from one well to the other, or from one torus to the other, with the potential $-V(q)$. In the case of the groundstate for the hydrogen-molecule ion, this trajectory coincides with the straight line connecting the two protons because of the symmetry of the two tori involved (Cf. Section 3.4). In general, however, there is no symmetry to indicate the best tunneling trajectory.

When the chemists calculate reaction rates where the system does not have enough energy to overcome the potential barrier classically, the trajectory has to be found by a numerical computation (cf. the work of Garret and Truhlar with their collaborators, 1983 and 1985). The conditions to be satisfied are quite different from the usual trajectory calculations; instead of known initial position and momentum, the initial and the final positions have to be located on the boundary of the classically forbidden region; but their exact locations and initial directions have to be found from the condition of minimal action. The same difficulty arises in phase space, when there is tunneling between different tori (cf. among the many efforts in this direction, Auerbach and Kivelson (1984), and the same authors with Nicole (1985), as well as Huang, Feuchtwang, Cutler, and Kazes 1989).

CHAPTER 15

Wave Functions in Classically Chaotic Systems

The state of a quantum-mechanical system at any fixed time t is uniquely described by its complex-valued wave function $\psi(q, t)$, where q represents one-half of a canonically conjugate coordinate system (p, q). Besides being square-integrable, and satisfying certain boundary conditions which depend on the problem at hand, the wave function $\psi(q, t)$ at a fixed time t is subject to few restrictions. Of course, when the time is allowed to vary, ψ has to satisfy Schrödinger's equation (13.1) with a well-defined real-valued potential $V(q, t)$.

Any symptoms of chaos are, therefore, found in the time dependence of $\psi(q, t)$. Wave functions of a fixed frequency $\omega = E/\hbar$, the usual eigenfunctions $\psi(q, t) = \phi(q) \exp(-i\omega t)$ of the stationary Schrödinger equation (13.6), can be expected to display whatever chaotic features can show up in the framework of quantum mechanics.

It is not clear what those features could be, however, because the eigenfunctions $\phi(q)$ are smooth in the strong mathematical sense of being analytic, i.e., expandable in a convergent power series with respect to the coordinates q. Also, the states of low energy E are uninteresting since they have few distinguishing characteristics such as maxima and minima, or complicated nodal lines where $\phi(q) = 0$.

Chaos in quantum mechanics, if there is any to be seen in the eigenstates, will show up only in the highly excited states. They bring us back to the limit of small Planck's quantum $\hbar$, and thus to the connection with classical mechanics. Since wave functions in general are hard to measure, and the highly excited ones are hard to calculate, the topic of this chapter is still wide open.

Many different approaches have been tried, nevertheless, some of which will be discussed here. No hard and fast rules are known so far; the numerical results and their interpretation have not yielded any simple criteria for the presence or absence of chaos. But they have led everybody to be keenly aware of what can be learned from looking at a wave function. The many manifestations of quantum mechanics have become much better appreciated, as well as the possibility of viewing the same situation from different vantage points (cf. the recent review of Eckhardt 1988).

15.1 The Eigenstates of an Integrable System

Since the Hamiltonian operator H_{op} in (13.11) is both linear and real, as long as there is no magnetic field, its eigenfunctions can always be made real-valued, and can be plotted without difficulty. Can we tell whether the system is integrable or not from looking at the plot? Pechukas (1972) addressed this issue and proposed a test which goes back to earlier work of Miller and Good (1953); as pointed out above, such a test is necessarily restricted to highly excited states.

An integrable system behaves very much like a separable one; in particular, the patterns of nodal lines where the wave function vanishes are expected to be similar. With n degrees of freedom, there are n families of (finitely many) roughly parallel $(n - 1)$- dimensional *nodal surfaces* where each surface intersects all the ones from the other families transversely. In two degrees of freedom, the nodal pattern looks like a distorted chess board. The quantum numbers are the numbers of nodal surfaces in each family.

As a simple example of a separable system, take two decoupled harmonic oscillators whose frequencies are not commensurate; their position coordinates are x and y. The eigenstates are products $\Phi_{\mu\nu}(x,y) = \phi'_{\mu}(x)\,\phi''_{\nu}(y)$, in terms of the appropriate Hermite functions; there are μ nodal lines x = constant and ν nodal lines y = constant. An integrable system with this structure should have eigenfunctions that can be expressed in the form

$$\overline{\Phi}(\xi\,,\eta) = \Phi_{\mu\nu}(T(\xi,\eta))\ , \tag{15.1}$$

where $(x,y) = T(\xi,\eta)$ is a one-to-one, sufficiently smooth transformation of the coordinates, independent of the quantum numbers μ and ν.

This transformation T satisfies a set of conditions that can be interpreted in terms of the Hamilton-Jacobi first-order partial differential equation (2.8) or (14.3). This idea, therefore, brings us back to Keller's construction in Section 14.2. The simple recipe above has to be adjusted when the nodal pattern in the position coordinates becomes more complicated, because the invariant tori in phase space have non-trivial projections, as in Figure 31.

Eigenstates of an integrable quantum system can probably be constructed with the help of some of the other methods that worked for classical integrable systems. Foremost among them are the canonical transformations, in particular the Birkhoff-Gustavson normalization scheme of Section 14.3. A start in this direction was made by Thomas (1942), and then by Eckhardt (1986); everything seems indeed set up for the generalization to quantum mechanics. The quantities ρ_k in Section 14.3 become the quantum number for the k-th oscillator as shown in (14.10).

More specifically, already Birkhoff used the complex quantities

$$a_k = (q_k + ip_k)/\sqrt{2} \quad , \; a_k^+ = (q_k - ip_k)/\sqrt{2} \quad , \qquad (15.2)$$

which now become the lowering and raising operators for the k-th oscillator. The classical procedures used the Poisson brackets to define the conditions for the generating functions W of the canonical transformations; now, the commutator with the Hamiltonian appears for unitary operators W, and their construction is entirely similar.

At the end, the starting Hamiltonian becomes a polynomial in the 'occcupation' numbers $N_k = a_k^+ a_k$, and the formula (14.11) follows. There are corrections in $\hbar^2$, however, which can not be obtained by the semiclassical approach of Section 14.3. By transforming back to the original coordinates, the corresponding eigenfunctions can be calculated. It would be of great interest to see where they are located, and how they are distributed in phase space; for this last notion, one needs the Wigner function which is the main topic Section 15.4.

15.2 Patterns of Nodal Lines

The simple checkerboard for regions of positive and negative values of the eigenfunction $\phi(x, y)$ gets destroyed when the system ceases to be integrable. The original arrangement may still be recognizable, but the lines separating black (positive) and white (negative) regions do not intersect any longer. Typically, these lines are long compared with the size of the available space, forming an involved zigzag curve; either they go from one boundary to another, or they are closed loops. Before trying to interpret this phenomenon, two preliminary remarks have to be kept in mind.

A theorem by Uhlenbeck (1976) states that it is a generic property of eigenfunctions to have non-intersecting nodal lines. The argument goes roughly as follows: at the point where two nodal lines intersect, not only does the wave function vanish, $\phi(x, y) = 0$, but so does its gradient because the partial derivatives are zero along each nodal line. Therefore, three simultaneous conditions have to be imposed on ϕ at the nodal intersection; but whereas a single condition defines a line in the (x,y) plane, and a double condition defines an isolated point, a triple condition cannot be satisfied in general.

It is very easy to destroy a nice checkerboard pattern, simply by adding two wave functions each of which all by itself produces such a pattern. As an example, from the first volume of Courant and Hilbert (1931 p. 396, 1953 p. 451), take the eigenfunctions of the Laplacian in the unit-square with the Dirichlet boundary condition, i.e., $\phi = 0$. The two functions,

$$\phi_1 = \sin(k\pi x)\sin(\ell\pi y)\ , \ \phi_2 = \sin(\ell\pi x)\sin(k\pi y)\ , \quad (15.3)$$

with $0 < k < \ell$, belong to the same eigenvalue, and each has a perfect checkerboard-like pattern of positive and negative values. And yet the sum $\phi_1 + (1+\varepsilon)\phi_2$ with $\varepsilon \neq 0$ does not have a single crossing of nodal lines.

Stratt, Handy, and Miller (1979) made a very thorough investigation of the nodal pattern for a variant of the Contopoulos Hamiltonian (cf. Chapter 8), due to Barbanis. Two harmonic oscillators of different frequencies are coupled by a third-order term αxy^2, where α is chosen such as to allow for about 135 eigenstates that are confined to the potential well near the origin. These eigenstates are represented as linear combinations of 420 products of Hermite functions, each similar to the first example in the preceding section.

The eigenstates are put into three classes, called regular, uncertain, and ergodic, simply by visual inspection; an example from each class is given in Figure 34. There is clearly room for personal bias; but the authors try to compare their assignments with other symptoms of chaos. We will only mention the spectral analysis of Section 14.4 where $x(t)$ in (14.13) now is the electric dipole-moment, and the averaging operation $< \ldots >$ is the quantum-mechanical expectation value. The averaged quantum-numbers are defined as

$$n_x + 1/2 = m\omega_x < x^2 > /\hbar\ , \ n_y + 1/2 = m\omega_y < y^2 > /\hbar\ . \quad (15.4)$$

Uncertain and ergodic states cannot be associated with weakly perturbed harmonic oscillators; on the basis of their averaged quantum-

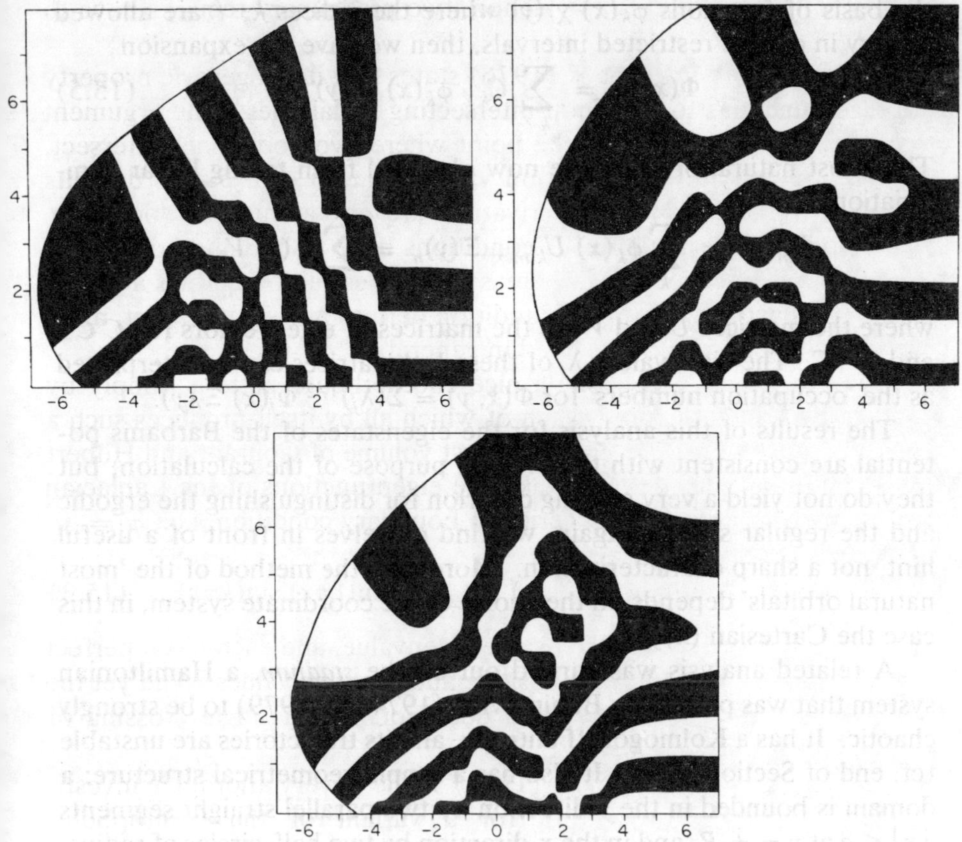

Figure 34 Nodal patterns of 104th, 121st, and 118th eigenstate in the Barbanis potential (non-linear coupling $\simeq x\,y^2$); Stratt, Handy, and Miller (1979) classify them as regular, uncertain, and ergodic.

numbers, both of their degrees of freedom carry about the same energy, indicating an effective resonance.

Stratt, Handy, and Miller also study the precise make-up of their eigenstates in terms of the underlying basis of 420 decoupled harmonic oscillator states. Nordholm and Rice (1974) had proposed that an ergodic eigenstate requires a wide distribution of uncoupled states in its representation, whereas a regular eigenstate is characterized by one, or at worst only a few, principal contributors. This argument was called into question, however, because it depends on the choice of the basis for the uncoupled states, which can be more or less appropriate to the particular Hamiltonian at hand.

A better method seems, therefore, to ask for the 'most natural orbitals' to represent a particular eigenstate $\Phi(x, y)$. If we started with the basis of functions $\phi_k(x)\ \chi_\ell(y)$ where the indices k, ℓ are allowed to vary in certain restricted intervals, then we have the expansion

$$\Phi(x,y) = \sum_{k\ell} C_{k\ell}\ \phi_k(x)\ \chi_\ell(y) \ . \qquad (15.5)$$

The 'most natural orbitals' are now obtained from taking linear combinations,

$$\Phi_m(x) = \sum_k \phi_k(x)\ U_{k\,m}\ ,\ \ \Xi(y)_n = \sum_\ell \chi_\ell(y)\ V_{\ell\,n}\ , \qquad (15.6)$$

where the matrices U and V are the matrices of eigenvectors for $C\,C^+$ and $C^+\,C$. The eigenvalues λ_j of these last matrices can be interpreted as the 'occupation numbers' for $\Phi(x,y) = \Sigma(\lambda_j)^{1/2}\ \Phi_j(x)\ \Xi_j(y)$.

The results of this analysis for the eigenstates of the Barbanis potential are consistent with the original purpose of the calculation; but they do not yield a very striking criterion for distinguishing the ergodic and the regular states. Again, we find ourselves in front of a useful hint, not a sharp characterization. Moreover, the method of the 'most natural orbitals' depends on the choice of the coordinate system, in this case the Cartesian (x,y).

A related analysis was carried out on the *stadium,* a Hamiltonian system that was proved by Bunimovich (1974 and 1979) to be strongly chaotic. It has a Kolmogoroff entropy, and its trajectories are unstable (cf. end of Section 10.3). It also has a simple geometrical structure: a domain is bounded in the y-direction by two parallel straight segments $|x| \leq a$ at $y = \pm R$, and in the x-direction by two half-circles of radius R centered in $x = \pm a$, $y = 0$; the potential is 0 inside the domain, and $+\infty$ outside. A classical particle makes a specular reflection on the walls of the stadium, while a wave function has to vanish there.

McDonald and Kaufmann (1979) were the first to calculate eigenstates for the stadium, and plot their nodal patterns. Their preliminary results appeared before the above work on the Barbanis potential, and left a strong impression on the aficionados of chaos in quantum mechanics. In particular, Figure 35 gives a vivid picture of a disorderly wave function, something that nobody had really seen or imagined before. Many more such pictures can be found in McDonald's Ph.D. thesis (1983), including the perspective drawings for the absolute values squared of the same eigenstates, like Figure 36; some of these were finally published by McDonald and Kaufmann (1988).

The stadium will come up several times later on; but at this point, a note of caution should be inserted about its ergodicity. The shape of the stadium is such as to allow for at least one class of almost stable trajectories; these correspond to the particle bouncing back and forth

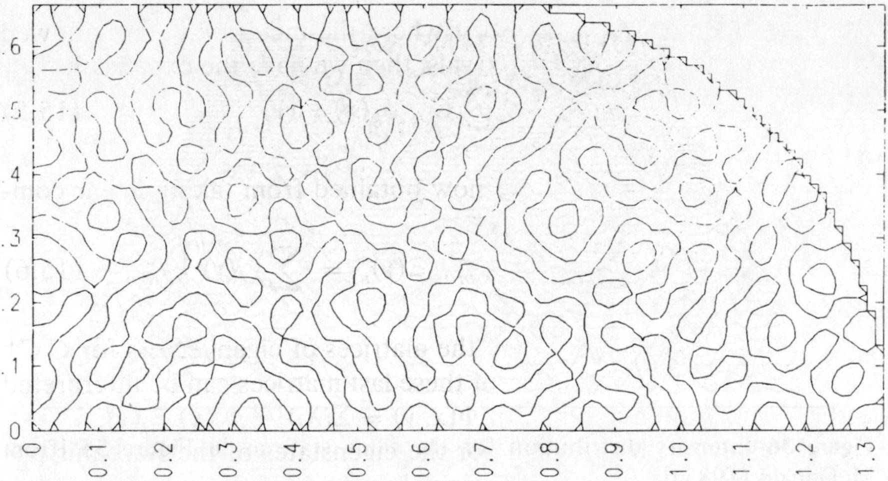

Figure 35 Nodal lines in the stadium (area normalized to π) for the odd-odd parity state with eigenvalue 100.297 [from McDonald (Ph.D. Thesis 1983)].

for a long time between the two parallel segments of the boundary. Correspondingly, an almost plane wave can slush back and forth in the y-direction; thus, in spite of the 'mathematically proven' classical chaos, there is a good fraction of regular eigenstates. They were constructed by Bai, Hose, Stefanski, and Taylor (1985) using what they called the 'Born-Oppenheimer adiabatic mechanism'.

The potential for the stadium is formally defined as a function $V(x\,;y)$ of y for a given value of the parameter x. The eigenstates for the one-dimensional Schrödinger equation in the y-direction at some fixed value of x are easy to obtain, because $V(x\,;y) = 0$ for $-y_0(x) < y < y_0(x)$, and $V(y\,;x) = +\infty$ for $|y| > y_0(x)$. The function $y_0(x)$ describes the boundary of the stadium in the upper half plane; let $F_n(x)$ be the n-th eigenvalue with the eigenfunction $\phi_n(x\,;y)$, both of which depend on the parameter x.

True to the Born-Oppenheimer philosophy, the one-dimensional Schrödinger equation in the x-direction with the potential energy $F_n(x)$ is now solved to yield the energy E_{nj} with the eigenfunction $\chi_{nj}(x)$. The total wave function for the stadium, therefore, becomes $\Phi_{nj}(x,y) = \chi_{nj}(x)\phi_n(x\,;y)$ with the energy E_{nj}. Figure 37 compares Φ_{nj} for $n = 22$ and $j = 2$ with the exact numerical wave function. About 10% of the exact wave functions have this quasiregular structure.

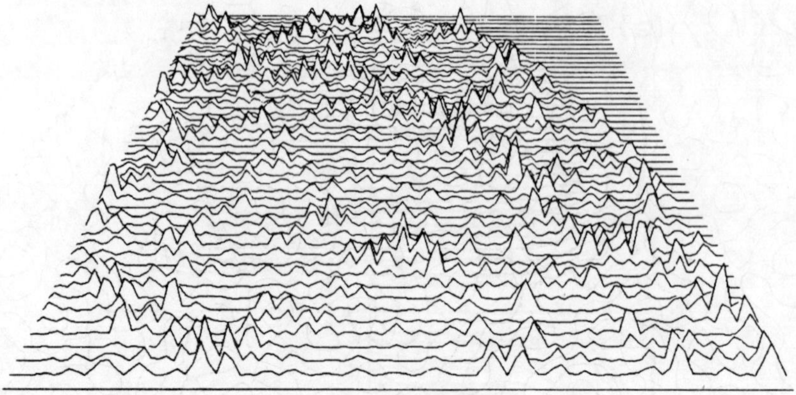

Figure 36 Intensity distribution for the same state as in Figure 35 [from McDonald (1983)].

15.3 Wave-Packet Dynamics

Wave functions with minimum uncertainty $\Delta p\, \Delta q$ are Gaussians both in position and in momentum space. Such functions are convenient as a basis from which to build many-particle correlated wave functions for complicated atoms and molecules, because overlap integrals and interaction matrix elements can be worked out algebraically. Therefore, the quantum chemists have developed the art of using Gaussian wave functions, even though atomic wave functions decay rather like a simple exponential with the distance from the nucleus.

Heller (1975) proposed a new application for Gaussians as a basis for constructing wave functions. He thought of a classical trajectory in phase space where the Hamiltonian in the immediate neighborhood of the moving point p_t , q_t at a given instant of time t can be expanded in powers of $(p - p_t)$ and $(q - q_t)$ to second order, just as in a harmonic oscillator. The wave function then becomes

$$\psi(q, t) = \exp\left[(i/\hbar)\left\{\alpha_t(q - q_t)^2 + p_t(q - q_t) + \gamma_t\right\}\right] , \quad (15.7)$$

where α_t is a complex symmetric matrix in as many rows and columns as degrees of freedom, and γ_t is a complex phase. The expectation values for the position and the momentum are simply $< p > = p_t$ and $< q > = q_t$; the matrix α_t gives the spread of the wave packet which is related to the approximate shape of the Hamiltonian near (p_t , q_t). The complex phase γ_t provides the necessary normalization, as well as the crucial phase angle.

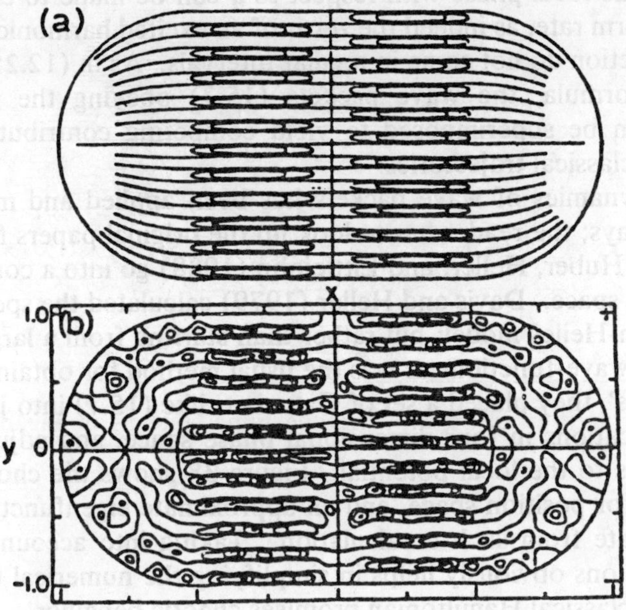

Figure 37 Contour plot of the 'adiabatic' wave function for $n = 22$, $j = 2$, and of the corresponding exact wave function of the stadium; the eigenvalues are 52.2405 and 52.2547 [from Bai, Hose, Stefanski, and Taylor (1985)].

The time-dependence of α_t, p_t, q_t, γ_t is obtained by inserting (15.7) into Schrödinger's equation (13.1). The potential energy is expanded to second order,

$$V(q,t) = V(q_t, t) + V_q(q - q_t) + (1/2)V_{qq}(q - q_t)^2 , \quad (15.8)$$

where the first and second derivatives, V_q and V_{qq} are to be evaluated at q_t and at the time t. In terms of the Hamiltonian $H(p_t, q_t) = p_t^2/2m + V(q_t)$, one gets for p_t and q_t the usual equations of motion (2.2), and moreover the conditions

$$\frac{d\alpha_t}{dt} = -\frac{2\alpha_t^2}{m} - \frac{V_{qq}}{2}, \frac{d\gamma}{dt} = i\hbar\frac{\text{trace}(\alpha_t)}{m} + p_t\frac{dq_t}{dt} - H(p_t q_t) . \quad (15.9)$$

The Hamilton-Jacobi equations of motion have been amplified to include the spreading Gaussian and its phase.

The second half of the last formula is of special interest: its right-hand side has an imaginary term $i\hbar$ trace(α_t), which serves to normalize the Gaussian, and a real part which fixes the change in phase angle. The rate of change is nothing but the classical Lagrangian in agreement with (2.1). Heller's scheme is, therefore, a generalization of the Van Vleck formula (12.20). Notice that the spread α of the wave packet can be kept real, if it started out that way; but by making it partly imaginary, the local phase with respect to q can be made to change at a non-uniform rate, as indeed the zeros of an excited harmonic-oscillator wave-function do not come in regular intervals. As in (12.25) for Van Vleck's formula, the wave packets (15.7) obeying the conditions (15.9) can be superimposed to yield competing contributions from different classical trajectories.

The dynamics of wave packets has been applied and modified in various ways; the reader has to look up the original papers for the details, e.g., Huber, Heller, and Littlejohn (1988) go into a complex valued phase space. Davis and Heller (1979) calculated the spectrum for the Hénon-Heiles model; but rather than starting from a large base of oscillator wave-functions as was the usual method for obtaining 'exact eigenstates', they placed a set of Gaussians like (15.7) into judiciously chosen locations in four-dimensional phase space, and adjusted their parameters to the local potential. Figure 38 shows the choice of the locations for position space, and an approximate eigenfunction for an excited state from such a calculation. Taking into account classical considerations obviously helps in simplifying the numerical task, even when the classical Hamiltonian promises chaotic behavior.

The original idea of spreading a Gaussian like (15.7) along a classical trajectory was implemented by the same authors (Davis and Heller 1981), again in the Hénon-Heiles model, but this time in the integrable region. After the appropriate torus for a particular quantum state had been found, the corresponding wave function was obtained. The resulting pictures of highly excited, but still integrable states are very impressive, e.g., Figure 39. These figures show some unexpected distributions of the amplitude, and prove that the method of wave-packet dynamics leads to the subtle interference effects which are the essence of quantum mechanics.

The frequency analysis of a classical trajectory in Section 14.4 was carried out by Davis, Stechel, and Heller (1980) for a wave packet (15.7) traveling according as (15.9), using again the Hénon-Heiles model for energies in the transition region between integrable and chaotic behavior. The dynamical variable $x(t)$ to define the correlation function (14.12) is the overlap between the starting wave function $\psi(q, 0)$ and the running $\psi(q, t)$, so that $x(t) = \int dq\, \psi^+(q, 0)\psi(q, t)$. Figure 40 shows the two starting wave functions in the surface of section, while Figures 41 and 42 give both the correlation function $C(t) = |x(t)|^2$ and the energy spectrum $I(\omega)$. In the author's words, the difference is profound; but Brumer and Shapiro (1980) see no difference in a similar calculation; for further studies along these lines cf. Feit and Fleck (1984).

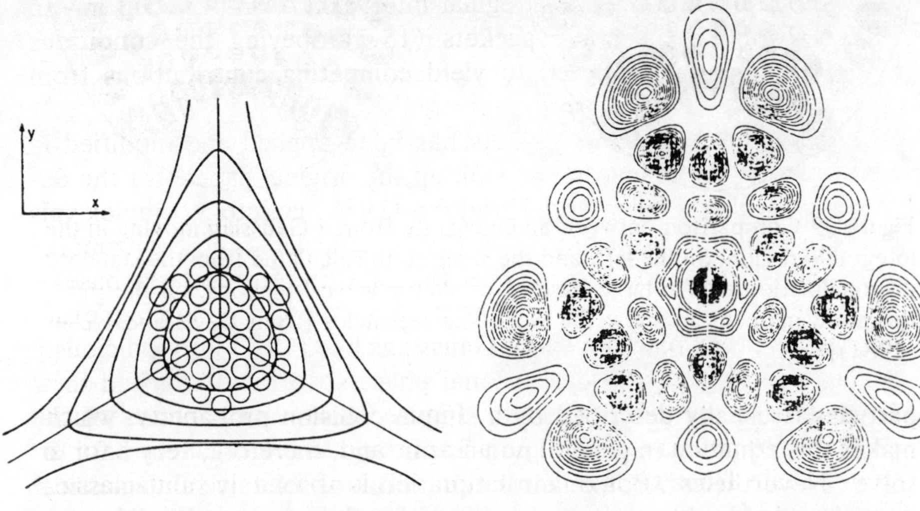

Figure 38 Choice of the locations in position space for a basis of Gaussians to calculate the eigenfunctions of the Hénon-Heiles model, and contour plot for the 62nd state, of energy 10.5905, when the dissociation occurs at 13.333 [from Davis and Heller (1979)].

Other investigations of this kind have been reported, usually with the help of different methods and trying out other dynamical systems such as the coupled-rotator model of Feingold and Peres (1985). While everybody now agrees on the most basic features, there is no detailed explanation for curves such as Figures 41 and 42. One would like to find some categories for frequency spectra, so that the traveling wave packets could be classified and related to the Hamiltonian which generated them. Such a thorough understanding, however, seems a distant goal at this time, offering a great opportunity for the reader.

15.4 Wigner's Distribution Function in Phase Space

Statistical mechanics studies large ensembles of identical particles that interact with one another such as in a gas or a liquid. The main tool is the probability in phase space, $f(p, q, t)\, d^n p\, d^n q$; it indicates the likelihood for a typical particle to have its momentum in the small volume $d^n p$ near p and its position in the small volume $d^n q$ near q. The distribution $f(p, q, t)$ satisfies *Boltzmann's equation,* a first-order partial differential equation which expresses Liouville's theorem on the invariant volume of phase space (cf. Section 7.2); the interaction between the

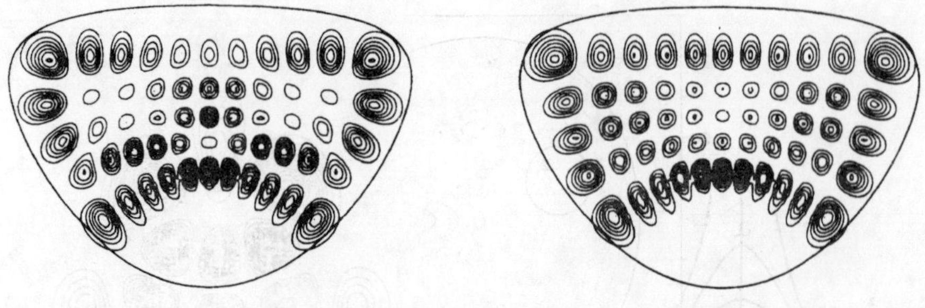

Figure 39 Comparison between an eigenstate from a Gaussian moving in the integrable region, on the left, and the same state calculated from the standard diagonalization over a large base of oscillator wave-functions, on the right; this state happens to show the worst discrepancies [from Davis and Heller (1981)].

particles is usually described by a simple collision probability, which makes this equation in general non-linear, and, therefore, very hard to solve. Nevertheless, Boltzmann's equation is absolutely fundamental; its author deduced from it an expression for the entropy of the ensemble, and showed that it increases with time. Boltzmann's formula for the entropy is none other than (10.1) which is used for the metric entropy of a dynamical system.

Wigner (1932) tried to derive the quantum-mechanical corrections to Boltzmann's equation. His expression for the distribution $f(p, q, t)$ in terms of the wave function $\psi(q, t)$ has become a very popular study object among people who are interested in quantum chaos. We shall introduce the reader very briefly to *Wigner's function* because it often yields a concise statement for the general appearance of wave functions and their relation to classical mechanics. Although Boltzmann's and Wigner's functions look and act in a very similar manner, they are bona fide representatives of classical and of quantum mechanics with all the implied differences. In particular, p and q are not independent variables in quantum mechanics, but are tied together by Heisenberg's uncertainty relations.

The following definitions and properties are valid for both time-dependent and stationary wave functions; neither the time t nor the energy E, whichever applies, will be mentioned explicitly. In order to prevent confusion with Boltzmann's $f(p,q,t)$, Wigner's function is called $\Psi(p, q)$. It is defined by the integral

$$(2\pi\hbar)^{-n} \int d^n Q \, \psi(q - Q) \, \psi^+(q + Q) \, \exp(-2ipQ/\hbar) \, . \tag{15.10}$$

Let the wave function $\phi(p)$ in momentum space corrresponding to $\psi(q)$ be defined by the Fourier transform $\phi(p) = (2\pi\hbar)^{-n/2} \int d^n q \, \psi(q) \, \exp(-ipq/\hbar)$, then

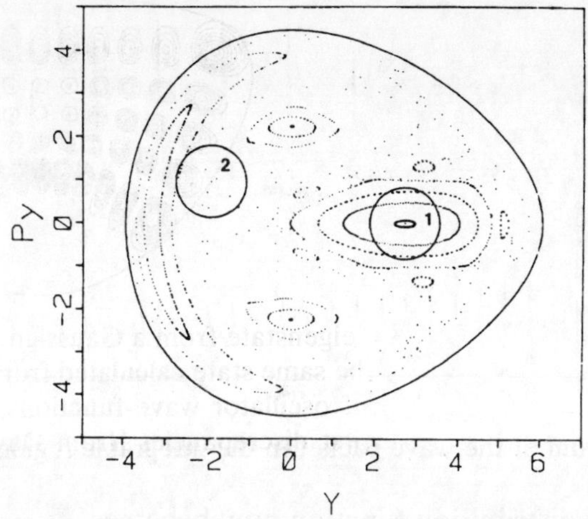

Figure 40 Surface of section for the Hénon-Heiles model at the energy 10.0 when dissociation occurs at 13.3333; two initial, Gaussian wave packets are placed in the integrable (1) and the ergodic (2) region; their spectrum is shown in Figures 41 and 42 [from Davis, Stechel, and Heller (1980)].

$$\int d^n q\, \Psi(p,q) = |\phi(p)|^2, \int d^n p\, \Psi(p,q) = |\psi(q)|^2 . \quad (15.11)$$

Therefore, if $\Psi(p,q)$ is interpreted as a probability in phase space in the same way as $f(p,q)$, then it yields at least the correct 'projection' into position as well as into momentum space.

Although Ψ is real, the main objection against such an interpretation comes from the fact that Ψ can be negative. Moreover, the function $\Psi(p,q)$ is badly redundant; it is written as if it depended on $2n$ variables, while it is defined on the basis of a single function $\psi(q)$ of only n variables. Equivalently, the Heisenberg uncertainty relation prevents us from specifying both position and momentum for a quantum system. This last objection was already met by Wigner by invoking what is now commonly called the *density matrix* in quantum mechanics.

A dynamical system in an experiment is not usually guaranteed to be in a particular, well-defined quantum state $\psi(q)$, a *pure state.* Rather, there is a *mixed state,* i.e., a collection of states $\psi_j(q)$, assumed to be mutually orthogonal for the discussion's sake, and a probability $\rho_j \geq 0$ for the system to be in the state $\psi_j(q)$; of course, $\Sigma\, \rho_j = 1$. Such a mixture is fully described by the density matrix

$$\rho(q'', q') = \sum_j \rho_j\, \psi(q'')\, \psi^+(q') \ ; \quad (15.12)$$

the pure state $\psi_k(q)$ is characterized by the choice $\rho_k = 1$ with $\rho_j = 0$ for $j \neq k$.

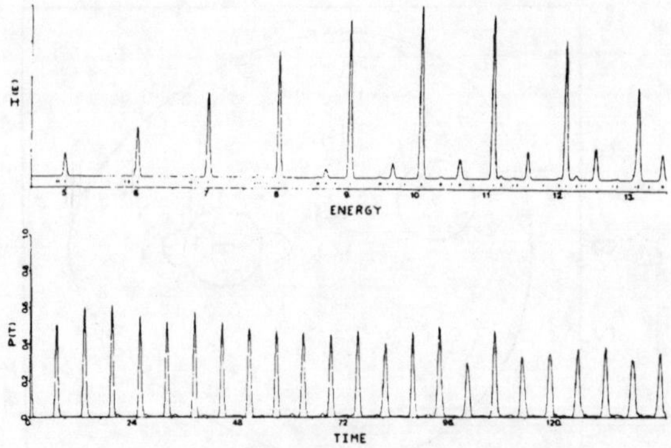

Figure 41 Spectrum of the wave packet in the integrable region.

The Wigner distribution function now becomes

$$\Psi(p, q) = (2\pi\hbar)^{-n} \int d^nQ \; \rho(q - Q , q + Q) \; \exp(-2ipQ/\hbar) \, . \quad (15.13)$$

The probabilities in momentum space or in position space alone are still given by the integrals (15.11); the redundancy in the information contained in Ψ has been removed. The difficulty now comes from the representation (15.12) of the density matrix; it is a Hermitian operator whose kernel is given by (15.12) and has all its eigenvalues between 0 and 1, with a trace equal to 1. Clearly, not every real function $\Psi(p, q)$ satisfies this requirement and can claim a constituent density matrix $\rho(q'' , q')$ for itself.

Berry (1977a), and later his collaborators, Ozorio and Hannay (1982, 1983), have explored in great detail what becomes of Wigner's function when the wave function $\psi(q)$ is associated with an invariant torus in phase space. The generalized classical approximation to Green's function is used, $G_c(q , I , E)$, where the starting coordinates are the invariant actions I, and q is the final position, as explained at the end of Section 12.2; with these replacements, formula (12.28) now yields the wave function $\psi(q)$ to be inserted into (15.10). The limit of small Planck's quantum has not been taken as yet, although the wave function has been approximated in the spirit of Van Vleck. It takes considerable work to establish what happens in the neighborhood of caustics, in particular when their projection into position space is not simple; the results can be phrased in the terms of *catastrophe theory*.

In the limit of $\hbar \to 0$, Wigner's function gets concentrated on the invariant torus that is associated with the eigenstate $\psi(q)$. If the invariant actions $I(p,q)$ are expressed as functions of the original momentum p and position q, this result can be written formally as

$$\Psi(p, q) = \quad \delta(I(p,q) - I_\psi)/(2\pi)^n \, . \quad (15.14)$$

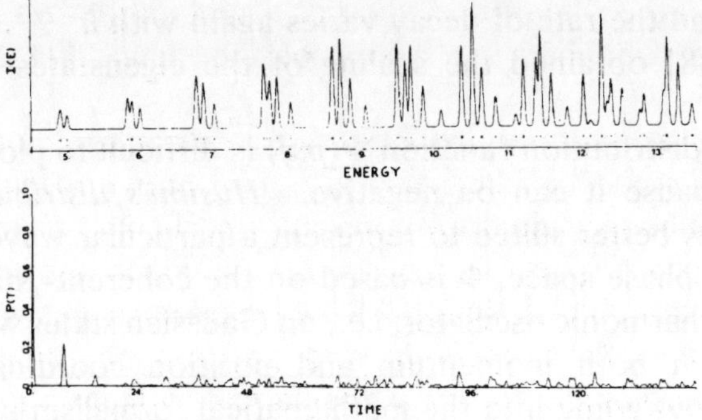

Figure 42 Spectrum of the wave packet in the ergodic region.

The symbols $I(p,q)$ and I_ψ represent vectors of as many components as degrees of freedom, and the Dirac δ-function is to be taken as n-dimensional; the values of the actions for the stationary state ψ have been called I_ψ. When $\hbar$ is small, but not zero, Berry shows that Wigner's function broadens; the size of the peak varies with $\hbar^{-2n/3}$. The 1/3-powers are typical for the Airy functions which arise when a wave function is investigated near its classical turning point. The simultaneous broadening of the δ-function to a width $\simeq \hbar^{2n/3}$ implies that Wigner's functions for different eigenstates in one dimension tend to overlap, because the space between the corresponding tori equals $2\pi\hbar$.

While these characteristics are fairly well understood in the integrable part of phase space, we have to rely on guesswork and numerical results about Wigner's function in the chaotic region. It is easy to conjecture that Ψ now spreads over the whole domain which is accessible to a particular trajectory. Numerical calculations to prove this view were first carried out by Hutchinson and Wyatt (1980) on the Hénon-Heiles model. The full four-dimensional phase space, not only a surface of constant energy, is required for plotting the results; but the usual surface of section gives already a clear picture. Wigner's function appears to be quite evenly distributed over the ergodic part. Also, it vanishes very abruptly outside the region where the potential energy $V(q)$ exceeds the available total energy E.

When the classical surface of constant energy divides into distinct chaotic regions which are separated by a KAM torus, Wigner's function is expected to remain confined to one or the other region. Again the boundary cannot be sharply defined in quantum mechanics; Geisel et al. (1986) and Radons et al. (1988) made calculations on the kicked rotator (cf. Section 9.8), a one-dimensional time-dependent system (one and one-half degrees of freedom in the language of Section 4.3), where a 'last' torus disappears when one of the parameters exceeds a critical value. Wigner's function decays exponentially across this classical limit, and the rate of decay varies again with $\hbar^{-2/3}$. Radons and Prange (1988) obtained the scaling of the eigenstates near the last KAM torus.

Wigner's distribution function $\Psi(p,q)$ is difficult to plot and hard to interpret because it can be negative. *Husimi's distribution function* $\Phi(p, q)$ seems better suited to represent a particular wave function or a mixture in phase space. Φ is based on the coherent- state representation of the harmonic oscillator, i.e., on Gaussian states with minimum uncertainty in both momentum and position coordinates (Husimi 1940). Without going into the mathematical formalism it may be sufficient to quote the formula which allows one to calculate Φ from Ψ,

$$\Phi(p,q) = \frac{1}{\pi\hbar} \int dp'\, dq' \exp\{ - \frac{(q-q')^2}{\hbar w^2} - w^2 \frac{(p-p')^2}{\hbar} \} \, \Psi(p', q') ,$$

where w is the width of the uncertainty in the position coordinate q. It is not obvious from this formula that $\Phi \geq 0$; but its interpretation is straightforward. It is also possible to calculate Ψ from Φ, although such a transformation is not easy to implement on a computer because the result depends very sensitively on the exact shape of Φ.

The reader is referred to the recent work of Saraceno and collaborators who have produced some striking pictures of Husimi distributions for non-integrable systems (cf. Leboeuf and Saraceno, preprint ITP). They also show how the underlying coherent-state formalism can be generalized to any system with a symmetry group, such as a rotator (cf. Kramer and Saraceno 1981). Adachi (1989) has studied a form of path integral in phase space with the help of coherent states; together with Toda and Ikeda (1988 and 1989), he has also tried to display the quantum-mechanical symptoms of chaos in phase space.

15.5 Correlation Lengths in Chaotic Wave Functions

The spread of Wigner's function over the accessible chaotic part of phase space can be expressed by the formula

$$\Psi(p,q) = \frac{\delta(E - H(p,q))}{\int dp \int dq \ \delta(E - H(p,q))} \ , \qquad (15.15)$$

for a wave function of energy E, and assuming that a typical classical trajectory actually covers the whole surface of energy E. The δ-function now is one-dimensional in contrast to (15.14); the distribution in phase space is the same as the microcanonical ensemble of statistical mechanics. All points on the surface of constant energy E are equally probable.

This distribution can be projected into position space as indicated in (15.11) to yield $|\psi(q)|^2$. If the Hamiltonian has the standard form $H(p,q) = p^2/2m + V(q)$, the integral over the momentum at a fixed position q gives

$$|\psi(q)|^2 = (E - V(q))^{-1 + n/2} \, \theta(E - V(q)) / \int dq \ .... \qquad (15.16)$$

where $\theta(x) = 0$ for $x < 0$, and $\theta(x) = 1$ for $x \geq 0$; the normalization integral in the denominator has not been written out. Chaos in time-independent Hamiltonian systems occurs only for $n \geq 2$; the wave function $\psi(q)$, therefore, fills out the whole classically allowed domain in position space, and has no singularities at its boundary; it even vanishes at the boundary when $n > 2$.

If Wigner's function (15.14) is projected according as (15.11) into position space for an integrable system, $|\psi(q)|^2$ is found to be restricted to the inside of the various caustics that are associated with the invariant torus I_ψ. As a function of the distance Δq from a particular caustic, $|\psi(q)|^2 \simeq \Delta q^{-1/2}$; the classical approximation to the wave function is singular near the caustics. In contrast to the chaotic wave functions, the regular ones have to accumulate intensity at the caustics because their domain is severely reduced from the classically allowed region.

The Fourier transformation in the definition (15.10) of Wigner's function can be undone to yield the spatial *autocorrelation function*

$$\begin{aligned} C(X;q) &= \psi(q + X/2)\psi^+(q - X/2) \\ &= (2\pi\hbar)^{-n} \int dp \exp(-ipX/\hbar) \ \Psi(p,q) \ . \end{aligned} \qquad (15.17)$$

To qualify as a proper correlation function, an average over the position coordinate q should be taken covering some neighborhood of diameter Δq, not too large to prevent the next step in the calculation, and not too small to reveal some untypical feature of the wave function ψ.

The local details, however, have already been eliminated in the microcanonical Wigner's function (15.16) which can, therefore, be directly inserted into (15.17). The integration over p is carried out in polar coordinates; the δ-function in (15.15) fixes the absolute value of the momentum, $|p| = \sqrt{2m(E - V(q))}$, leaving only the integral over the angles, written as an integral over the unit vector Ω,

$$C(X;q) = \int d\Omega \exp[i\Omega X\sqrt{2m(E - V(q))} /\hbar] / \int d\Omega \,. \quad (15.18)$$

This result can be expressed using the ordinary Bessel functions,

$$C(X;q) = \Gamma(n/2) \frac{J_{-1+n/2}(|X|\sqrt{2m(E - V(q))} /\hbar)}{[|X|\sqrt{2m(E - V(q))} /2\hbar]^{-1+n/2}} \quad (15.19)$$

which was first obtained by Berry (1977b). The oscillations in (15.19) have the local de Broglie wavelength $\hbar/\sqrt{2m(E - V(q))}$.

The wave function $\psi(q)$ can be conceived as built by superposing the phases from all the classical trajectories which pass through a particular neighborhood; but these phases are uncorrelated because between consecutive returns to the same neighborhood, the trajectory wanders almost randomly around the energy surface. Various authors have, therefore, speculated that ψ is a *Gaussian random function* of q with a spectrum given by the averaged Wigner's function and the spatial correlation (15.19).

Shapiro, Ronkin, and Brumer (1988) have calculated the autocorrelation function

$$F(\delta) = \int dq\, \psi(q + \delta)\, \psi^+(q) \,, \quad (15.20)$$

for different eigenstates of the stadium (cf. the end of Section 10.3). With $n = 2$ degrees of freedom, the potential $V(q) = 0$ inside the stadium, and its area normalized to 1, the average of (15.19) becomes $F(|\delta|) = J_0(|\delta|\sqrt{2mE})$. The agreement of this simple formula with the numerical calculations for some of the higher excited states is quite striking. Nevertheless, the oscillations for large δ are more pronounced than the zero-order Bessel function indicates.

15.6 Scars, or What Is Left of the Classical Periodic Orbits

Although the results in the preceding section give a rather convincing explanation for the general shapes of chaotic wave functions, at least in terms of some global averages such as the correlation functions, Heller, O'Connor, and Gehlen (1989) tried to get a more detailed picture of the eigenstates for the stadium. The calculation was pushed to the 10,000-th eigenstate, all of them antisymmetric with respect to reflection on the x-axis as well as the y-axis. There are roughly 100 nodes in each direction of position space, so that even rather elaborate patterns can show up quite visibly. Figure 44 shows the absolute value squared inside the stadium, black when $|\psi|^2$ exceeds a certain threshold. Ten consecutive eigenstates in order of increasing energy are plotted.

The results are striking. Every eigenstate has an intensity pattern showing what Heller calls *scars,* i.e., narrow linear regions with an enhanced intensity which stands out clearly and appears to be coming from classical periodic orbits. It is reasonable to argue that a particular periodic orbit will show up sometimes in this manner, provided it is not too unstable; but Figure 44 seems to demonstrate more than such an obvious comment.

The stability of an unstable periodic orbit is characterized by its exponent χ as defined in Section 6.4. Equivalently, a frequency ω and a Lyapounoff exponent λ can be used; the distance of neighboring trajectories from the periodic orbit increases with time t as $\exp(\lambda t)$, as shown in Section 10.6. Scars of a periodic orbit will appear if

$$\chi \;=\; 2\pi\,\lambda/\omega \;<\; 2\pi\ . \tag{15.21}$$

Such a criterion is, of course, not logically precise. Nevertheless it is remarkably generous, since it allows the appearance of a very long complicated orbit like the one mentioned in Figure 27 of Section 11.5; its stability exponent $\chi = 6.875$, and neighboring trajectories, therefore, drift away by a factor $\exp(6.875) \simeq 1000$ at every turn around the periodic orbit.

Heller's argument (Heller 1986 and 1987) for the criterion is best explained with the help of Figure 43. A Gaussian wave packet (15.7) is launched along a classical periodic orbit at time $t = 0$. The overlap $<\psi|\psi(t)> = \int dq\psi^+(q,\ 0)\ \psi(q,\ t)$ has the appearance of Figure 43a as a function of time: tight Gaussian peaks repeat at intervals corresponding to the period of the orbit $\tau = 2\pi/\omega$; but they decay exponentially with half the Lyapounoff exponent λ. The Fourier spectrum of the overlap $S_T(E) = (2\pi)^{-1}\int dt\, \exp(iEt/\hbar)\ <\psi(0)|\psi(t)>$, ex-

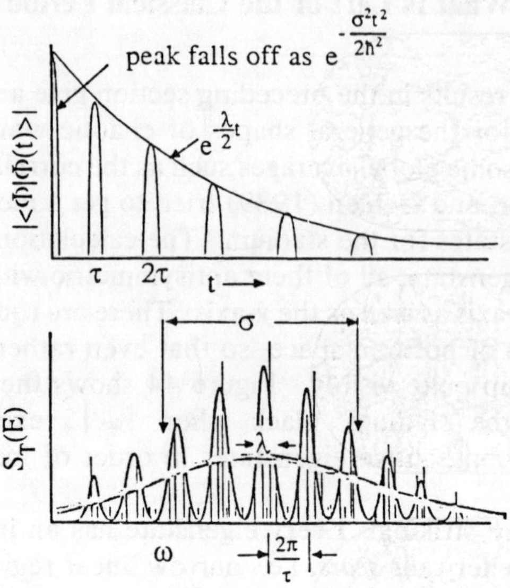

Figure 43 Schematic plot of the overlap function in time (a) and in energy (b) for a Gaussian wave packet, in order to justify the scar condition (15.21) [from Heller (1986)].

tended over the finite time interval from $-T$ to $+T$, has the complementary shape given in Figure 43b.

If the original wave packet $\psi(q, 0)$ is expanded in the eigenstates ϕ_n of the system, one notices that the function $S_T(E)$ is further resolved into narrow peaks of widths $\hbar/T$, each corresponding to an energy level E_n; the height of the peak gives the intensity I_n with which ϕ_n is participating in $\psi(q, 0)$; of course, these intensities add up to 1. Since they are concentrated in bands of width $\hbar\lambda$, which occur at intervals of $\hbar\omega$, the intensities are enhanced by a factor ω/λ compared to a completely random distribution. We find, therefore, a privileged set of eigenstates that have a marked preference for the particular periodic orbit at the beginning of this argument. This privilege becomes the more exclusive the larger the ratio ω/λ.

The reader should study both the many pictures and the additional explanations in the papers by Heller and his coworkers. The identification of *ridges* in the eigenstates with particular periodic orbits is not always obvious, because there may be families of such orbits that cover the stadium in an increasingly more involved, yet similar manner, as shown in Figure 45. As $\hbar \to 0$, they do not show up more strongly, because their *depth,* i.e., contrast with the general background, again depends mostly on χ, while the number of nodes goes up inversely with $\hbar$.

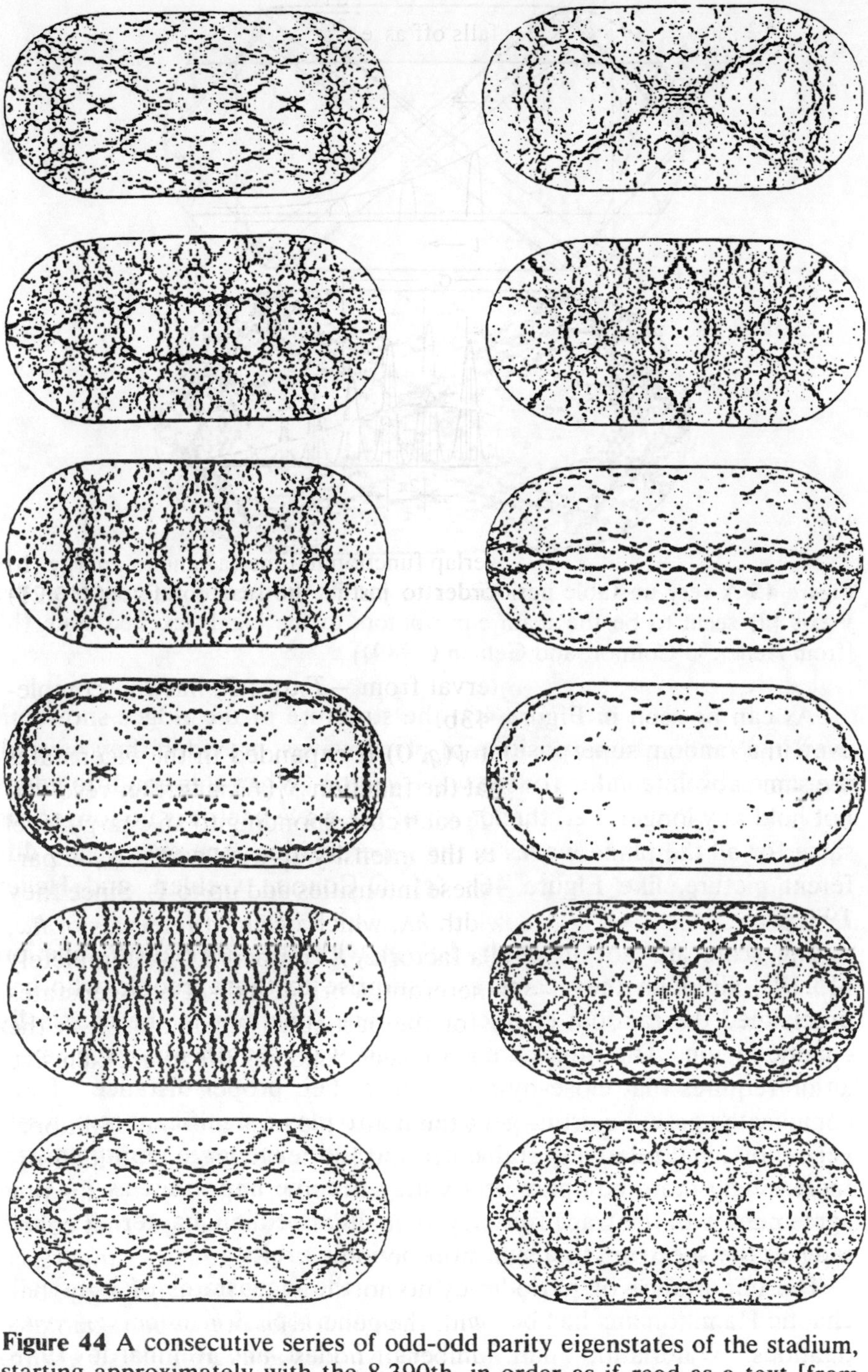

Figure 44 A consecutive series of odd-odd parity eigenstates of the stadium, starting approximately with the 8,390th, in order as if read as a text [from Heller, O'Connor, and Gehlen (1989)].

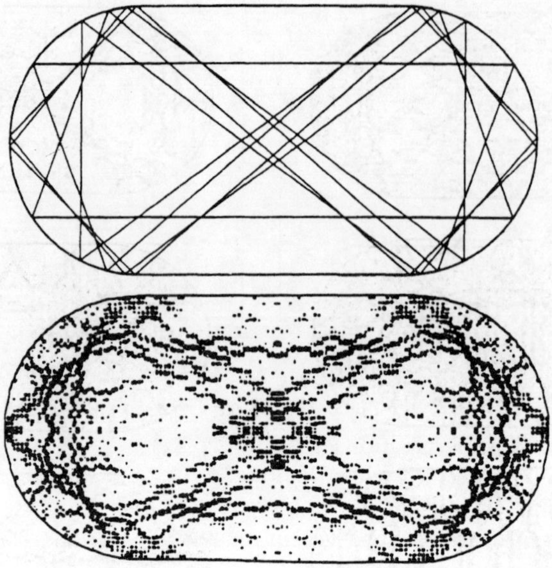

Figure 45 A very unstable periodic orbit and its brother by reflection on the y-axis (a) seem to be the main contributors to the highly excited state (b) [from Heller, O'Connor, and Gehlen (1989)].

As can be seen in Figure 46a the structure of the ridges shows up even in a random superposition of plane waves, provided they have all the same absolute value for the wave vector. When this condition does not hold any longer, i.e., the length of the wave vector is no longer the same for all the plane waves in the superposition, one gets a very different picture, like Figure 46b (cf. O'Connor, Gehlen, and Heller 1987).

The scars are not in contradiction with the correlation function (15.19), although they came as a surprise to those people who had proclaimed the random character of wave functions in the classically chaotic region. The ridges are about one wave length wide; the correlation requires that close-by ridges keep their proper distance. Thus, complicated periodic orbits get a chance to play a significant role, while very short, and relatively stable ones would leave large empty spaces. Increasingly more complex orbits may become important as $\hbar \to 0$, in agreement with the trace formula to be discussed in Chapter 17, which reduces the spectrum to a sum over periodic orbits.

The importance of periodic orbits in the eigenstates of classically chaotic Hamiltonians has become a popular topic for numerical investigations. Waterland, Yuan, Martens, Gillilan, and Reinhardt (1988) have looked at the quantum surface of section, i.e., the plot of the Husimi distribution $\Phi(p,q)$ in a surface of section for the corresponding

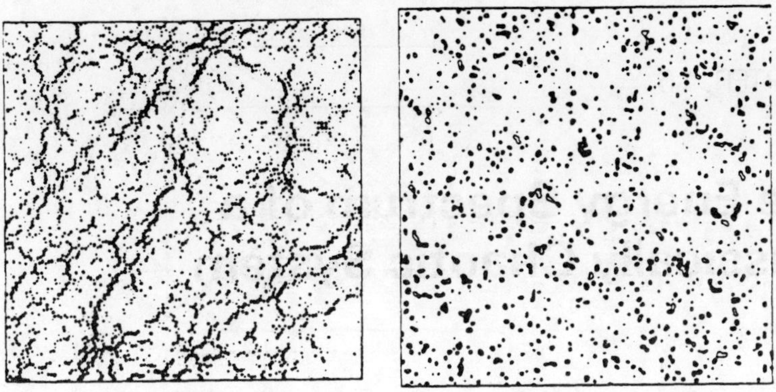

(a) (b)

Figure 46 (a) A superposition of 10,000 plane waves of random direction, amplitude, and phase shift when the magnitude of the wave vector is the same for all the plane waves. (b) A speckle pattern, which is produced in the same manner; but with various wave-vector magnitudes. The field is about 60 wave lengths on the side [From O'Connor, Gehlen. and Heller (1987)].

classical system. Most remarkably, they find not only large intensities near one particular, short periodic orbit, but moreover a pronounced concentration near the stable and unstable manifolds for the x^2y^2 potential. Eckhardt, Hose, and Pollak (1989) present the pictures for many eigenstates in this potential, and associate them with groups of periodic orbits called 'channels', where simple quantization rules like (14.1) apply. Founargiotakis, Farantos, Contopoulos, and Polymilis (1989) compare the eigenstates for a Hénon-Heiles type potential with the classical structure of intersecting stable and unstable manifolds of Figure 24.

CHAPTER 16

The Energy Spectrum of a Classically Chaotic System

The quantum mechanics for a dynamical system in a compact space yields a set of discrete energy levels, rather than a continuous range of possible energies as in classical mechanics. Obviously, the discrete set of numbers, called the spectrum, can have many more characteristic features and contains much more information than the continuous, classical interval. The details of the spectrum change from system to system; but certain properties are quite general, and some seem to be directly correlated with the nature of the chaos in the corresponding classical system.

This chapter is mainly concerned with the statistical distribution of the energy levels. The average number of levels per unit interval of energy is directly related to the volume in classical phase space, according to a famous theorem of Weyl which goes back to the very beginning of quantum theory. The more detailed information, however, such as the spacing between neighboring levels, has only been recognized very recently as connected with the chaotic nature of the trajectories.

After deriving Weyl's formula, we will discuss various measures for the fluctuations in the spectrum. A particularly useful model is based on Wigner's idea of choosing a Hamiltonian at random from a sample that satifies certain basic requirements. The correlations in the resulting Gaussian ensembles of random matrices are found to match very closely the empirical correlations among energy levels in classically chaotic systems. This phenomenon has not been explained satisfactorily as yet.

The exceptional properties of integrable systems can be explained; but the change in the characteristics of the spectrum with the transition to chaos, and the distribution of energy levels in the chaotic regime, can only be understood with the help of mostly qualitative arguments. Nevertheless, the investigation of spectral statistics has revealed a very important connection between the classical and the quantal behavior of dynamical systems.

16.1 The Spectrum as a Set of Numbers

Planck started quantum mechanics in 1900 when he assumed that the energy of a harmonic oscillator comes in discrete steps, rather than varying continuously as classical mechanics requires. His magic formula, $E = h\,\nu$, or equivalently, $E = \hbar\,\omega$, gives the smallest increment in energy which the oscillator can experience if its natural frequency is $\nu = \omega/2\pi$. The word 'quantum' to describe this new phenomenon was used to catch the essence of physics at the atomic scale. Some 25 years later, it appeared that the wave nature of matter is a more fundamental attribute than the discrete steps in its allowed energy; but the idea of 'quantum jumps' has caught the imagination of many people, and has become a figure of speech to dramatize certain unusual events in almost any field of human activity.

The simple positive multiples of $\hbar\omega$ were postulated by Planck, in order to explain the distribution of optical frequencies emitted by a hot body. It was soon realized, however, that the *spectrum*, i.e., the allowed values of the energy, could be much more complicated. Bohr found the set of numbers $E = \text{Ryd}/n^2$, with n integer and positive, for the energy levels of the hydrogen atom. ('Ryd' is the natural unit of energy in atomic physics, and is equal to 13.59 electron-volts.) He also explained the *combination principle:* any frequency ω seen in an optical experiment can be interpreted as the difference $\omega = (E_k - E_\ell)/\hbar$ between two energy levels, E_k and E_ℓ, in the atom or molecule.

Therefore, the word 'spectrum' as applied to a particular small dynamical system has sometimes a different meaning. It designates the set of frequencies that are seen in an experiment and which are characteristic of the system. An additional effort is necessary to disentangle all of them into a set of energy differences.

Given the many frequencies that such a system emits or absorbs, it is no mean task to find all the energy levels E_k. A complicated atom like iron emits and absorbs thousands of optical frequencies, and a reasonably simple molecule such as acetylene, $H - C - C - H$, tens of thousands if one includes the infrared (cf. Chen et al 1988); they can not be labeled with a set of quantum numbers in the usual way (cf. the work of Abramson et al. 1985, and of Pique et al. 1987, as well as the review of Jost and Lombardi 1986).

An experiment offers us in general a large, seemingly indiscriminate set of energy differences and no hint where to start looking for the underlying set of energy levels. This situation is typical of classically chaotic systems, because the all-important *selection rules* for the interpretation of experimental spectral data are missing. They follow directly from the symmetries in the system; but there are none in a chaotic system, beause otherwise there would be some constants of motion leading to integrability and invariant tori in phase space.

Experiments or numerical calculations may give us a set of energy levels, $E_0 \leq E_1 \leq E_2 \leq ...$, and no sensible numbering scheme, except the increasing value of the energy. By way of contrast, the reader should recall how energy levels in simple atoms, like sodium with a single electron in the outer shell, are described by the appropriate *quantum numbers:* the principal quantum number, usually called n; the orbital angular momentum ℓ, also given in the old spectroscopic notation s, p, d, f, g, etc.; the magnetic number m, and so on.

The hydrogen atom in a strong magnetic field (cf. Chapter 18) gives a prime example where this scheme breaks down completely. A direct connection with the classical periodic orbits, however, offers a novel and effective method to understand some of the apparent, but confusing regularities. This idea will be discussed in the next two chapters, whereas at present we will view the spectrum as merely a set of increasing numbers. We will not attempt to give detailed physical arguments for individual energy levels and particular Hamiltonians.

The interesting feature in Planck's spectrum for the harmonic oscillator, as far as this chapter is concerned, is not the magic formula $E = \hbar\omega$ giving the quantum jumps, but the completely regular distribution of its energy levels. If we *scale* the spectrum by dividing with $\hbar\omega$, then we end up with the simple sequence of integers 0, 1, 2, 3, ..., about the most boring one could imagine.

The world is full of more interesting sequences whose consecutive steps are of roughly the same size, and yet they contain all kinds of non-trivial information in their slight irregularities. A famous example are the stops on the number 1 subway line which follows Broadway in New York City for most of its course; they are designated by the streets the line crosses, from 14-th, 18-th, 23-rd, 28-th, on up to 238-th, and the last stop at 242-nd. Contrary to these indications, the average distance between stations is 8 1/7 city blocks, leading to a slow ride at the beginning and at the end, as well as long walks to the nearest station in the middle section of the line.

Scientifically minded people from grade school on up are fascinated by the *prime numbers,* positive integers p whose only divisors are 1 and p itself. They are the exceptions among all the integers, and show up at irregular intervals, 2, 3, 5, 7, 11, 13, 17, 19, ..., 503, 509, 521, 523, 541, 547, 557, 563, 569, 571, 577, 587, 593, 599, ...; notice the crowding near 570 as opposed to the gap around 530.

The *Prime Number Theorem,* first proved by Hadamard and de la Vallée-Poussin in 1896, states that the average density in the neighborhood of $x > 0$ is given by $1/\log x$, provided x is sufficiently large. Another way of expressing this result is to say that the numbers $p/\log p$ have an average density of 1; 'log' is the natural logarithm. On this scale, the distance between the neighbors 523 and 541 is 2.6794, while the distance between 569 and 571 is only .2655, a factor of 10 less.

This last example illustrates two basic ingredients into the discussion of this chapter. First, different sequences of energy levels can be compared provided they have been reduced to the same scale. This reduction can be carried out as soon as the average number of levels in an interval of fixed length ΔE is known in the limit of large E. Second, there can be extreme variations in the local distribution, even after the reduction has been carried out. The main question then is whether these variations can be characterized, and whether the occurrence of qualitatively different behavior can be associated with chaos or its absence in the corresponding classical system.

16.2 The Density of States and Weyl's Formula

The spectra of different dynamical systems can only be compared if they are scaled so as to have an average density of 1 per unit interval, for large values of the energy. Since the energy levels are the eigenvalues of the stationary Schrödinger equation (13.6), the asymptotic distribution of the large eigenvalues has to be found. This purely mathematical problem has a long and distinguished history, which started with the physicists Rayleigh, Lorentz, and Debye trying to understand the thermodynamic properties of electromagnetic radiation in a cavity and the lattice vibrations of a solid body.

The heat content in either case was known to be proportional to the volume; therefore, a specific heat could be defined, independent of the shape of the cavity or the solid body. Each vibration acts like an isolated harmonic oscillator of frequency ω, carrying an average energy $\bar{n}\hbar\omega$ where Planck's distribution law gives $\bar{n} = 1/[\exp(\hbar\omega/kT) - 1]$. Many of the higher frequencies contribute to the specific heat at moderate temperatures. Their frequency distribution can, therefore, not depend significantly on the shape of the cavity or solid body, but only on the volume.

This result was first checked directly for simple rectangular shapes and spheres; the general statement was then proposed by the physicists as a challenge to the mathematicians. Hermann Weyl proved the first theorem of the required kind in 1912, on the basis of Hilbert's theory of integral operators. Other proofs and more general propositions have been found since, although not nearly to the extent needed to justify all the applications in quantum mechanics.

The mathematicians concentrate mainly on eigenvalue problems related to the Laplacian in a fixed domain D; that takes care of the vibrations in cavities and solids, but does not address Schrödinger's equation, except for the various billiard models. The discussion by Courant and Hilbert in their first volume is based on comparing different shapes, volumes, and boundary conditions. As a rule, the wider domains and the softer boundary conditions, e.g., vanishing normal derivative *(Neumann)* rather than vanishing amplitude *(Dirichlet),* give consistently lower frequencies. Thus, both lower and upper limits on the frequencies for a given eigenvalue problem can be found, in terms of already known spectra.

A different line of reasoning starts from the partial differential equation for heat flow in a body with a well-defined coefficient of heat transport, or equivalently, with the *diffusion equation.* Rather than making Planck's constant imaginary, the time in Schrödinger's equation (13.1) is replaced by $t = -i\tau$ with real $\tau \geq 0$. The expansions like (13.9) for the propagator remain valid and are now interpreted in terms of the probability $P(q''q'\tau)$ for a particle to diffuse from its starting position q' to its final position q'' in the given time τ. The further argument has been presented by Kac (1966b) in a famous article with the title: "Can you hear the shape of a drum?" Weyl's original theorem gives only the dependence of the spectrum on the volume, but not on the shape.

In a purely formal manner, we replace *it* in the exponent by τ in the expansion (13.9). Then we set $q'' = q' = q$, and argue that $|\phi_j(q)|^2$ averages out to $1/A$, where A stands for the area of D. Therefore,

$$P(q, q, \tau) \simeq A^{-1} \sum_{j=0}^{\infty} \exp(-E_j \tau/\hbar) = A^{-1} \int_0^{\infty} e^{-E\tau/\hbar} dN(E) , \qquad (16.1)$$

where the sum over the energies has been written as a Laplace integral; $N(E)$ is the number of eigenstates of energy $\leq E$. The density $\rho(E) = dN(E)/dE$ consists of a sequence of Dirac δ-functions in the case of a discrete spectrum; the number of energy levels $N(E)$ looks like an ascending staircase. The expression (16.1) is now interpreted as arising from a diffusion process for short times τ.

As a first step, we note that a particle starting in q', inside D, takes some time to reach the boundary. Until then, the formula (12.21) for the free particle is applicable; since $q'' = q' = q$, the exponential becomes 1. If we consider the case of two dimensions, we are left with $P(q, q, \tau) \simeq m/2\pi\hbar\tau$, which is to be compared with the last expression in (16.1). Since both sides are valid for small values of τ, their equality should tell us something about the density $\rho(E)$ for large values of E.

Some subtle mathematics gets involved at this point in the form of the *Tauberian theorems* from the theory of Laplace transforms. We can simply check that the two expressions become equal if one replaces $\rho(E)$ in (16.1) by

$$\overline{\rho} = \frac{A}{4\pi} \frac{2m}{\hbar^2} , \tag{16.2}$$

which is already *Weyl's theorem,* if we interpret $\overline{\rho}$ as the average density of energy levels.

The next step lets the diffusing particle feel the effect of the boundary for a short while. The average over $q'' = q' = q$ now has a term from the boundary where the particle does not diffuse freely, but gets reflected or absorbed according as the boundary condition. This process can at first be treated as if it occurred in one dimension where everything can be worked out in detail; later on, the curvature of the boundary can be taken into account. The new term is proportional to the length of the boundary, and inversely proportional to $\sqrt{\tau}$. The average density $\overline{\rho}$ has to be corrected correspondingly, again with the help of the Tauberian theorems, or simply by checking the Laplace integral.

If even the next term in the diffusion process near the boundary is included, either with the Dirichlet condition $\psi = 0$ (upper sign), or with the Neumann condition $\partial\psi/\partial n = 0$ (lower sign), the average number $\overline{N}(E)$ of energy levels below the energy E becomes

$$\overline{N}(E) = \frac{A}{4\pi} \frac{2mE}{\hbar^2} \mp \frac{L}{4\pi} \sqrt{\frac{2mE}{\hbar^2}} + K , \tag{16.3}$$

where L is the length of the perimeter, and K is a number that carries information about the topological nature of the domain D, and the curvature of its boundary. The second and third term depend on the

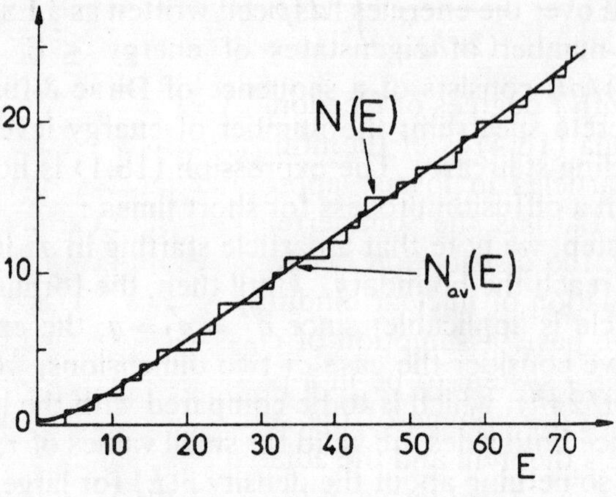

Figure 47 Cumulative density of eigenvalues $N(E)$ and its average $\overline{N}(E)$ for the stadium with straight line and radius of curvature equal to one [from Bohigas and Giannoni (1984)].

boundary conditions, as well as on the shape. Figure 47 shows the staircase-like function $N(E)$ for the billiard inside the stadium, as well as the corresponding smooth function $\overline{N}(E)$ according as (16.3).

The explanations in the preceding few paragraphs are meant to give an idea of the physical arguments which go into Weyl's formula and its extension to lower energies. The reader will find a much more informative account in two remarkable papers by Balian and Bloch (1970 and 1971). These authors deal with the asymptotic distribution of the eigenvalues for the Laplacian in a three-dimensional domain; they work out the surface as well as the curvature term, and they include the general linear boundary condition $\partial\phi/\partial n = -\kappa\phi$ where $\partial/\partial n$ is the derivative along the normal to the bounding surface. Maxwell's equations are covered, as well as waves on a Riemannian surface; the underlying space is allowed to have any number of dimensions.

Weyl's formula (16.2) has a remarkable interpretation in terms of classical mechanics. Since the potential energy $V = 0$ inside the D, the kinetic energy for the particle bouncing around is constant, and the absolute value of its momentum $|p|^2 = 2mE$. The total volume of phase space with energy less than E is given by $\Omega(E) = \pi\, 2mE\, A$. The first term in (16.3) becomes, therefore, $\overline{N}(E) \simeq \Omega(E)/(2\pi\hbar)^2$. Quite generally, for a Hamiltonian system with f degrees of freedom, *in the limit of large E, the number of eigenstates with energy less than E equals the number of cells of size h^f contained in the volume of phase space with energy less than E.* The average density of eigenstates per unit interval of energy is given by

$$\bar{\rho}(E) = \frac{1}{(2\pi\hbar)^f} \int d^f p \, d^f q \, \delta(E - H(p,q)) \qquad (16.4)$$

in a system with f degrees of freedom.

The formula (16.4) is of the utmost importance for treating many interacting particles in lowest approximation, such as the *Thomas-Fermi model* of atoms, the Sommerfeld theory of metals, Bose-Einstein condensation, the distribution of electron energies in β-ray emission, and the calculation of nuclear binding energies. It also provides a rationale for the basic assumption of classical statistical mechanics: the probability for a system to be in a particular subset of phase space is proportional to the volume of the subset. Of course, it is consistent with Liouville's theorem and the adiabatic principle.

The second and third terms in (16.3) extend the validity of Weyl's formula (16.2) to lower energies; but they give no information about local accumulations of eigenvalues. The remainder of this chapter is concerned with finding the appropriate statistical measures for the fluctuations, $N(E) - \overline{N}(E)$.

16.3 Measures for Spectral Fluctuations

Instead of studying the values of the energy levels E_k, we will look at the values $N_k = \overline{N}(E_k)$, where $\overline{N}(E)$ gives the number of levels below E in the limit of large E. The function $\overline{N}(E)$ is either taken directly from (16.3) in the case of a billiard in a compact domain, or it is calculated from the volume of phase space as indicated in (16.4). This reduction was already illustrated when the prime numbers were mentioned; $\overline{N}(x) \simeq x/\log x$ so that the points $p/\log p$ on the real axis, rather than the prime numbers p themselves, are investigated. The reduced energies will be called E_k again.

The simplest, and perhaps most telling, statistical measure is the *spectral distribution* $Q(x)\,dx$, which is the probability of finding an energy level in the interval $\alpha + x \leq E < \alpha + x + dx$ if there is an energy level at $E = \alpha$. A related measure, called the *spacing of energy levels,* is the probability $P(x)\,dx$ for finding two consecutive energy levels a distance x apart, the lower at $E = \alpha$, and the upper in the interval $\alpha + x \leq E < \alpha + x + dx$; of course, $\int P(x)\,dx = 1$. For the harmonic oscillator, $P(x)$ is simply a δ-function peak at $x = 1$. The opposite extreme has a completely random distribution of energy levels with average density 1. The function $Q(x) = 1$ since the probability for finding an energy level is now independent of the already known level at $x = 0$.

The functions $Q(x)$ and $P(x)$ are related in the following way. We choose a segment of (integer) length L, and divide it into small intervals, all of the same length ε. We place L markers at random, independently of one another, with the probability $Q(x)\,dx$, into the small intervals. The first marker above 0 will hit any particular small interval with probability $\varepsilon Q(\xi)/L$, and miss any other small interval with probability $1 - \varepsilon Q(\xi)/L$ where ξ is some coordinate inside the small interval in question. Now we choose a contiguous interval of length x, and require the probability that none of the markers fall within x, while there is a marker in $\alpha + x \leq E < \alpha + x + dx$. Clearly, we have to form the product of all the $1 - \varepsilon Q(\xi)/L$ for $0 < \xi < x$, and multiply with $\varepsilon Q(x)/L$. In the limit of small ε and/or large L, we find that $P(x) = Q(x) \exp[-\int d\xi Q(\xi)]$. In particular the random distribution of energy levels, $Q(x) = 1$ yields the *Poisson distribution* $P(x) = \exp(-x)$.

The spacings of energy levels will presumably have a distribution between the rigid ladder of the harmonic oscillator and the exponential decay. Quite unexpectedly, the integrable systems are found to have a Poisson-like spacing of energy levels. The classically chaotic systems, however, tend toward a rigid spacing, although their statistics is much more complicated as we shall see.

The staircase character of $N(E)$ is more pronounced, the more the spectrum deviates from the rigid spacing. A striking case is the set of the eigenvalues λ of the Laplacian on the sphere, $\Delta\phi + \lambda\phi = 0$; the area $A = 4\pi$, so that Weyl's formula yields a level density 1 for large λ. The spectrum is given by the values of the total angular momentum in units of $\hbar$, so that $\lambda = \ell(\ell + 1)$ with the multiplicity $2\ell + 1$ and $\ell \geq 0$.

A measure for the deviation from equal spacing, called the *rigidity,* was introduced by Dyson and Mehta (1963) who defined the function

$$\Delta_3(L; \alpha) = \frac{1}{L} \operatorname*{Min}_{A,B} \int_{\alpha}^{\alpha+L} [N(E) - AE - B]^2 dE \ , \quad (16.5)$$

where the constants A and B will give a best local fit to $N(E)$ in the interval $\alpha \leq E < \alpha + L$. The variable α is used to define the various averages $< ... >$ over part or all of the spectrum, while the length L gives an indication of the window over which the spectrum is viewed. $\bar{\Delta}_3(L)$ is the average over α.

The harmonic oscillator gives the least possible value with $\Delta_3 = 1/12$. The motion of a particle on a sphere, or equivalently, the symmetric rotator yield a rigidity which varies as α increases, and will go through maxima given by $L^2/3$. A completely random spectrum, with the Poisson spacing, has an average rigidity $\overline{\Delta}_3 = L/15$ independent of α.

For completeness sake, we shall list some further standard measures. They are based on the number $n(\alpha,L)$ of energy levels in the interval from α to $\alpha + L$. The *variance* is given by the standard expression

$$\Sigma^2(L) = < [n(\alpha,L) - < n(\alpha,L) >]^2 > \quad . \qquad (16.6)$$

The corresponding third and fourth power averages, after division by Σ^3 and Σ^4, give the *skewness* $\gamma_1(L)$, and the excess or *kurtosis* $\gamma_2(L)$.

16.4 The Spectrum of Random Matrices

Nuclear physics, in contrast to atomic and molecular physics, studies the interaction between particles whose forces of attraction and repulsion are not well understood. Protons and neutrons are not even point-like objects on the nuclear scale of 10^{-13} cm. The constituents of an atom or a molecule, the nuclei and the electrons surrounding them, however, have vanishing dimensions on the atomic scale of 10^{-8} cm; the electrostatic forces between them, with minor corrections for magnetic moments, are all we need to write down a complete Hamiltonian.

In the face of this uncertainty in nuclear physics, Wigner turned the problem around, and asked for the spectrum of an object whose Hamiltonian is unknown, or more precisely, a Hamiltonian that is drawn at random from a large collection, subject only to the general restrictions of quantum mechanics. Surprisingly, the result is not trivial. Wigner compared it to the distribution of the planets whose origin and order around the Sun is also poorly understood at this time. Nevertheless, the planets show a striking regularity in their distances from the Sun, and in their masses. Empirical 'laws' like the geometric progression in the major axes, known as the Titius-Bode law, might arise from almost any initial distribution of clouds in the solar neighborhood (a detailed review was recently given by Nieto 1972).

The detailed discussion of random Hamiltonians becomes rather technical. Bohigas and Giannoni (1984) have written a good introduction (cf. also José 1988). Some of the basic ingredients can be explained without too much formal mathematics. As usual, the upper index + on a matrix indicates the Hermitian transpose (or conjugate), $(T^+)_{jk} = (T_{kj})^+$, while the upper index + on a number indicates its complex conjugate. The Hamiltonian is assumed to be a Hermitian matrix H with a finite number K of rows and columns, i.e., $(H_{kj})^+ = H_{jk}$.

A first basic consideration concerns the invariance of the Hamiltonian with respect to *time reversal,* or in more practical terms, the presence or absence of a magnetic field B. If the Hamiltonian stays the same under time-reversal, or has an equivalent symmetry, then it can always be reduced to a real symmetric matrix by an appropriate unitary transformation without changing the eigenvalues (for an exhaustive discussion of this question, cf. Berry and Robnik 1986). That leaves $K(K+1)/2$ parameters to be chosen at random; but a reasonable measure to define the probability of each choice has to be determined.

The argument is no more sophisticated than finding the correct definition of the distance of a point (x, y, z) from the origin in a Cartesian coordinate system. Since the distance is not supposed to change when the coordinate axes are rotated around some arbitrary direction, it can only be some function $f(r)$ where $r^2 = x^2 + y^2 + z^2$. Similarly, the probability for finding a particular matrix H should be independent of the set of basis vectors that was chosen to obtain the matrix elements H_{jk}. A change of basis requires a transformation $H' = T^+HT$ with a K by K (real) orthogonal matrix T, satisfying the condition $T^+T = I$.

There exist a total of K different functions F_n of the matrix elements H_{jk} which remain the same under any orthogonal transformation T. These invariants were already mentioned in the discussion of the Toda lattice in Section 3.6. The two simplest were written down explicitly, $F_1 = \Sigma_j H_{jj}$ and $F_2 = \Sigma_{jk} H_{jk} H_{kj}$. For $\ell > 2$, however, F_ℓ becomes a rather complicated polynomial of degree ℓ. In defining our ensemble of random matrices, we are free to fix the value of any or all of the functions F_ℓ; but only F_1 and F_2 will be given in advance, while the others will be left to chance, because they are directly involved in the spacing of the eigenvalues.

If the matrix H has been diagonalized, its diagonal elements $\lambda_1, \lambda_2, ..., \lambda_K$ are the eigenvalues, so that $F_1 = \sum_\ell \lambda_\ell$, and $F_2 = \sum_\ell \lambda_\ell^2$ By fixing F_1 in advance, we prescribe the average of the eigenvalues; and by fixing F_2 in advance, we prescribe the sum of their squares; but their spacings have not been fixed except through their mean value and their overall spread. The product $d\Omega = \Pi\, dH_{jk}$ where $j \leq k$ is easily shown to be invariant under orthogonal transformations T and defines the appropriate measure in the space of the $K(K+1)/2$ matrix elements H_{jk}.

At this point, we have what could be called a microcanonical ensemble, if we associate F_2 with the energy, in analogy to the Toda lattice. Most calculations, however, are easier if we use the argument of Maxwell when he first wrote down the velocity distribution of the particles in a gas: the probability density which multiplies the measure $d\Omega$ is required to be the product of $K(K+1)/2$ functions $f(H_{jk})$ where each depends only on the particular matrix element H_{jk}. Thus, we finally get

$$\exp\left[-A \sum_{jk} (H_{jk})^2 - B \sum_{j} H_{jj} - C \right] \prod_{\ell \leq n} dH_{\ell n} \qquad (16.7)$$

for the probability of the matrix H, the typical expression for a canonical ensemble; C is the normalization constant.

The *Gaussian orthogonal ensemble* (GOE) is defined by (16.7) with $B = 0$. The answer to various questions is found by carrying out the necessary integrations. They become quite tricky, however, and the reader has to consult the review by Rosenzweig (1963) or the book by Mehta (1967) for the details. As a preliminary step, the integration is carried out while keeping the eigenvalues $\lambda_1, \lambda_2, \ldots, \lambda_K$ fixed and arbitrary. The result is remarkably simple and clear-cut,

$$\exp\left[-A \sum_{j} (\lambda_j)^2 - C' \right] \prod_{k<\ell} |\lambda_\ell - \lambda_k| \prod_{n} d\lambda_n , \qquad (16.8)$$

with a new constant of normalization C'. The main result is the repulsion between the eigenvalues: the probability density goes to zero, linearly as their distance, every time two eigenvalues λ_ℓ and λ_k get close to each other.

The probability $Q(y)\,dy$ for any eigenvalue λ to be found in the range $y \leq \lambda < y + dy$ is given, for large values of K, by

$$Q(y) = \begin{cases} \dfrac{2}{\pi\sigma^2}\sqrt{K\sigma^2 - y^2} \quad \text{for } y < \sqrt{K\sigma^2} ; \\ 0 \quad \text{for } y > \sqrt{K\sigma^2} ; \end{cases} \qquad (16.9)$$

the spread σ is related to the constant A in (16.7) and (16.8) by $\sigma^2 = 1/A$. This function represents a half-circle of radius $\sigma\sqrt{K}$, a result that Wigner (1957) had already conjectured.

The main result of this whole theory is the probability $P(x)\,dx$ for finding two adjacent eigenvalues with a spacing between x and $x + dx$. There is no simple expression in closed form, but there is the *Wigner's surmise* that

$$P(x) \simeq \frac{\pi}{2}\, x \exp[-\pi x^2/4] . \qquad (16.10)$$

This function, although not exact, is within a few percent of the exact result; since most experimental and numerical data come in the shape of rather coarse histograms, Wigner's surmise is amply sufficient. (The exact slope of the $P(x)$ at $x = 0$ is $\pi^2/6 = 1.6449$ rather than

$\pi/2 = 1.5708$; in other words, the percentage difference in the initial slope is $\pi/3 - 1 = .0472$!)

The spacing distribution (16.10) is in stark contrast to the Poisson distribution $P(x) = \exp(-x)$ of the preceding section, which arises when the eigenvalues are thrown at random into a finite interval. Already the nuclear resonances in uranium U^{238} under neutron bombardment, which were obtained in the 1950s, show very clearly the level repulsion as given by (16.10). Notice that both probabilities are not only normalized, but their mean is also 1, i.e., $\int x\, P(x)\, dx = 1$. Their variances, however, are very different; for the Poisson distribution, $\int (x-1)^2 P(x)\, dx = 1$, whereas for Wigner's surmise this integral $= 4/\pi - 1 = .5224$.

The rigidity (16.5) for the GOE is

$$\bar{\Delta}_3(L) = \frac{1}{\pi^2}\,(\log L - .0687)\,, \qquad (16.11)$$

instead of $L/15$ for the random distribution of eigenvalues in the interval of length L, or $1/12$ for equal spacing. The strong tendency toward a harmonic-oscillator-like spectrum is quite striking in the GOE.

16.5 The Density of States and Periodic Orbits

Percival (1973) first coined the words "regular" and "irregular" as appplied to the spectrum of a quantum-mechanical system. The criterion for labeling energy levels in this manner is their sensitivity to perturbations. Pomphrey (1974) showed that this idea worked well for the Hénon-Heiles model, when the perturbation was a change in the non-linear coupling parameter; Pullen and Edmonds (1981) got similar results for a fourth-order potential. He also found that the relative number of regular levels in some limited energy range was roughly the same as the proportion of volume in phase space that is occupied by invariant tori.

Since then, closer inspection of the spectrum in the Hénon-Heiles model has shown that most energy levels can be labeled with ordinary quantum numbers all the way into the chaotic region (cf. Section 14.3). Doubts about the simple distinction between a regular and an irregular part in the spectrum have also been voiced by Weissman and Jortner (1981). They investigated the evolution of wave-packets with time, and concluded that there were two entirely different limiting types: rapid dephasing and nearly quasiperiodic behavior. But they could not identify a clearcut transition from one regime to the other, and compare it to the transition from invariant tori to chaos in the classical system as illustrated in Figure 15.

A regular spectrum will, therefore, simply be defined as given by a formula like (14.11): there are as many quantum numbers $n_1, n_2, \ldots$ as degrees of freedom, and a smooth function Γ of as many actions $I_1, I_2, \ldots$ to yield the whole spectrum of interest. The Maslov indices, and the consequent replacement of n_j by $n_j + \beta_j/2$, depend on the problem at hand, and do not change any of the discussion; they will be left out of the further manipulations. The energies are then the values of a smooth function Γ on a regular grid of lattice distance $\hbar$ in the positive 'quadrant'; Γ does not necessarily come from a calculation based on classical orbits. It can be a fit to the empirical data, or it could result from a theory like the Birkhoff-Gustavson normalization of Section 14.3 which seems sometimes valid even in the classically ergodic region of phase space.

The density of states can then be written as

$$\rho(E) = \sum_{\mathbf{n}} \delta(E - \Gamma(\mathbf{n}\,\hbar)) \, , \tag{16.12}$$

where $\mathbf{n}$ is the vector of the integers $n_1, n_2, \ldots$. The regular lattice in $\mathbf{n}$ immediately invites the theorist to write $\rho(E)$ as a Fourier sum over a reciprocal lattice indexed by a vector $\mathbf{M}$ of integers, exactly as in solid-state physics; the δ-function peaks in (16.12) are then generated by a superposition of plane waves of wave-vector $\mathbf{M}$.

This route was taken for the first time by Berry and Tabor (1976, 1977a, 1977b), in order to calculate the spacing probability $P(x)$ for such a regular spectrum. We shall follow these authors fairly closely, not only because of the final result, but also because their arguments give a good introduction to the more general considerations of the next chapter. Again, we shall only indicate the important steps, and leave to the reader the task of working through the mathematical details.

Poisson's formula carries out the transformation from the *direct lattice* of quantum numbers $\mathbf{n}$ to the *reciprocal lattice* $\mathbf{M}$, whose interpretation will be given shortly. For simplicity's sake, let us consider the variable s in one dimension, $-\infty < s < +\infty$, and some well-behaved integrable function $g(s)$; its Fourier transform is called $\chi(\sigma) = \int ds\, g(s) \exp(-2\pi i s \sigma)$ with $-\infty < \sigma < +\infty$. A short calculation yields

$$\sum_{n=-\infty}^{+\infty} g(n) = \sum_{\nu=-\infty}^{+\infty} \chi(\nu) \, . \tag{16.13}$$

As an example, take the Gaussian $g(s) = \exp(-\pi s^2/\varepsilon^2)$ and its Fourier transform $\chi(\sigma) = \varepsilon \exp(-\pi \sigma^2 \varepsilon^2)$. The width ε of the peak in $g(s)$ is reciprocal to the width $1/\varepsilon$ of $\chi(\sigma)$, and the convergence of the two series in (16.12) behaves accordingly.

The summation over each quantum number n_j is now converted with the help of (16.13) into a summation over **M**,

$$\rho(E) = \frac{1}{\hbar^f} \sum_{\mathbf{M}} \int d^f I \, \delta(E - \Gamma(I)) \, e^{2\pi i (\mathbf{M},\mathbf{I})/\hbar} , \qquad (16.14)$$

where f is the number of degrees of freedom. The reader may want to replace the Dirac δ-function in (16.12) and (16.14) by $\exp[-\pi(E - H(I))^2/\varepsilon^2]$, where both the energy E and the width ε are parameters, while the actions I take on integer multiples of $\hbar$. Formula (16.14) is then the limit of ε going to zero.

The first term in (16.14), $\mathbf{M} = 0$, is the classical density of states as given in (16.4),

$$\rho_0(E) = \hbar^{-f} \int d^f I \, \delta(E - H(I)) \; . \qquad (16.15)$$

In this section, however, we are interested in the deviations from this smooth asymptotic density of states. Thus, we will discuss only the fluctuations $\tilde{\rho}(E)$, i.e., the terms in (16.14) where $\mathbf{M} \neq 0$. They all have a non-trivial factor $\exp(2\pi i(\mathbf{M},\mathbf{I})/\hbar)$ in the integrand and will be treated by the stationary-phase method of Section 12.5. The validity of this method depends on the remainder of the integrand varying slowly compared to the exponential; therefore, the integers **M** are assumed to be large. Since the δ-functions are very sharp, the Poisson formula requires many terms.

The stationary phase method requires finding the values of the actions I, where the first derivatives of the exponent $M_1 I_1 + \ldots + M_f I_f$ with respect to the I's vanish, provided $H(I) = E$. With the help of a Lagrange multiplier $1/\omega_0$, and the formula $\partial H/\partial I_j = \omega_j$, it follows that $\omega_j = M_j \omega_0$ as in Section 6.2. The individual frequencies ω_j are integer multiples of the common frequency ω_0, as in Section 6.2. Therefore, *the formula (16.14) for the density of states at the energy E can be interpreted as a sum over the classical periodic orbits at E.*

This central idea will be further explored in the next chapter for the general case of a classically chaotic system. Right now, the formulas for the integrable case will be worked out in more detail, so that the contrast with the next chapter can be seen more clearly. Although the presentation in this chapter is largely due to Berry and Tabor, the author (Gutzwiller 1970) was the first to discuss such a formula for an integrable system with two degrees of freedom. In one degree of freedom, the reduction to periodic orbits is trivial, and has been effectively practiced for a long time. The author's work on the hydrogen atom

(Gutzwiller 1967) is an example, as is some of the work of Norcliffe and Percival (1968).

The integration extends over the constant energy surface near the stationary point where $\omega_j = M_j \omega_0$. The exponent (**M**, **I**) in (16.13) is expanded to second order in the displacements δI_k around the stationary value $\mathbf{I}_0$, while the energy is kept fixed. It is natural to use a local coordinate system in action space where the first action points along the normal to the surface of constant energy, while the $f-1$ remaining actions are tangential. In order to do the integration over the $\delta(E - H(I))$ in (16.14), we also replace the first of the new actions by the local variation in energy δE. The Jacobian for these trivial transformations of the integration variables is $1/\omega_1$; the exponent now becomes simply $2\pi i |\mathbf{M}| \delta I_1/\hbar$, where $|\mathbf{M}|$ is the absolute value of the vector **M**. The constant energy surface is then given by $\delta I_1 = -\Sigma' H_{jk}\, \delta I_j\, \delta I_k / H_1$; the indices on H indicate the corresponding derivatives with respect to the actions I, and the prime on Σ restricts the summation to $j,k > 1$.

Each term in (16.14) is now in the standard form (12.23) for a multidimensional Fresnel integral. The fluctuating part of the density of states becomes

$$\begin{aligned} \tilde{\rho}(E) &= \frac{1}{\hbar^f} \sum_{\mathbf{M} \neq 0} \frac{1}{\omega_1} (\hbar\omega_0)^{\frac{f-1}{2}} \frac{\exp^{2\pi i(\mathbf{M},\, \mathbf{I}_0)/\hbar}}{\sqrt{\det'(H_{jk})}}\,, \\ &= \sum_{\mathbf{M} \neq 0} A(E, \mathbf{M}) \exp^{2\pi i(\mathbf{M},\, \mathbf{I}_0)/\hbar}\,, \end{aligned} \tag{16.16}$$

where the actions at the stationary point have been distinguished by the index 0 from the local action coordinates in the determinant and the prefactor. The second line defines the coefficients $A(E, \mathbf{M})$, which multiply the phase factors $\exp[iS(E, \mathbf{M})/\hbar]$. Notice that $\omega_1 = |grad_I\, H|$ and that the $\det'(H_{jk})$ is a measure for the curvature of the constant-energy surface in action space. The prime indicates again that only the local actions $\delta I_2, ..., \delta I_f$ are involved.

The reader should try to apply the formula (16.16) to the simplest case of an integrable system, the motion of a particle in a rectangular box with periodic boundary conditions,

$$H(I_1, I_2, I_3) = \frac{I_1^2}{2m\, a_1^2} + \frac{I_2^2}{2m\, a_2^2} + \frac{I_3^2}{2m\, a_3^2}\,, \tag{16.17}$$

where a_1, a_2, a_3 are the length, width, and height of the box. The behavior of the various terms in (16.16) for large values of the integers $M_1, ..., M_f$ is typical for any other shape of a container. The qualitative features of (16.16) appear in this special example with particular clarity; but they are valid for more general integrable systems.

16.6 Level Clustering in the Regular Spectrum

Formula (16.16) for the fluctuations in the density of states $\tilde{\rho}(E)$ will be derived in the next chapter from a different starting point. We will then inquire how this sum over all the periodic orbits manages to yield the sharp δ-function peaks in (16.12). In this section, however, we will ask whether (16.16) can tell us something about the correlations in $\tilde{\rho}(E)$. The method is motivated by the formulas (14.12) and (14.13), although certain modifications will have to be made ultimately, in order to put the whole thing together as Berry and Tabor did.

The variable x in (14.12) whose autocorrelation we are trying to find, is the density $\tilde{\rho}$, and the time variable t is replaced by E. Formula (14.13) requires that we calculate the Fourier transform $\tilde{\sigma}(2\pi/\hbar\omega)$ of $\tilde{\rho}(E)$ with respect to E; the variable of this transform will be called $2\pi/\hbar\omega$ for the lack of a better name. The absolute square $|\tilde{\sigma}|^2$ gives us the 'power spectrum' $\Pi(2\pi/\hbar\omega)$ of $\tilde{\rho}(E)$, i.e., the Fourier transform of the autocorrelation $< \tilde{\rho}(E)\tilde{\rho}(E+\varepsilon) >$ according as (14.12).

The formula (16.16) can be interpreted as the Fourier representation of $\tilde{\rho}(E)$, if we consider the exponent of each term $2\pi i(\mathbf{M}, \mathbf{I}_0)/\hbar$ as a function of E. Therefore, the squares of the absolute values for the coefficients $A(E,\mathbf{M})$ in (16.16), if properly summed up, are directly involved in the autocorrelation of $\tilde{\rho}(E)$.

Quite generally, the phase of a particular term in (16.16) changes with the energy E as the period $(2\pi/\hbar)(\mathbf{M}, \partial\mathbf{I}_0/\partial E) = 2\pi/\hbar\omega_0$, calculated for fixed $\mathbf{M}$. Since $1/\omega_0$ increases as $|\mathbf{M}|$, if the energy and the frequency ratios are kept constant, terms in (16.16) with different M are completely out of phase (incoherent), as E varies in the neighborhood of E_0, unless they share the same value of ω_0. Our job is then to sum up the squares $|A(E, \mathbf{M})|^2$ of the amplitudes in (16.16) which belong to the same value of the energy E near E_0, and to the same value of the frequency ω_0 near ω.

If the reader finds the formal manipulations at this point discombobulating, the following steps are recommended. Take $|A(E, \mathbf{M})|^2$ in (16.16), multiply with the necessary δ-functions, $\delta(E - E_0)$ and $\delta((2\pi/\hbar\omega_0) - (2\pi/\hbar\omega))$, where ω is the required constant value of ω_0. Now consider the integral over the $f+1$ variables $E, M_1, ..., M_f$; replacing the sum over the M 's by an integral is justified, because each term has the small factor ω_0^{f-1}. Since $\omega_j = M_j\omega_0$, however, and the energy E can be considered as a well-defined function of $\omega_1, ..., \omega_f$ (cf. the latter part of Section 14.4), we now switch to the integration variables $\omega_0, \omega_1, ..., \omega_f$; without the help of the local coordinates, the Jacobian is given by $(\omega_1 \partial E/\partial\omega_1 + ...)/\omega_0^{f+1}$. At this point, the integral over ω_0 is carried out with the help of its δ-function, and we are left with an integral over all frequency space,

$$\frac{1}{2\pi\hbar^f}\int d\omega_1 ... d\omega_f \, \frac{\omega_1 \partial E/\partial\omega_1 + ...}{\omega_1^2} \, \frac{\delta(E - E_0)}{\det'(H_{jk})} \, ,$$

where we have already used the local coordinates in the denominators, as in (16.16).

The last step converts everything to the local variables, and again to the energy in order to get rid of $\delta(E - E_0)$. In particular, $d\omega_1 = dE/(\partial E/\partial\omega_1)$, and the additional terms in the numerator vanish. The integral over E is now done, which leaves us with an integral over the surface of constant energy E_0 in frequency space.

The determinant in the denominator is nothing but the Jacobian $\partial(\omega_2, ..., \omega_f)/\partial(I_2, ..., I_f)$, which can be brought into the numerator by inverting it. Thus, we get rid of the frequency space, and are now left with an integral over the constant-energy surface in the action space, as in (16.15). The final result is

$$\int d^f M \, \delta(\frac{2\pi}{\hbar\omega_0} - \frac{2\pi}{\hbar\omega}) \, |A(E, \mathbf{M})|^2 = \frac{1}{2\pi}\rho_0(E) \, . \quad (16.18)$$

The left-hand side is the 'power spectrum' $\Pi(2\pi/\hbar\omega)$ of the density $\tilde{\rho}(E)$, i.e., the Fourier transform (in terms of the variable $2\pi/\hbar\omega$) of the autocorrelation function $< \tilde{\rho}(E)\tilde{\rho}(E + \varepsilon) >$. The result of our labors is given by the right-hand side, which turns out to be surprisingly simple, since it is independent of $2\pi/\hbar\omega$; the energy occurs only because all our calculations apply to a particular interval of energy around E.

Such a relation is called a *sum rule;* it sets the sum of the absolute squares in an expansion like (16.16) equal to a quantity which is easy to calculate like (16.15). Sum rules usually have a simple physical origin, not unlike Weyl's formula (16.2), which says that a particle needs a minimum time to reach the boundary of a domain by diffusion. The above sum rule, due to Berry and Tabor, has meanwhile found its explanation by Hannay and Ozorio de Almeida (1984) and will be discussed again in the next chapter.

Since the average density of states $\rho_0(E)$ in (16.15) is, in general, not a constant, a fake energy U has to be defined such that the density with respect to U becomes 1. Therefore,

$$U(E) = \hbar^{-f}\int dI^f \, \Theta(E - H(I)) \, , \quad (16.19)$$

where $\Theta(x) = 0$ for $x < 0$, and $\Theta(x) = 1$ for $x \geq 0$. The modified spectrum is defined by the fake Hamiltonian $W(I) = U(H(I))$, which replaces $H(I)$ everywhere in the preceding derivation, including the formulas (16.14), (16.15), (16.16), and (16.18). The new density will be called $\rho'(U)$ and its mean $\rho'_0(U) = 1$. The spectral density (14.12) for the autocorrelation of $\rho'(U)$ reduces then simply to the value $1/2\pi$, independent of both the frequency ω_0 and the energy U_0.

The last step in finding the autocorrelation function for an integrable spectrum relates directly the 'power spectrum' Π with the spectral distribution $Q(x)$ (cf. Section 16.3). The reader is asked to check the paper by Berry and Tabor (1977b) for the proof of the general formula,

$$Q(x) = 1 + \int_{-\infty}^{+\infty} d\kappa \, e^{i\kappa x} \, [\Pi(\kappa) - \frac{1}{2\pi}] ,$$

where we have abbreviated $\kappa = 2\pi/\hbar\omega$. Because of (16.18) and $\rho_0 = 1$, the function $Q(x) = 1$; therefore, the energy levels are spaced according as the Poisson distribution (again cf. Section 16.3). In a very qualitative argument we could say that the energy levels are randomly distributed because the 'power spectrum' (16.18) does not single out any preferred frequencies.

Is the Poisson distribution for the energy levels in an integrable system correct? The first numerical check was carried out by Casati, Chirikov, and Guarneri (1985) on the simplest possible case, a particle in a rectangular box, the two-dimensional version of (16.17). If the ratio γ of the square of the sides, a_1^2/a_2^2, is rational, there are many degeneracies among the energy levels; they are explained by one of the highly developed as well as non-trivial parts of number theory, the quadratic forms of integers. To get away from these accidental complications, γ is chosen to be irrational.

The authors put $\gamma = \pi/3$ and calculated 100,000 levels from the simple formula $E = \gamma m^2 + n^2$. With this kind of statistics, even seemingly minor deviations from the expected result quickly indicate some major discrepancies. The most important discovery was the *saturation* of the rigidity; instead of increasing as $L/15$ with L, as indicated after formula (16.5), the rigidity $\overline{\Delta}_3(L)$ becomes effectively constant rather abruptly for $L > L_c$; the exact value of the critical length L_c depends on the sample chosen, i.e., the range of α in (16.5).

According to Casati, Guarneri, and Valz-Gris (1984), the saturation is explained by the underlying structure for the energy spectrum of an integrable system, namely a simple grid of points. The number of levels between two energies E and $E + \varepsilon$ is quite random as a function of E, as long as ε is small, i.e., L is sufficiently small in (16.5), say of the order 15. If L becomes much larger, however, the elliptic annulus between E and $E + \varepsilon$ in the plane of actions (I_1, I_2) contains a good portion of the regular grid structure, and the random character gets lost. The rigidity then approaches the constant value corresponding to the harmonic oscillator.

Seligman and Verbaarschot (1987) and Verbaarschot (1987) have studied various statistical measures for the spectrum of independent non-linear oscillators. Using the ideas of Berry (1985), some of which were discussed above, and others that will be brought up in the next chapter, they succeeded in explaining the saturation of the rigidity analytically, in addition to making extensive numerical checks. Nevertheless, their arguments apply only to *scale-invariant* systems where the potential energy is a homogeneous polynomial in the position coordinates, while the reasoning of Berry and Tabor is more general. The deviations from Poisson statistics for the spacing of energy levels in integrable systems need further work to be fully understood.

16.7 The Fluctuations in the Irregular Spectrum

The evidence for a connection between nuclear energy levels and random matrices remained inconclusive for almost 25 years after the pioneering work of Wigner. Haq, Pandey, and Bohigas (1982) then analyzed the whole body of high-quality data, 1407 resonances corresponding to 30 sequences of 27 different nuclei. They investigated both the spacing of the energy levels $P(x)$ and the rigidity (16.5) as a function of L, both shown in Figure 48. "Astonishingly good agreement" was found between the experimental data and the Gaussian Orthogonal Ensemble (GOE), although there are no parameters to fit. Further statistical analysis is given by Bohigas, Haq, and Pandey (1985).

Bohigas, Giannoni, and Schmit (1984a) applied the same statistical technique for the first time to an artificial spectrum. They chose Sinai's billiard, a unit-square with a circular hole of radius R in the center, and hard reflections on the boundary. Its classical behavior is known to be strongly ergodic; there exists a good asymptotic formula for its level distribution, (16.3), including the terms in the length of the boundary and its curvature. Fluctuations are, therefore, easily separated from the smoothed density of levels. Berry (1981) had earlier established an efficient algorithm for finding the eigenvalues of the Laplacian in this domain, with the help of the Kohn-Korringa-Rostoker method from band calculations in solid-state physics.

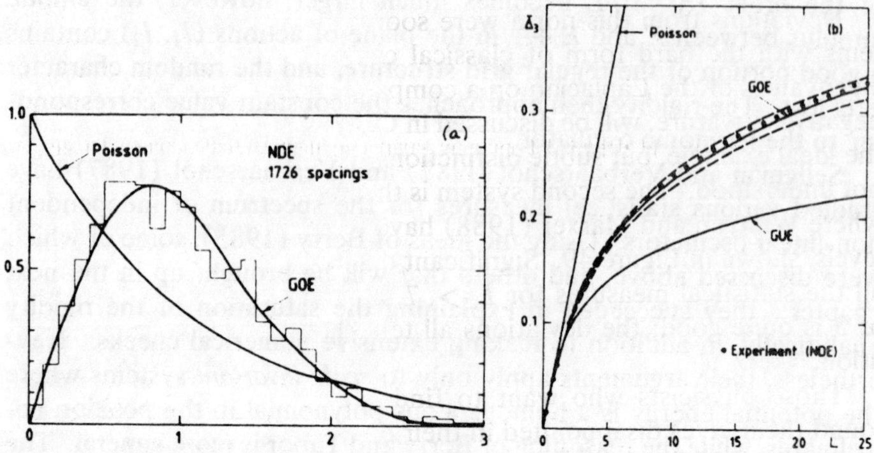

Figure 48 The spacing distribution $P(x)$ and the rigidity $\bar{\Delta}_3$ for the composite of 30 energy-level sequences from 27 different nuclei [from Bohigas, Haq, and Pandey (1983)].

Since the domain allows for a group of symmetries with eight elements, the results have to be segregated for each particular symmetry of the wave functions with respect to reflections on the median lines and the diagonals of the square. If the eigenfunctions of the Laplacian are antisymmetric with respect to every one of these reflections, the wave functions vanish on the symmetry lines, corresponding to Dirichlet boundary conditions. The spectrum is the same as for the reduced domain that consists of a right-angled isoceles triangle from which a pie-shaped piece has been removed on one of the corners. The deviations from the ideal GOE curves can be accounted for by the finite sampling, i.e., the finite upper limit to the reliable eigenvalues.

The success of the GOE statistics was further confirmed by data on *atomic spectra* of rare-earth metals by Camarda and Georgopoulos (1983), as well as in simple molecules such as NO_2 by Haller, Koppel, and Cederbaum (1983), at least as far as Wigner's surmise (16.10) for the spacing of levels (further data are given by Zimmermann et al. 1987 and 1988).

Large-scale calculations of energy levels were carried out for the stadium by Bohigas, Giannoni, and Schmit (1984b). They computed not only the rigidity $\bar{\Delta}_3$ and the spacing $P(x)$, but also the variance Σ^2 as given in (16.6), as well as the skewness and the excess. The agreement with GOE is unexpectedly close and complete. It led to the speculation that GOE statistics is a "universal" characteristic of the energy levels in classically chaotic systems.

Deviations from this norm were soon found, however, in two systems with the hard form of classical chaos. The first of these, the eigenvalues of the Laplacian on a compact smooth surface of constant negative curvature, will be discussed in Chapter 19; it should have been the ideal example, but subtle distinctions seem to be at work which are not understood. The second system is the Anisotropic Kepler Problem where Wintgen and Marxer (1988) have recently obtained some 5000 levels, shown in Figure 49. Significant deviations from GOE appear in all the statistical measures for $L > 7$, although the fit for L up to 2 or 3 is quite good; the deviations all tend toward the Poisson distribution.

Those physicists who want to find universally valid simple rules found themselves disappointed in their hopes, whereas those who look for a rich variety of different behavior felt encouraged. The search is now on, both for reasons explaining the departures from GOE, and for getting better numerical data to see where GOE breaks down in spite of the classical chaos. Some features might then be discovered whereby different types of quantum chaos can be recognized, if not explained.

16.8 The Transition from the Regular to the Irregular Spectrum

Since the statistical properties are so clearly different for the regular and the irregular spectra, the question about the transition from one to the other arises immediately. The theory of random matrices gives no clue concerning the connection between the two types of statistics. While there are good arguments to associate regular spectra with classically integrable systems (cf. Section 16.5), and irregular spectra with the classically chaotic ones (cf. the next section as well as the next chapter), the intermediate regime is poorly understood. Nevertheless, a number of numerical studies have been undertaken, and some will be mentioned in this section to give the reader an idea of the extent to which this general problem has been investigated.

The computations of this kind require a continuous sequence of dynamical systems which go from integrable at one end to chaotic at the other end. A particularly obvious example comes from the conformal maps of the unit-circle, such as the simplest quadratic map proposed by Robnik (1984), $w = Az + Bz^2$ where $w = u + iv$ and $z = x + iy$ are complex variables, and the ratio B/A is the parameter starting from 0 at the integrable end of the sequence. The image of the unit-circle is the domain for a billiard with reflecting walls; its area is

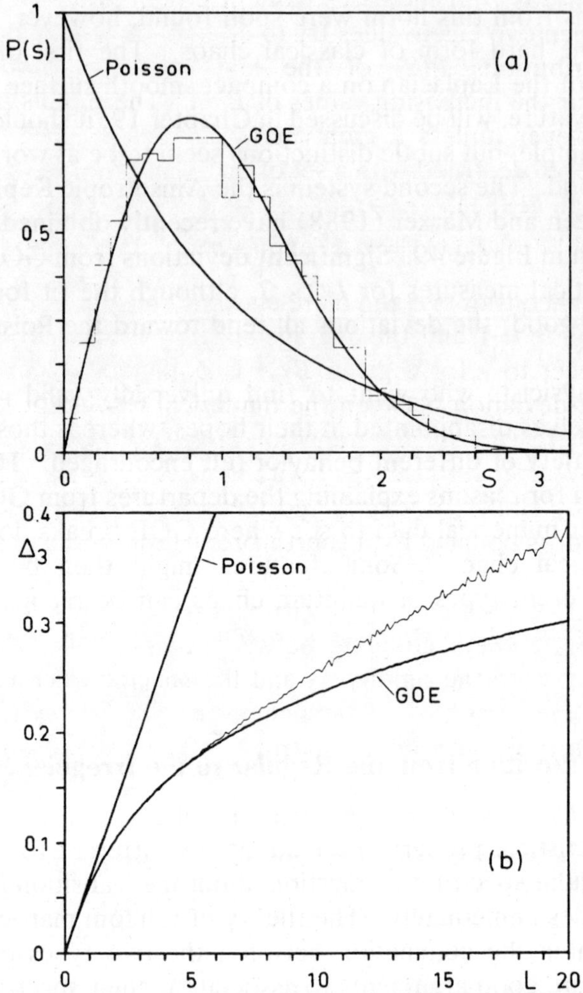

Figure 49 Nearest spacing $P(s)$ and spectral rigidity Δ_3 for the Anisotropic Kepler Problem with the mass ratio 5/4; the continuous lines represent these functions for the Poisson distribution and the Orthogonal Gaussian Ensemble, as discussed in Section 16.4 [from Wintgen and Marxer (1988)].

$\pi(A^2 + B^2)$, which is normalized to 4π so as to yield the asymptotic density 1 of eigenvalues for the Laplacian according to Weyl's formula.

This scheme suggests a relatively simple set of basis functions in which to expand the eigenfunctions: indeed, if we use the variable z as coordinate in the transformed unit circle, Schrödinger's equation becomes $\Delta\phi + \lambda J(x, y)\phi = 0$, where $J = |dw/dz|^2$ and z is limited to the unit-circle. If $J = 1$, the eigenfunctions are a well-known set of Bessel functions, which can serve as a basis set when $J \neq 1$; the Laplacian acting on this set is easy to handle. The continuous distortion of the eigenfunctions, as B/A increases from 0, can be watched in detail, in particular when two eigenvalues try to cross each other.

The distributions $P(x)$ of the level spacings are plotted as histograms for the increasing values of B/A. These plots are not very striking, however, unless a great many levels have been computed. They are, therefore, fitted with a set of distributions that was suggested by Brody et al. (1981),

$$P_\nu(x) = a\, x^\nu \exp(-b\, x^{\nu+1}) \ , \tag{16.20}$$

where the real constants a and b are chosen to insure both the normalization $\int P_\nu(x)dx = 1$ and the average spacing $\int xP_\nu(x)dx = 1$. A best fit with respect to ν for a given B/A is obtained by minimizing the mean-square deviation between the numerical results for the spacings and P_ν. There is a steady transition from the Poisson distribution $\nu = 0$ to the Wigner surmise $\nu = 1$ with increasing B/A. Similar results were found by Haller, Koeppel, and Cederbaum (1984) for two harmonic oscillators coupled by a fourth-order term in the position coordinates.

Seligman, Verbaarschot, and Zirnbauer (1984 and 1985) investigate a more elaborate coupling between two non-linear oscillators; they work out both the rigidity $\bar{\Delta}_3$ and the spacing $P(x)$ as a function of the coupling strength. The spacings are fitted with a modified Gaussian ensemble of random matrices: the off-diagonal matrix elements H_{jk} in the ensemble are subject to a cut-off factor $\exp(-(j-k)^2/\sigma^2)$. Poisson statistics requires $\sigma = 0$, whereas $\sigma = \infty$ for GOE. There is a smooth transition from σ=0 to $\sigma \simeq 1$, as the coupling strength increases.

As a last example, we mention the work of Ishikawa and Yukawa (1985a and b) who study the billiard in the oval given by $r = r_0(1 + \cos 2\phi)$ in polar coordinates. The eigenfunctions are expanded in the Bessel functions $J_n(kr)\exp(in\phi)$ where $E = \hbar^2k^2/2m$, so that the matrix elements $M_{jn}(k)$ can be expressed analytically. The authors compute a great many different statistical measures, including the ones discussed above; they also calculate Lyapounoff exponents (cf. Section 1.6) and the resulting entropy, i.e., the average over (10.6) for many different classical trajectories. The qualitative conclusions remain the same as in the previous examples.

Berry and Robnik (1984) pointed out a peculiar difficulty that affects the numerical results in this section. As long as the classical system is not fully ergodic, i.e., a typical trajectory does not go through the whole energy surface, the phase space is divided into distinct parts, which do not communicate classically. Experience with the Wigner distribution functions (cf. Section 15.4) shows that the highly excited eigenfunctions respect this classical break-up of phase space; such a dynamically isolated region may be integrable, or nearly integrable, i.e., be covered with many remaining KAM tori; or it may be fully ergodic, but separated from another fully ergodic region by a last KAM torus; or finally the most common case, two regions may belong to different exact symmetries in the Hamiltonian, and such symmetries divide the spectrum into two entirely independent parts.

In all these cases, each dynamically distinct region of phase space contributes a set of energy levels with its own statistics, such as Poisson or GOE; but the superposition of two or more such independent contributions makes the result look as if the energy levels were totally uncorrelated. E.g., the existence of a single KAM torus may have the effect of faking a Poisson distribution for the spacing $P(x)$. Spectral statistics has to be viewed with great caution and has to be supplemented with a careful examination of the classical system. But even then, it may be hard to separate the parts of the spectrum that one may vaguely attribute to one or the other region of phase space. Further analysis along these lines is presented by Wintgen and Friedrich (1987a) and by Hönig and Wintgen (1989).

16.9 Classical Chaos and Quantal Random Matrices

If the Gaussian Orthogonal Ensemble (GOE) is assumed, precise statements about the statistical nature of the spectrum can be made, some of which are confirmed to an unexpected degree of accuracy in Hamiltonian systems with classical chaos, as shown in Section 16.6. Can we understand this empirical result?

Two kinds of explanations have been offered so far: the first gives an argument like that of Berry and Tabor in Section 16.5 where at some crucial moment an explicit statement about the classical behavior is used and is then transformed into a precise statistical proposition, e.g., the regular grid in the space of classical actions yields the Poisson distribution for the spacings. The second kind of explanation is more intuitive and general, but gives no more than a vague connection between a particular symptom of classical chaos and some characteristic feature of the spectrum.

The first kind is clearly preferable in the long run, but the only useful approach so far for classically chaotic systems is due to Berry (1985). It will be discussed in the next chapter, because it starts with the trace formula; it ends up calculating the rigidity, and yields the same value (16.11) as the GOE. The physics is hidden in a sum rule due to Hannay and Ozorio de Almeida (1984), whose analog for integrable systems is responsible for the result of Berry and Tabor (cf. Section 16.5). Work by Balazs and Voros (1987) involving the zeta-function is of the same kind. Progress along this direction is badly needed.

Zaslavski (1977) also starts from the trace formula; he gets a rather explicit expression for the spacing probability $P(x)$ which involves the Kolmogoroff entropy, basically the average of (10.6) over the energy surface. Such a relation would be very desirable; but Casati, Valz-Gris, and Guarneri (1980) see no evidence for such a direct connection in their calculation of eigenvalues for the stadium. Ishikawa and Yukawa (1985a and b) correlate the entropy directly with the parameter ν in the spacing distribution (16.17), but they make no effort to explain their result quantitatively.

A more formal argument is given by Alhassid and Levine (1986) who consider the transition strengths of a quantum system; $y = |x|^2$ where $x = \langle f | T | i \rangle$ is the transition amplitude from the initial state i to the final state f through the observable T. The aim is to find the most likely distribution of $Q(x)dx$, given the sum rule to be obeyed by the transition strengths, such as the sum over the final states $\Sigma | \langle f | T | i \rangle |^2 = \langle i | T^+ T | i \rangle$. The likelihood for a particular distribution $Q(x)$ is defined in terms of its 'entropy', $\int dx\, Q(x) \log Q(x)$, which is then minimized under the constraint of the sum-rule. Additional constraints to limit the spread of the transition strengths from their mean eventually lead to distributions similar to (16.20). A related argument is advanced by Wilkinson (1988) who thinks of the matrix elements in the quantum Hamiltonian as executing a random walk as a function of the perturbation.

The distribution $P(x)$ of the level spacings in chaotic systems is narrowly peaked, because *level crossings are avoided* as the strength of the coupling between different degrees of freedom increases. Presumably, since the eigenfunctions of the Hamiltonian don't have any simple pattern, their overlap integrals don't vanish, and degeneracies are avoided. The energy levels try to stay away from one another as best they can, and thereby account for the strong peak in $P(x)$ at $x = 1$. Gomez, Zakrzewski, Taylor, and Kulander (1989) illustrate this point with examples such as the hydrogen atom in a strong magnetic field, as well as the molecule H_3^+. Heiss and Sannino (1989) consider quite generally the orientation of the eigenvectors in Hilbert space, and their rotation as function of the perturbation parameters, to explain the level repulsion.

Graffi, Paul, and Silverstone (1986) have related this phenomenon directly to the overlap of classical resonances that was discussed in

Section 9.4, and applied this argument to a time-dependent Hamiltonian system with one degree of freedom; but further work is needed to explain avoided crossings in conservative chaotic examples. Reichl and Lin (1986), as well as Lin and Reichl (1987 and 1988) have worked out the idea of a quantum-resonance overlap to explain the transition from regular to irregular spectra, and related manifestations of chaotic behavior in quantum systems; but their examples are time-dependent and one-dimensional, and will not be discussed here.

The idea that the overlap between many different states leads to a GOE type spectrum has been formalized by Pechukas (1983); his method was further developed by Yukawa (1985). We will follow the latter. The Hamiltonian is written as $H = H_0 + tV$, where H_0 belongs to an integrable system, and V is the coupling whose strength t increases from 0 as if it were the time. The eigenvalues of H are $x_n(t)$, and the corresponding eigenfunctions are $\phi_n(t)$. The matrix elements for the perturbation are $V_{mn}(t) = < \phi_m(t) \mid V | \phi_n(t) >$. These quantities are shown by simple differentiation (perturbation theory) to satisfy a number of 'equations of motion',

$$\frac{dx_n}{dt} = p_n \, , \quad \frac{dp_n}{dt} = 2 \sum_{m \neq n} \frac{f_{nm}^2}{(x_n - x_m)^3} \, , \tag{16.21}$$

$$\frac{df_{mn}}{dt} = (x_n - x_m) \sum_{\ell \neq (n,m)} \frac{f_{n\ell}\, f_{\ell m}}{|x_n - x_\ell \mid | x_\ell - x_m \mid} \left(\frac{1}{x_n - x_\ell} + \frac{1}{x_m - x_\ell} \right) ,$$

where $f_{mn} = | x_m - x_n | \, V_{mn}$ for $n \neq m$, and $x_n > x_m$ for $n > m$.

It takes a moment of thought to realize that these equations describe a perfectly acceptable classical dynamical system in a phase space of coordinates (x_n, p_n, f_{mn}) where the indices n and m range over the number of dimensions in the assumed Hilbert space. The Liouville volume is preserved, as are the 'energy' E and the total coupling strength Q,

$$E = \frac{1}{2} \sum_n p_n^2 + \frac{1}{2} \sum_{n \neq m} \frac{| f_{nm} |^2}{(x_n - x_m)^2} \, , \tag{16.22}$$

$$Q = \sum_{n \neq m} | f_{nm} |^2 .$$

There are infinitely many 'constants of motion', because increasing the coupling parameter t generates an orthogonal transformation in the Hilbert space whose invariants, like the trace of various operators, stay the same. Indeed, Nakamura and Lakshaman (1986) have shown that the system (16.22) is completely integrable.

The problem is now shifted toward understanding a classical dynamical system of infinitely many degrees of freedom with as many constants of motion. One could resort to statistical mechanics, and consider a stationary distribution $(1/Z)\exp(-\beta E - \gamma Q)$ where Z is the normalization constant, alias partition function. The 'inverse temperature' β and the 'chemical potential' γ can be determined by prescribing the mean values $E_0 = \langle E \rangle$ and $Q_0 = \langle Q \rangle$. The variables p_n and f_{nm} are integrated out, so as to yield the distribution for the eigenvalues x_n of the Hamiltonian H. The result is exactly the probability distribution (16.8) for the eigenvalues in the GOE, and one is tempted to recognize a mathematical justification for the random matrix model; but the GOE now appears to be almost too generally applicable to be of much value in the characterization of chaotic features in quantum mechanics.

CHAPTER 17

The Trace Formula

The main topic of this chapter gets its name from a spiritual ancestor, the Selberg Trace Formula, which was discovered by the Norwegian mathematician Atle Selberg in the 1950s. Whereas the original trace formula claims the numerical equality of two rather unequal-looking functions, the descendant makes this claim only in the limit of Planck's quantum going to zero, or equivalently, in the limit of a wavelength small compared to the size of the container. While the degree of validity for the new formula has been reduced, its domain of application has been greatly enlarged.

Instead of a well-defined and crisp formula, we now have a basic approach to making a connection between a function of time or of energy computed in quantum mechanics, on the left-hand side of the $\simeq$ sign, and another such function calculated in classical mechanics on the right-hand side. The exact formal expressions on the right depend somewhat on the behavior of the classical dynamical system, whether regular, softly, or harshly chaotic; but the left side is independent of this classical predicament. The main idea is to use the knowledge of the classical behavior to compute the right-hand side, and then switch sides so as to draw conclusions about the quantum mechanics of the system.

The author was the first to devise this general scheme (Gutzwiller 1970 and 1971) as a way to answer Einstein's question: how can classical mechanics give us any hints about the quantum-mechanical energy levels when the classical system is ergodic? Similar issues were addressed by Balian and Bloch (1972 and 1974), although in the somewhat reduced context of waves inside a cavity; Colin de Verdière (1973) and Chazarain (1974) established the exact mathematical propositions for the free motion of a particle on a Riemannian surface; Berry and Mount (1972) wrote an early review of the field. Dashen, Hasslacher, and Neveu (1974) were the first to apply the main ideas to high-energy physics.

The trace formula and its variants rely on the knowledge of the classical periodic orbits, in order to understand the quantal spectrum. This approach has seen a number of significant accomplishments:

among others, the energy levels have actually been calculated in this manner for the Anisotropic Kepler Problem (Gutzwiller 1980 and 1982), a system with hard chaos (cf. Chapter 11); the level repulsion in the spectrum of a classically chaotic system was explained on this basis by Berry (1985); and perhaps the most important, and satisfying for both experimental and theoretical physicists, the correlations in the spectrum of hydrogen in a strong magnetic field near the ionization threshold has been reduced to a set of classical periodic orbits (cf. 18.6). Considering the scope of these ideas, the present chapter can only provide some general guidance to the perplexed who are not used to thinking in terms of periodic orbits.

17.1 The Van Vleck Formula Revisited

Van Vleck's formula (12.20), and its obvious generalization (12.25), gives the classical approximation $K_c(q'' q' t)$ to the quantum-mechanical propagator $K(q'' q' t)$. No effort was made in Section 12.5 to prove this fact; it was only shown that the convolution formula (12.22) holds, provided the integral is evaluated by the stationary phase method, and the phase loss of $\pi/2$ at a conjugate point is taken into account.

The discussion in Section 13.6 was designed to convince the reader that the path integral (13.14) for $K(q'', q', t)$ can be worked out in the neighborhood of a classical trajectory, if we take only the second variation into account. The corresponding integral over these quadratic fluctuations yields exactly the Van Vleck expression.

Another, more straightfroward argument comes from inserting (12.20) directly into (13.4). The right-hand side does not vanish any longer; but it can be written in a fairly compact form, after some non-trivial rearrangements of the terms. The calculation can just as well be done with the general expression (13.2) for the kinetic energy operator in a Riemannian space. The resulting right-hand side of (13.4) becomes

$$\frac{\hbar^2}{2m} \frac{1}{\sqrt{C}} (\Delta'' \sqrt{C}\,) K_c(q'' q' t) , \qquad (17.1)$$

where the Laplacian acts only on the amplitude factor $\sqrt{C}$ in the Van Vleck propagator $K_c(q'' q' t)$.

The factor of K_c in (17.1) depends not only on q'', but also on q' and t ; although it contains second derivatives with respect to q'', its variation is independent of $\hbar$, except for the factor $\hbar^2/2m$ in front. Thus the whole right-hand side (17.1) can be interpreted as an extra potential energy, comparable to $V(q'')$ on the left, but whose value is proportional to $\hbar^2$. Therefore, as Planck's quantum goes to 0, the classical propagator K_c tends toward a solution of Schrödinger's equation (13.4).

Green's function $G(q'' q' E)$ is obtained from the propagator $K(q'' q' t)$ through the Fourier-Laplace transform (12.27) with respect to the time. It transforms all other quantities from the time domain into the energy domain, e.g., K_c goes into G_c, provided the stationary phase method is used consistently. The right-hand side of the inhomogeneous Schrödinger equation (13.11) acquires a term exactly as (17.1), with D and G_c replacing C and K_c,

The ideal use of the sums over the classical trajectories, (12.25) for K_c and (12.28)for G_c, would be to transform either one into a form resembling as closely as possible (13.9) for K and (13.10) for G. The author was able to carry out this program in the exceptional case of the bound states for the hydrogen atom, in momentum rather than position coordinates (cf. Section 12.8). Although the expression (12.31) is not the correct Green's function in momentum space, it has only poles for the real part of $E < 0$; they are at the right place, and have the correct residues.

The same kind of calculation was also carried out for the bound states in a more difficult, but still separable problem. The pure Coulomb potential was replaced by a spherically symmetric, screened electrostatic field (Gutzwiller 1969). The trajectory from $q' = (r', \theta')$ to $q'' = (r'', \theta'')$ lies on an invariant torus in phase space that is characterized by the total angular momentum L and the energy E. When projected into position space, the invariant torus has to cover the starting point q' and the end point q''; at fixed E, the angular momentum L can vary only over a limited range.

The detailed construction of the possible trajectories is typical for integrable systems with more degrees of freedom. Each trajectory between the fixed endpoints is characterized by two integers, ν to count the complete circuits between extremal points in the radial direction, and λ for each completion of 2π in the angular motion. The angle around the origin increases by $2\pi + \gamma$ when the radial distance r goes through one cycle between the extremal values, $r_1 < r_2$, as in the discussion of Section 6.1. For the screened Coulomb potential, $\gamma > 0$, yielding a precession of the trajectories (by contrast, Figure 10 shows the situation for $\gamma < 0$, called regression of the orbit). The angle γ is not a simple function of E and L, however, since it vanishes for sufficiently large values of L when the trajectory stays completely on the outside or on the inside of the screening cloud.

The angle α covered before reaching the extremum r_2 for the first time, and the angle β after the last passage through r_2 must be computed; the details are worked out in the paper mentioned above. Given the energy E, the angular momentum L is fixed by the ratio λ/ν, with some minor adjustments to accommodate the start and the finish of the trajectory. Both the action integral S and the amplitude $\sqrt{D}$ must be calculated for each trajectory.

As in Section 16.5, but moving in the opposite direction, the two summations, over ν and over λ, are now transformed with the help of Poisson's formula (16.13); the new integer variables are called n and ℓ. A pole is associated with each admissible pair (n, ℓ), and its residue is a product of two wave functions; the angular part of each is the classical approximation to the Legendre function of order ℓ, while the radial part is the usual WKB approximation for the n-th radial eigenfunction in the assumed potential and the given angular momentum $(\ell + 1/2)\hbar$. The half-integer for the angular momentum can be directly traced back to the count of conjugate points; it would be an integer, if the problem had been treated in two dimensions.

The classical Green's function is, therefore, shown to yield the usual approximation to the spectrum, as well as the wave functions, for integrable systems. Nothing new has been achieved beyond the results of Section 14.1 and 14.2; but we can now move forward into the unknown territory of chaotic dynamical systems, with the full assurance that our method works well in all the cases that have been treated so far without the benefit of Van Vleck's formula for K_c and its offshoot G_c.

17.2 The Classical Green's Function in Action–Angle Variables

The classical Green's function G_c was explained in the preceding section, and hints were given toward its explicit calculation, for a particle in a spherically symmetric potential. The motion was separated into a radial and an angular component, but action-angle variables were not used. As mentioned already in Section 6.1, they are artificial, because they vary linearly with time, and perhaps more significantly, they are not directly accessible to measurement. The four angular variables in the three-body problem, e.g., Moon-Earth-Sun, can be associated with the four angles that define the relative positions in the system as described in Section 4.4; but making the connection between the real and the artificial angles is the main job of celestial mechanics, and it is very difficult.

Nevertheless, the action-angle formalism is helpful when discussing the general properties of integrable systems. Therefore, we will assume in this section that the canonical transformation from the physical coordinates in phase space to the action-angle variables has been completed. The Hamiltonian is given by the function $H(I_1, ..., I_f)$ as explained in Section 3.3. The starting point of a trajectory is given by the angles $w'_1, ..., w'_f$, and the endpoint by the angles $w''_1, ..., w''_f$. In order to compute the time elapsed t, the energy E has to be specified; then, a relatively tricky procedure is necessary to determine the actions $I_1, ..., I_f$.

Since $w''_j - w'_j = \omega_j t$, the frequency ratios are known. Equation (3.3) tells us that $\omega_j = \partial H/\partial I_j$; the direction of the normal to the surface of constant energy E in action space is defined thereby. If we assume, as always in these dicussions, that the determinant of the second derivatives $\partial^2 H/\partial I_j\,\partial I_k$ does not vanish, the direction of the normal will uniquely specify the point on the energy surface. The action integral $S = I_1(w''_1 - w'_1) + \ldots\ + I_f(w''_f - w'_f)$ can then be calculated to provide the exponent in the classical Green's function (12.28).

The amplitude requires the second derivatives of the action S with respect to w'', w', and E. If we set $w_j = w''_j - w'_j$, the second derivatives with respect to w are just as good. For the first derivatives, the same argument that leads from (6.5) to (6.6), now yields $\delta(I_1 w_1 + \cdots + I_f w_f) = I_1 \delta w_1 + \cdots + I_f\,\delta w_f$. The next derivative is more complicated, because a change δw_j entails both a change $\delta\omega_k$ in all the frequencies and a change δt in the time elapsed, as the energy stays fixed at E. If the derivative is taken with respect to E for constant w's, however, the products $\omega_j t$ have to remain fixed, while E varies.

Although these operations are not difficult, they require close attention; the author has found the function $F(\omega_1, ..., \omega_f)$ useful, which was defined in connection with formula (14.15) and is the dual of the Hamiltonian $H(I_1, ..., I_f)$. The determinant D of (2.7) becomes

$$D(w''\,w'\,E) = \frac{(-1)^{f+1}\det(F_{jk})}{t^{f-1}\sum \omega_m F_{mn} \omega_n}, \tag{17.2}$$

where the indices on F indicate derivatives with respect to the frequencies ω. The matrix of the second derivatives of F is the inverse of the matrix of the second derivatives of H. If we use local coordinates in the action space, as in Section 16.5, so that the direction of the normal to the energy surface becomes the I_1-axis, (17.2) reduces to $(-1)^{f+1}/t^{f-1}\omega_1^2 \det'(H_{jk})$, where the prime on the determinant means leaving out the first row and column. The classical Green's function $G_c(w''\, w'\, E)$ in action-angle variables becomes

$$\frac{2\pi}{(2\pi i\hbar)^{(f+1)/2}} \sum_{\text{class.traj.}} \frac{t^{(1-f)/2}}{\omega_1 \sqrt{\det'(H_{jk})}} \exp[i \frac{\mathbf{I}(\mathbf{w}'' - \mathbf{w}')}{\hbar} - i\phi], \quad (17.3)$$

where the phase ϕ counts all the phase losses of $\pi/2$ along the trajectory. As a rule, each increase of one of the angles w by π implies a well-defined number of conjugate points.

The classical trajectories in (17.3) can be visualized in an f-dimensional space of coordinates $w_1, \ldots, w_f$ similar to Figure 19. A rectangular grid of of cells is defined by the points whose coordinates are the multiples of 2π; the starting point w' is put into the first cell of the 'positive quadrant'. The endpoint w'' has a representative in each cell of the grid; it is obtained by adding a multiple of 2π to one of the components of w''. The straight line connecting w' with one of the w'''s represents a trajectory; its slope gives the direction of the frequency vector $(\omega_1, \ldots, \omega_f)$, or equivalently, the direction of the normal to the energy surface $H(I) = E$ in the action space. Each cell is numbered by f integers ranging from $-\infty$ to $+\infty$. Thus, the trajectories in (17.3) are characterized by an f-tupel of integers, exactly as the trajectories in the spherically symmetric potential of the preceding section were numbered by the two integers ν and λ.

The classical Green's function (17.3) is related to the quantum-mechanical expression (13.10), but the correspondence between the two expressions is not obvious. As we will show shortly, even the motion of a particle in a square with periodic boundary conditions leads to serious discrepancies between the classical and the quantum-mechanical Green's functions.

17.3 The Trace Formula for Integrable Systems

Since the full-fledged summation over all classical trajectories from the starting point q' to the endpoint q'' is so difficult to evaluate, we will settle for considerably less. The trace in (13.10) is taken, namely

$$g(E) = \int dq^f\, G(q\, q\, E) = \sum_{j=0}^{\infty} \frac{1}{E - E_j}, \quad (17.4)$$

where the energy E is assumed to be a complex number in the neighborhood of the real axis.

The information on the right-hand side can be obtained equally well, if we use the decomposition

$$\frac{1}{E + i\varepsilon - E_j} = P\left(\frac{1}{E - E_j}\right) - i\pi \, \mathrm{sign}(\varepsilon) \, \delta(E - E_j) \quad , (17.5)$$

into a principal-part integral P and a δ-function. Thus, we can write the discontinuity in the trace as the density of states $\rho(E)$,

$$\frac{i}{2\pi} \int dq^f \lim_{\varepsilon \to 0} [G(q\, q\, E + i\varepsilon) - G(q\, q\, E - i\varepsilon)] = \sum_{j=0}^{\infty} \delta(E - E_j) \quad (17.6)$$

where $\varepsilon > 0$. This form of the trace is sometimes more convenient because the integral over q in (17.4) may diverge for small q when G is replaced by G_c.

In taking the trace of (17.3), the endpoint w'' is identified with the starting point w', modulo 2π in every component. Therefore, we have $w''_j - w'_j = 2\pi M_j$, where M_j can be any integer. The integral is taken over the starting point, which is then allowed to sample the whole invariant torus by roaming through the first cell of the grid in w-space. The integrand does not depend on w' as long as w'' stays always the same multiples of 2π away from w'. The corresponding trajectories are periodic orbits that wind around the invariant torus M_j times in the w_j coordinate.

The integral over w' contributes a factor $(2\pi)^f$ to (17.3); the time t in (17.3) is the period of the periodic orbit, and its frequency is, therefore, $\omega_0 = 2\pi/t$. With these minor manipulations, and remembering (17.6), the trace of (17.3) becomes identical with (16.16), at least as far as the terms with $\mathbf{M} \neq 0$ are concerned.

Since the original formula (12.28) is applicable for any pair of points q'' and q', there is naturally a term where $q'' \to q'$ when the trace is calculated. For that exceptional term, it is necessary to expand G_c in powers of $|q'' - q'|$ with the help of (2.5). The Green's function is singular, however, and special tricks are necessary in order to obtain a finite expression; e.g., the difference between the Green's function and its complex conjugate is used, exactly as in (17.6). This exceptional term in (12.28) leads to Weyl's formula for the asymptotic distribution of the eigenvalues; indeed, it is involved in the argument of Mark Kac, which we sketched in Section 16.2. All further discussion will deal only with the terms with $\mathbf{M} \neq 0$ which are responsible for the fluctuations in the spectrum.

The formula (16.16) has, therefore, been derived from the trace of (17.3), which in turn resulted from the classical Green's function (12.28). Does the expression (16.16) indeed have δ-function like singularities with respect to the energy? We could simply backtrack from (16.16) through (16.14) to (16.12); a stationary-phase integral has to be 'inverted', i.e., each term in (16.16) is recognized as the result of a stationary-phase evaluation of the corresponding term in (16.14).

More directly, however, we can ask why there would be any singularities in the sum over periodic orbits (16.16) as a function of the energy E. Let us consider *contiguous terms,* i.e., two terms where only one of the integers M_j is different, and differs only by 1. Constructive interference takes place when the difference between the exponents $2\pi i(\mathbf{M},\ \mathbf{I}_0)/\hbar$ is a multiple of 2πi. The change in the action S is given by (6.6) in the limit of large integers M; constructive interference between contiguous terms requires, therefore, that each action I_k be a multiple of $\hbar$, exactly as in the Bohr-Sommerfeld quantization (14.1). Had we included the count of conjugate points and the consequent phase-loss in the derivation of (16.16) from the trace of G_c, i.e., had we started from (17.3), this simple criterion would have yielded the quantization condition (14.4), including the Maslov indices.

The author (Gutzwiller 1970) has investigated more closely how the classical trace, i.e., the trace calculated from the classical Green's function, depends on E in a number of simple integrable systems. As pointed out in Chapter 12, Van Vleck's formula (12.20) gives the exact quantum-mechanical propagator for the free particle in three dimensions (12.21), and the stationary-phase method in the Fourier integral (12.27) gives the same result (12.29) as the exact integration over time. Therefore, the classical trace coincides with the exact trace for the three-dimensional box with periodic boundary conditions. The Cartesian coordinates q are the angle variables in this case, and the formula (17.3) represents no more than the *method of images* in the solution of Schrödinger's equation. The boundary conditions are enforced by matching each source with its image upon reflection on one of the edges.

The two-dimensional box, however, does not work out so well, because the expression for G_c does no longer coincide with the solution of the inhomogeneous Schrödinger equation (13.11) for the free particle. G_c is only the asymptotic form for large distances $|q'' - q'|$, whereas the exact solution is a Bessel function. The method of images is still applicable, however, and is expressed in (17.3). The classical density of states, as computed from (17.6) with (16.16), shows not only δ-functions in the energy at the correct places, but also a continuous part whose average is 0. Thus, even in the most favorable cases, the classical trace may contain some very unphysical parts.

Rather than to start from the classical trace as the author did, Keating and Berry (1987) started from the expression (16.12), and inserted the expression for the energy levels in a rectangle, the same as used by Casati et al. (1985; cf. Section 16.6). The number of states below E, $N(E) = \int^E \rho(E)\, dE$, can be transformed by Poisson's formula (16.13). The resulting exact expression, obtained without the help of the stationary-phase method, can still be interpreted as a sum over periodic orbits. Each term has a spurious singularity, however, and one could be fooled if the infinite sum is truncated too early; but upon carrying the summation far enough, these false singularities disappear, and the steps in the function $N(E)$ appear rather miraculously at the right places.

In both of these two elementary examples the particle moves in a domain that can be made to *tile or pave* the infinite Cartesian space, i.e., copies of the domain are obtained by translation or other symmetry operations, and they cover completely the infinite space without overlap; Keller and Rubinow (1960) have given other, non-trivial examples. This common feature allows Schrödinger's equation to be solved by the method of images: Green's function for the domain is obtained from Green's function for the whole space by taking the sum over all the copies of either the starting point or the endpoint. That is exactly the content of (17.3). Difficulties arise, however, if Green's function for the open space is approximated, as in the two-dimensional Cartesian space where the Bessel function was replaced by its asymptotic expansion.

There are remarkable cases where the classical approximation to the trace still gives the correct quantum-mechanical result, although the classical Green's function G_c differs from the quantum-mechanical G. Selberg's trace formula in two dimensions, for the spectrum of a particle on a surface of constant negative curvature, belongs to this category; its derivation, however, is no more than an application of the method of images (cf. Section 19.5). To compound the mystery further, G_c coincides with G in the three-dimensional hyperbolic space, just as G_c in three-dimensional Cartesian space is given by (12.29) which happens to be the solution of the inhomogeneous Schrödinger equation (13.11) for $V(q) = 0$.

17.4 The Trace Formula in Chaotic Dynamical Systems

Van Vleck's formula (12.20) and its offsprings, in particular the classical Green's function G_c in (12.28), are assumed to be valid approximations for chaotic Hamiltonian systems. The difficulty now lies in finding all the classical trajectories from the starting point q' to the endpoint q'', either in the given time t or at the given energy E, for a chaotic dynamical system. Apparently, nobody has made sufficient progress in this problem to publish any results. The construction of quasiclassical wave functions, starting either from the classical propagator K_c or from Green's function G_c, seems out of the question for the time being.

The next best thing is to take the trace, say of G_c; this process is now quite different from the calculation in the preceding section. Each term in the sum over classical trajectories is treated separately and remains separate in all the further developments. There is no simple way to compare different trajectories, as in the preceding section where the notion of contiguous orbits was used. The presence of sharp spikes, like δ-functions in the density of states $\rho(E)$, will remain a problem without a general solution. The final result, called the trace formula, can only be tested in special examples like the Anisotropic Kepler Problem. These preliminary remarks are meant to put the rest of this chapter into perspective, and to help the reader spot areas where progress can be made.

The general argument was given by the author (Gutzwiller 1971), and we will follow his development fairly closely; Littlejohn (1990) gives a more general approach using Lagrangian manifolds. Most of the formal manipulations have already been carried out in the earlier chapters, so that we are left with relatively little tedious work at this point. Let us call $g_c(E)$ the integral (17.4), where G has been replaced by G_c as defined in (12.28). The starting point q' and the endpoint q'' coincide in the integrand, $q'' = q' = q$, i.e., if we pick any particular term in the sum over the classical trajectories, that trajectory has to close. There is no need, however, for the trajectory to close smoothly; the starting momentum p' may be different from the final momentum p''.

The integration in (17.4) requires that the common starting and endpoint q vary continuously over all of the position space that is accessible at the stipulated energy E. The particular trajectory gets deformed thereby in a continuous manner, but it maintains its basic shape. In the spirit of the stationary-phase method, the variation of the exponent $i\, S(q''\, q'\, E)/\hbar$, as the point $q'' = q' = q$ varies, is now of crucial importance.

The formula (2.4) gives the derivatives of the action S with respect to q'' and q'; therefore,

$$\frac{\partial S(q\,q\,E)}{\partial q} = \left(\frac{\partial S(q''q'E)}{\partial q''} + \frac{\partial S(q''q'E)}{\partial q'} \right)_{q''=q'=q} = p'' - p' . \quad (17.7)$$

If the trajectory does not close smoothly, i.e., $p'' \neq p'$, the integration over q tends to reduce the term by the destructive interference between trajectories of the same kind, provided the action integrals S vary as much as $2\pi\hbar$. If the trajectory closes smoothly, however, i.e., $p'' = p'$, the contributions from different starting = end points q add up in phase. Thus, *the calculation of the trace $g_c(E)$ is reduced to a summation over all periodic orbits.*

Each term in the sum (12.28) or (12.25) is now characterized by a particular periodic orbit. Contrary to the situation in an integrable system, the only parameter that can be varied continuously while holding on to this special orbit, is either the energy E or the time t, to be called the period from now on. *The periodic orbit is isolated.*

The integration over q will be carried out for a system with three degrees of freedom, as in the discussion of periodic orbits of Sections 6.3 and 6.4; all the basic features are already present in this special case. We use the results of Chapter 7 on the surface of section, to construct a special system of canonical coordinates. The coordinate q_1 runs along the particular periodic orbit, while q_2 and q_3 belong to the surfaces of section, which are always transverse to the periodic orbit. For convenience, the origin for the coordinates $(q_2\,,q_3)$ in each surface of section is the intersection with the periodic orbit.

As long as $(q_2\,,q_3) = 0$, the action integral $S(q\,q\,E)$ keeps the same value $S(E)$, independent of q_1. When $(q_2\,,q_3) \neq 0$, however, the closed trajectory is no longer a periodic orbit. Its action integral gets a correction to second order,

$$\frac{1}{2}\left(\frac{\partial^2 S}{\partial q''\partial q''} + 2\,\frac{\partial^2 S}{\partial q''\partial q'} + \frac{\partial^2 S}{\partial q'\partial q'} \right)_{q''=q'=\bar{q}} \bullet\ \delta q\,\delta q \ , \quad (17.8)$$

where $\bar{q} = (q_1, 0, 0)$ and $\delta q = (0, \delta q_2, \delta q_3)$; the summation over the indices 2 and 3 is implied.

The integration over q_2 and q_3 is done by stationary phase; only the variation in the phase is important, while the amplitude is computed on the periodic orbit. With the help of (12.23) and (12.24), the integral over q_2 and q_3 becomes

$$\frac{2\pi\hbar \exp(\pm i\frac{\pi}{4} \pm i\frac{\pi}{4})}{\left(\det' \left| \frac{\partial^2 S}{\partial q'' \partial q''} + 2\frac{\partial^2 S}{\partial q'' \partial q'} + \frac{\partial^2 S}{\partial q' \partial q'} \right|\right)^{1/2}} , \tag{17.9}$$

where the double signs refer to the eigenvalues of the quadratic form in (17.8), and the prime on the determinant reminds the reader that only the derivatives with respect to the variables 2 and 3 are involved.

The integration over q_1 now becomes relatively simple. The action integral $S(\bar{q}\bar{q}E)$ does not depend on q_1; we are left with the amplitude in (12.28), i.e., the determinant $D(q''\,q'\,E)$ of (2.7). It is simplified with the help of the same argument that led to (2.9); take the derivative of (2.8) with respect to E, keeping q'' constant; then notice that in our special coordinates $\partial H/\partial p = \dot{q} = (|\dot{q}_1|, 0, 0)$. As in (2.10), we have

$$\frac{\partial^2 S}{\partial E\, \partial q_{1'}} = -\frac{1}{|\dot{q}|} , \quad \frac{\partial^2 S}{\partial E\, \partial q_{1''}} = \frac{1}{|\dot{q}|} , \tag{17.10}$$

or (2.10), which yields

$$D(q''\,q'\,E) = \frac{1}{|\dot{q}|^2} \det' \left| \frac{\partial^2 S}{\partial q' \partial q''} \right| . \tag{17.11}$$

The expression (17.9) and the square root of D from (17.11) are now combined to yield the integrand for the integration over q_1, i.e., along the periodic orbit. If we go through Section 6.3 and 6.4, we find that the result can be expressed directly in terms of the function $F(\lambda)$, which is defined by (6.11). Actually, we need its value for $\lambda = 1$, which becomes

$$F(1) = \begin{cases} -4\,\mathrm{Sinh}^2(\chi/2) & \text{for a direct hyperbolic orbit ,} \\ 0 & \text{for a direct parabolic orbit ,} \\ 4\sin^2(\chi/2) & \text{for an elliptic orbit ,} \\ 4 & \text{for an inverse parabolic orbit ,} \\ 4\,\mathrm{Cosh}^2(\chi/2) & \text{for an inverse hyperbolic orbit ,} \end{cases} \tag{17.12}$$

for each degree of freedom transverse to the periodic orbit. We have omitted the loxodromic case because no example will be mentioned in this book. Notice that the possible values of $F(1)$ range continuously from $-\infty$ to $+\infty$; they are the *residues* in the work of Greene (1979), as mentioned in Sections 6.4 and 9.8.

Elliptic periodic orbits are very important in systems with soft chaos where we have to deal with small islands of stability; an interpretation of this phenomenon due to Miller (1975) will be discussed in Section 6. Direct parabolic periodic orbits are the hallmark of integrable systems; $F(1) = 0$ indicates that the periodic orbit is not isolated, because the square-root of $F(1)$ goes into the denominator of the integrand.

The two hyperbolic cases are characteristic of hard chaos, but they occur in soft chaos as well. Indeed, if anything, they are typical of chaotic systems, and our discussion will be focused on them. Our two prime examples, the anisotropic Kepler problem and the motion on a surface of constant negative curvature, have only direct hyperbolic orbits.

The integral along the periodic orbit, over q_1, now becomes easy; the only quantity in the integrand depending on q_1 is the velocity $|\dot{q}|$ in the denominator. Although the orbit may consist of the system going around it more than once, the integral arises from integrating the trace over all position space; every point should be covered only once. Therefore, $\int dq_1/|\dot{q}|$ is the simple (primitive) period, to be called T_0 from now on, in contrast to the full period T or the action integral S, which may be multiples of the corresponding primitive quantities.

Conjugate points are associated with a change in sign of $D(q''\,q'\,E)$; their effect is contained in the multiple signs (17.9). Since all our examples have two degrees of freedom, the discussion will be restricted to this case. We assume that an odd number of conjugate points in a hyperbolic orbit leads to the inverse hyperbolic case of (17.12); the only change in the formula below is then to replace $\mathrm{Sinh}(\chi/2)$ by $\mathrm{Cosh}(\chi/2)$. In this connection, it is comforting to know that according to Littlejohn (private communication) the number of conjugate points always increases with time, provided the kinetic energy as a function of the momentum is positive definite.

As in the transition from the classical propagator K_c to the classical Green's function G_c in Section 12.7, the count of conjugate points may change by 1 in going from G_c to g_c. A careful discussion of the multiple signs in (17.9) shows the following: the stable (or unstable) manifold rotates in phase space by a multiple of π, before closing the periodic orbit. This multiple ℓ is most easily recognized by the number of times the stable manifold is 'vertical', i.e., oriented in the local p-direction (cf. Creagh, Robbins, and Littlejohn 1990).

Putting all the pieces together then yields the *trace formula:* the classical approximation $g_c(E)$ to the quantum-mechanical trace $g(E)$ is given by the sum over the periodic orbits (po),

$$g_c(E) = \frac{1}{i\hbar} \sum_{po} \frac{T_0}{2\mathrm{Sinh}(\chi/2)} \exp\left[\, i\left(\frac{S}{\hbar} - \ell\,\frac{\pi}{2}\right)\right]. \quad (17.13)$$

Each term in the sum over periodic orbits consists of a coefficient $A = T_0/2i\hbar\,\mathrm{Sinh}(\chi/2)$ and a phase factor. The primitive period T_0, the instability exponent χ, and the action integral $S = \int p\,dq$ taken over the periodic orbit, are continuous functions of the energy E. As in the integrable case (16.16), the superposition of these smooth classical functions of E is expected to yield at least an approximate spectrum for the quantum-mechanical energy levels.

The mode of reasoning in this section has been extended by several authors to quantum-mechanical objects of a more complicated type, which can still be reduced to a trace with the propagator or Green's function. Indeed, one could think of $\int d^f q'' d^f q' \, F(q' q'') \, G(q'' q' E)$, where F is the kernel of some Hermitian operator. When Green's function is replaced by its classical form (12.28), and the integration is carried out by stationary phase, closed classical trajectories are bound to arise quite naturally, although the exact kind of closure will depend on the operator F and its classical analog. Wilkinson (1987) has discussed some of the details; the important application of this idea will come up in the next chapter, in connection with the transition matrix elements in optical emission or absorption.

17.5 The Mathematical Foundations of the Trace Formula

The trace formula for classically chaotic Hamiltonian systems seems to be the only general tool available at this time to establish a quantitative connection between the classical and the quantal regime. Before trying to use this tool, however, we have to gain trust in its reliability. The classical trace $g_c(E)$, as written down in formula (17.13), has to be understood more clearly as the limit for small Planck's quantum of the quantum-mechanical trace (17.4).

This section will offer, therefore, various comments on research concerning the trace formula that appeared shortly after the author's publication of the results in the preceding section. Most of the work to be reported was done quite independently, and reflects slightly different interests and intentions, although the results are all closely related to one another.

Section 16.2 toward the end mentioned the results of Balian and Bloch to derive higher-order corrections to Weyl's formula. The third installment of their work (Balian and Bloch 1972) uses the same approach as the two earlier ones; but now, formulas for the fluctuations in the spectrum are obtained which depend on finding the classical periodic orbits, just like (16.16) and (17.13). The argument is quite different, however, and well worth considering.

The propagation of waves inside a cavity is reduced to multiple reflections: a spherical wave is 'started' in the interior at q' at a given energy E with the wave function (12.29), which is the correct Green's function for a free particle in three dimensions. When this original wave reaches the boundary, a reflected wave gets superimposed whose form is again (12.29); but the starting point q' is at the boundary, and the amplitude depends on the value of the original wave as it reaches the boundary, as well as on the stipulated boundary conditions. The reflected wave generates a second-generation reflected wave in exactly the same way the original wave caused the first-generation reflection, and so on.

This manner of speaking, as if the consecutive reflections were following one another in time, is only valid by analogy. The fixed energy E is, of course, equivalent to a fixed frequency, and all the reflected waves form a permanent superposition for all times. If E is made complex by adding a positive imaginary part, i.e., $E \to E + i\varepsilon$, the amplitude (12.29) decays exponentially with the distance $|q'' - q'|$, and the consecutive reflections get weaker. The expansion in multiple reflections converges and can be broken off at some convenient order depending on the value of $\hbar/\sqrt{2m\varepsilon}$ compared to the size of the cavity. This argument of Balian and Bloch for $\varepsilon > 0$ is very similar to the discussion in Section 16.2, and leads to the smoothed-out spectrum, i.e., Weyl's formula and its extensions.

In order to get the fluctuations, however, ε has to be reduced below a critical value, and all the higher-order reflections have to be included when writing down Green's function for the cavity. The trace $g(E)$ in (17.4) now becomes a sum over all closed sequences of straight-line trajectories, bouncing off the walls of the cavity. Each term has an exponential factor, $\exp[i\sqrt{2mE}\,(r_{01} + r_{12} + \dots + r_{n0})/\hbar]$, where $r_{j,j+1}$ is the distance between the j-th and the $(j+1)$-th reflection points on the walls, and the index 0 refers to the arbitrary starting = end point. The integration over these $n + 1$ points is now done by stationary phase and leads to the periodic orbits.

The geometrical details are worked out much more explicitly than the rather formal discussions of the preceding section. The parallelepiped and the sphere are done in closed form, to nobody's surprise, since they are integrable; but no chaotic systems are treated in detail. In a subsequent paper, Balian and Bloch (1974) discuss Schrödinger's equation using the same idea of multiple reflections by starting from the integral equation for Green's function in terms of Green's function for the free particle.

The program of Balian and Bloch was carried out by Berry (1981) for Sinai's billiard (cf. Section 10.3). He first calculated the quantum-mechanical spectrum using the Kohn-Korringa-Rostoker method from solid-state physics. The bounded region is translated so as to cover the whole plane, and Schrödinger's equation is then solved as a multiple-scattering problem in the resulting infinite grid. Berry was able to rearrange the final expression for the density of states so as to make it look exactly like the multiple reflections from the boundary of original domain.

Selberg's trace formula (cf. Chapter 19) is essentially equivalent to the statement that $g_c(E)$ as given by (17.13) is equal to the trace (17.4), for the motion of a particle on a compact surface of constant negative curvature. The proof relies on the high degree of symmetry of such surfaces, as does Berry's use of multiple scattering in the Sinai billiard (cf. Section 16.7). Colin de Verdière (1973) is probably the first mathematician to prove a theorem similar to (17.13) in a case where the symmetry arguments cannot be used. He still needs a compact surface of (non-constant) negative curvature, however, and he derives his theorem for the heat equation, i.e., he replaces the time t in Schrödinger's equation by a negative imaginary quantity $-i\tau$ as in the discussion of Section 16.2.

In a second paper, Colin de Verdière (1973) proves a proposition like (17.13) for the motion on a compact Riemannian surface without any strong assumption about its curvature; the price for this new result, however, is a rather subtle condition concerning the asymptotic relation between $g(E)$ and $g_c(E)$. The proof first constructs the equivalent of the propagator $K(q'' q' t)$, called the *parametrix,* and then studies the space of all curves connecting the starting point and the endpoint; the stationary-phase method is used eventually. A physicist can only express his awe in front of such a difficult, technical development.

In a similar vein, Chazarain (1974 and 1980) investigates the wave equation $(\partial^2/\partial t^2 - \Delta)u(q, t) = 0$ on a Riemannian surface, and the corresponding propagator, say $K(q'' q' t)$. The construction of K uses Hörmander's theory of *Fourier integral operators,* based on the notion of *bicharacteristics,* which are none other than the classical trajectories in phase space. Taking the trace of K, and using again the stationary-phase method allows the detailed study of the singularities of the trace on the time axis.

The trace is singular with respect to t on a set of isolated points, $0 \le t_1 \le t_2 \le \ldots$. with only a finite number of consecutive = signs; each time t_j is the period of a periodic orbit. Between these singular values of the time, the trace is infinitely differentiable. Near them, the trace can be expanded in an exponential multiplying a power series. Although it seems difficult to extract the details of this last statement from the formal mathematical context, these results seem to be about the best a physicist can ever hope for. Unfortunately, the author is not aware of any special, but non-trivial examples worked out along these lines.

Another mathematically rigorous approach is due to Albeverio, Blanchard, and Hoegh-Krohn (1982) who deal directly with Schrödinger's equation in the usual form, although some mild restrictions are imposed on the potential energy. These authors apply their own method to treat the kind of oscillatory integrals that are the characteristic feature of Feynman's path integral. Like Colin de Verdière and Chazarain, they work mostly with the so-called Theta-function $\Theta(t)$, the trace of the propagator $K(q'' q' t)$; cf. also Voros (1987).

17.6 Extensions and Applications

The trace formula is easy to write down in the two extreme cases of classical mechanics: the formula (16.16) of Berry and Tabor is valid for integrable systems without degeneracies, and the formula (17.13) of the author is good for systems with hard chaos. In both cases, the main idea is to start from Green's function or the propagator, and then reduce their trace to a sum over all periodic orbits to get an approximate spectrum. When dealing with soft chaos, however, or various special situations in the two extreme cases, the formulas (16.16) and (17.13) have to be modified, while holding on to the basic idea.

The derivation of (17.13) from (17.12) suggests that a similar formula holds for isolated elliptic orbits, provided $\mathrm{Sinh}(\chi/2)$ in the denominator is replaced by $\sin(\chi/2)$. The author (Gutzwiller 1971) already noticed the trouble with this modification when χ is a multiple of 2π. He proposed to remedy the difficulty by interpreting the angle χ as part of the count of conjugate points; the denominator is eliminated, and the exponent in (17.13) gets a term $-i\chi/2$ which includes the contribution $-i\ell\pi/2$ from the conjugate points. The sum in (17.13) is then restricted to a primitive periodic orbit and all its multiples; the resulting geometric series has well-defined poles, where the action integral over the primitive periodic orbit $S_0(E) = (2n\pi + \chi/2)\hbar$.

Voros (1975) and Miller (1975) had a better proposal. They expanded

$$\frac{T_0}{2i\sin(\chi/2)} = T_0\,(e^{-i\chi/2} + e^{-3i\chi/2} + e^{-5i\chi/2} + \cdots). \qquad (17.14)$$

The first term is the same as in the author's argument; the additional terms generate poles whenever $S(E) = [2n\pi + (m + 1/2)\chi + \ell\pi/2]\hbar$

for $m \geq 0$. Very ingeniously, $(m + 1/2)\hbar$ is interpreted as due to the quantization of a harmonic oscillator whose elongation is transverse to the periodic orbit. Without going into the details of his reasoning, it may be enough to note that the surface of section around a stable periodic orbit does indeed look like that of a harmonic oscillator. The main energy level E_n gets broadened into a vibrational band $E_n + (m + 1/2)\hbar\omega$ with $\hbar\omega = \chi/T_0$.

The same reasoning can be applied to systems with more than two degrees of freedom, and results in more vibrational frequencies transverse to the periodic orbit. The neighborhood in phase space is assumed, in effect, to be isolated from the rest, as if it were a small integrable domain. This view led Miller (1972) to quantize some simple many-body systems: e.g., in helium-like atoms, two electrons on opposite sides of the nucleus in a 'symmetric stretch' define a periodic orbit; similarly, three identical particles with the same pairwise potentials, oscillating around their equilibrium positions in an equilateral triangle. The quantization of these isolated periodic orbits yields good energies when compared with exact quantum-mechanical calculations.

The issue of isolated stable periodic orbits was taken up again by Richens (1982) who pointed out that Miller's approach was equivalent to replacing the surface of constant energy in action space by its tangent plane. The formula of Berry and Tabor has to be refined in this situation because the curvature in the denominator of (16.16) would be 0, and yet the physical interpretation is not difficult.

The neighborhood of a periodic orbit was analyzed in Section 6.3 by expanding the action integral to second order in the displacements. The classification in Section 6.4 and in (17.13) ignores the complications which arise when the relevant quadratic forms have vanishing eigenvalues. Ozorio de Almeida and Hannay (1987) push the required expansions to higher order, and use the Birkhoff normalization (Sections 8.4 and 14.3) to find the trajectories. Different types of resonances must be distinguished, as the difference ε between some eigenvalues in the quadratic form (17.8) becomes degenerate. Since the trace formula is valid in the limit $\hbar \to 0$, the necessary modifications depend on the way ε and $\hbar$ go to zero relative to each other.

The same technique is used by Ozorio de Almeida (1986) to study the modifications of the trace formula (16.16) for integrable systems, when small perturbations of order ε destroy the structure of invariant tori in phase space. If $\varepsilon/\hbar$ remains small, the trace formula remains correct as if the system was still integrable; the density of states is insensitive to the intricacies of the classical orbit structure. If $\varepsilon/\hbar$ is large, however, the periodic orbits have to be treated as isolated, and (17.13) applies.

Although conservative systems with one degree of freedom are of no interest for the study of chaotic behavior, another idea of Miller (1979) is worth mentioning, because it can be generalized to two and more degrees of freedom. The classical trace $g_c(E)$ is easy to get in terms of the periodic orbits for a *double-well potential,* if we are willing to accept imaginary action integrals, $i\int dq\,[2m(V(q) - E)]^{1/2}$, whenever $V(q) > E$. The corresponding exponential is real, whereas so far we had a purely imaginary dependence. The effect of various potential shapes on the splitting of nearly degenerate energy levels comes out very clearly.

Another twist was given to the trace formula by Tabor (1983) when he applied it to an *area-preserving map* in the spirit of Section 8.6. Percival (1979) had shown how to define the required action integrals so that all of the machinery in the preceding section can be used to derive the formulas (17.12) and (17.13). Greene (1979) had developed a scheme for locating periodic orbits in the standard map (9.14), as explained in Section 9.8, with the help of the residues (17.12). The stable periodic orbits lead rather quickly to well-defined peaks in the density of states; it looks, however, as if many more of the unstable ones have to be found before their contribution leads to any recognizable structure. Similarly, Wintgen (1988) has used relatively few periodic orbits to get the level density for the hydrogen atom in a magnetic field.

The *scars* of periodic orbits in the quantum eigenstates (cf. Section 15.6) have been analyzed recently with the help of the trace-formula idea. Bogomolnyi (1988) goes as far as (17.8) in the derivation of the trace formula, i.e., he identifies the two endpoints $q' = q'' = q$ in Green's function, and restricts himself to the neighborhood of a periodic orbit; but he does not carry out the integration over the transverse coordinates q_2 and q_3. Therefore, he gets the oscillatory part in the square of the wave function because the quadratic terms (17.8) now remain in the exponent. All kinds of valuable information about the scars in terms of the monodromy matrix and the phase along the periodic orbit is obtained in this manner.

Berry (1989) uses a somewhat different strategy: the classical Green's function (12.28) is used to calculate *Wigner's distribution* $\Psi(p, q)$ as the integral (15.13) where the density matrix $\rho(q'', q')$ is replaced by $G_c(q'', q')$. The double integration again is carried out with the help of stationary phases in the neighborhood of a periodic orbit, although all expansions are now made in phase space rather than position space. The calculations get quite involved because the expansions must be carried to third order in the displacements δq and δp. The trace formula (17.13) and Bogomolnyi's density in position space are then obtained quite naturally by performing the required averaging in phase space. The nature of scars can now be understood against the full background of classical mechanics.

17.7 Sum Rules and Correlations

The trace formula (16.16) for an integrable system was shown in Section 16.6 to yield the correlations among the energy levels. The crucial step was to form the sum of the absolute squares of the coefficients $A(E,\mathbf{M})$ in (16.16), and then reduce this sum to the density of states $\rho_0(E)$ within a factor 2π.

An analogous sum rule, for the coefficients in the trace formula (17.13) of a dynamical system with only isolated periodic orbits, was found by Hannay and Ozorio de Almeida (1984) (cf. also Hannay 1985); Berry (1985) then derived the spectral rigidities from the trace formula. These developments will be discussed in this section, without attempting to present all the formal arguments. Finally, a number of related papers will be mentioned that took off from the remarkable results of these two papers.

It would take a lot of space to do justice to all of the fine points that are hidden in the article of Hannay and Ozorio de Almeida. The main line of the argument, however, can be seen in the following simplified manner. In order to visualize the reasoning, the various δ-functions will have a finite width ε; $\mathbf{r}$ stands for the full complement of coordinates in the $2f$-dimensional phase space; $\mathbf{r}_t$ designates a point which arises from the initial point $\mathbf{r}_0$ after the time t. The time average $< \ldots >_t$ is defined as

$$< \delta(\mathbf{r} - \mathbf{r}_t) >_t \; = \lim_{T \to \infty} \frac{1}{2T} \int_{-T}^{+T} \delta(\mathbf{r} - \mathbf{r}_t)\, dt \;, \qquad (17.15)$$

where a short, but fixed interval of time near 0 is excluded from the integration for technical reasons.

As long as T is kept finite, the formula (17.15) represents a tube of width ε which surrounds the trajectory toward the past and toward the future. The integral of (17.15) over all phase space yields 1. In the two extreme cases of mechanics, the following assumptions are made for an arbitrary trajectory,

$$< \delta(\mathbf{r} - \mathbf{r}_t) >_t \; = \begin{matrix} \delta(H(\mathbf{r}) - H(\mathbf{r}_0))/\Omega(H(\mathbf{r}_0)) \text{ for ergodic ,} \\ \delta(\mathbf{I}(\mathbf{r}) - \mathbf{I}(\mathbf{r}_0))/\Omega(\mathbf{I}(\mathbf{r}_0)) \text{ for integrable} \end{matrix} \qquad (17.16)$$

systems; $\mathbf{I}$ is the vector of conserved action integrals, and Ω is the normalizing volume for the numerator. The upper formula is the *ergodic hypothesis:* the typical trajectory fills the whole energy surface. We will continue only with this part of (17.16); the lower part would lead to the earlier sum rule, cf. Section 16.5.

At this point, we proceed somewhat differently from Hannay and Ozorio. To make explicit use of (17.16), let us choose a particular surface of section Σ in the surface of constant energy. The $f-1$ pairs of canonically conjugate coordinates in Σ are collectively called $\mathbf{s}$; they are complemented by the additional conjugate pair (E, t) as explained in Section 7.5. A trajectory starting in $\mathbf{s}_0$ intersects Σ consecutively in the points $\mathbf{s}_1, \mathbf{s}_2, \ldots$ at the future times $t_1, t_2, \ldots$, and in the points $\mathbf{s}_{-1}, \mathbf{s}_{-2}, \ldots$, at the past times $t_{-1}, t_{-2}, \ldots$.

The measure on the surface of constant energy E is now expressed in terms of the function $\bar{\delta}(\mathbf{s}-\bar{\mathbf{s}}) = \delta(\mathbf{s}-\bar{\mathbf{s}})\,\tau(\bar{\mathbf{s}})$ on Σ, where $\tau(\bar{\mathbf{s}})$ is half the time for the trajectory to get from the intersection $\mathbf{s}'$ preceding $\bar{\mathbf{s}}$, and on to the first intersection $\mathbf{s}''$ with Σ after $\bar{\mathbf{s}}$. Instead of the function $\delta(\mathbf{r}-\mathbf{r}_t)$ in (17.15) and (17.16), we now use the more explicit, but somewhat disorderly looking function $\sigma(\mathbf{s}, t; \mathbf{s}_0) = \Sigma\, \delta(t-t_n)\, \bar{\delta}(\mathbf{s}-\mathbf{s}_n)$.

If $\sigma(\mathbf{s}, t; \mathbf{s}_0)$ is integrated over $\mathbf{s}$ with some fixed $\mathbf{s}_0$, one obtains a function of t with sharp peaks every time the trajectory starting in $\mathbf{s}_0$ at $t=0$ intersects Σ. The time average of this function, as defined in (17.15), yields 1 because every peak is weighted with the time interval $(t_{n+1}-t_{n-1})/2$.

The ergodic hypothesis (17.16) for ergodic systems now becomes

$$< \sigma(\mathbf{s}, t; \mathbf{s}_0) >_t \; = \; \tau(\mathbf{s})/\bar{\tau}(\Sigma) \; , \qquad (17.17)$$

where $\bar{\tau}(\Sigma)$ is the average time between intersections with Σ. In terms of the volume $\omega(\Sigma)$ of the surface of section, and $\Omega(E)$ for the surface of constant energy, one has $\bar{\tau}(\Sigma) = \Omega(E)/\omega(\Sigma)$. In this formulation, the different parts of Σ are weighted according as the times τ to next intersection. The surface of section Σ is used only to give a well-defined coordinate system on the energy surface; but the measure on Σ is chosen such as to yield the same measure as in (17.16), independent of Σ.

The statement (17.16) excludes a set of measure 0 that contains any initial point $\mathbf{r}_0$ belonging to a periodic orbit, and similarly, (17.17) excludes such a set; but those are exactly the points we need. Therefore, the ergodic hypothesis will be reformulated - strengthened would perhaps be a better term - using $\sigma(\mathbf{s}, t; \mathbf{s}_0)$ on Σ. The trick of Hannay and Ozorio is to identify $\mathbf{s}$ and $\mathbf{s}_0$, i.e., to consider $\sigma(\mathbf{s}_0, t; \mathbf{s}_0)$ as a function of $\mathbf{s}_0$. If there is a periodic orbit with the period t at the given energy E, and the point $\mathbf{s}_0$ is near it, this function differs from zero, and will be larger the closer s_0 happens to be. If s_0 is allowed to roam around in phase space, all the periodic orbits of period t will be picked up.

Since the periods of periodic orbits are a discrete set of increasing numbers, $T_1 \leq T_2 \leq \ldots$, including repetitions, we can write

$$\int d^{f-1}p\, d^{f-1}q\, \sigma(\mathbf{s}, t; \mathbf{s}) = \sum_j B_j\, \delta(t - T_j) \ , \quad (17.18)$$

with the intensities B_j. They will turn out to be closely related to the coefficients for the periodic orbit with number j in the trace formula (17.13).

The ergodic hypothesis (17.17) is now generalized: *the time average of $\sigma(s, t; s)$ as a function of s is the same as the time average (17.17), which was given in (17.17) as distributed on the surface of section with the relative weight $\tau(s)$.* Another way of phrasing this idea is to say that most of the very long periodic orbits are almost indistinguishable from the ordinary trajectories.

The hidden difficulty with this idea comes from the great proliferation of periodic orbits in chaotic systems. As explained in Section 10.5, their number increases exponentially with their length according as their topological entropy, cf. (10.10). The length was originally defined as the value of the action integral S; the mean of the ratio S/T can be obtained by integrating $(\mathbf{p}, \dot{\mathbf{q}})$ over the energy surface; it is $f\, W(E)/\Omega(E)$ where $W(E)$ is the volume of phase space below the energy E. Formula (17.18) in conjunction with the generalized ergodic hypothesis (17.17), and the definition (17.15) for the time average, now becomes

$$\sum_{|T_j| < T} B_j \underset{T \to \infty}{=} 2\,T \ . \quad (17.19)$$

In words, if we assign to each periodic orbit the appropriate fraction of the energy surface, the sum of these fractions for the periods in the interval from T to $T + dT$ adds up simply to dT. The 'appropriate fraction' for a particular periodic orbit decreases as its instability increases; indeed, we see from the definition of σ that it is then difficult to get $\mathbf{s}$ and $\mathbf{s}_0$ close to each other when t is large.

It remains to establish the connection between the intensities B_j and the coefficients A_j in (17.13) for the periodic orbit j, which correspond to the $A(E,\mathbf{M})$ in (16.16) for integrable systems. Since the δ-functions have a small width ε, the point $\mathbf{s}_0$ in the surface of section can vary by some small displacements $\delta\mathbf{s}_0$. The corresponding change in the argument of $\sigma(\mathbf{s}_0, t; \mathbf{s}_0)$ is given by $(I - \partial\mathbf{s}_t/\partial\mathbf{s}_0)\delta\mathbf{s}_0$, where I stands for the $2(f-1)$ by $2(f-1)$ identity matrix, while the matrix of partial derivatives is exactly the same as in (6.9). The integration over δs_0 yields $B_j = T_{0j} / \det(I - \partial s_t / \partial s_0) = T_{0j} / F(1)$ where $F(1)$ is defined in (6.11); it occurs again in (17.12), as well as in the trace formula (17.13), so that $B_j = \hbar^2 |A_j|^2 / T_{0j}$, in terms of the primitive period T_{0j}

Before inserting this relation into (17.19), a few comments are required. The behavior of the sum for large T is dominated by the long periodic orbits; very few of these are repetitions of shorter ones, so the distinction between primitive and ordinary period is not necessary. Upon forming the sum over $|A_j|^2 / T_{0j}$ for large periods, the denominator T_{0j} can be replaced by its limiting value T and taken outside the sum. For ergodic systems, therefore, the terms in the trace formula (17.14) grow as

$$\sum_{T_j < T} |A_j|^2 \underset{T \to \infty}{=} 2\, T^2 / \hbar^2 \; . \tag{17.20}$$

(For integrable systems, the analogous sum increases only linearly with T.) This *sum rule* is now used for an argument similar to the one given in Section 16.6.

The main energy dependence in both trace formulas, (16.16) and (17.13), resides in the exponent, $S(E)/\hbar$; the rate of change of this exponent with E is $T/\hbar$ since $T = dS/dE$. The coefficient A_j in (17.13) can be directly interpreted as the Fourier transform of $x(t)$ in (14.13), with the following replacements: $t \to E$, $\omega \to T/\hbar$, and $x(t) \to \rho(E)$. According to (14.12), the asymptotic behavior of $I(\omega)$ for large ω determines the short-time behavior of $C(t) = \langle x(0) x(t) \rangle$. In the present interpretation, the correlations for small energy differences are such that the probability $P(x)\,dx$ for finding two adjacent levels at a distance between x and $x + dx$ goes to zero as $x \to 0$. For integrable systems, the analogous sum rule implies that $P(x) \to 1$ as $x \to 0$.

These results have been greatly strengthened by the work of Berry (1985). His procedure is more direct; he inserts the trace formulas (16.16) and (17.13) directly into the definition (16.5) for the *rigidity* $\Delta_3(L/\rho_0, \alpha)$, where the range of integration has been relabeled so that L becomes the average number of states. If L is small compared with the base $\alpha\rho_0$, the energy dependence in the various terms can be approximated by expanding around α; again, this dependence is important only in the exponents of (17.13). The analytical formulas show that Δ_3 depends only on the periodic orbits of period greater than $2\rho_0\hbar/L$, and, therefore, on the sum rule (17.20). If L is safely below an upper limit $L_{\max}$, to be specified shortly, the rigidities come out as $L/15$ for integrable systems, and as (16.10) for ergodic systems.

The limit L_{max} arises from the term in the trace formula with the shortest period T_{min}, presumably the first term in (16.16) or (17.13). This period gives the fluctuation in the level density with the longest correlation, and it is a very special feature of the particular dynamical system. It makes itself felt over distances of the order of $L_{max} \simeq h\rho_0/T_{min}$ in the energy spectrum. Since the ρ_0 increases as h^{-f}, the limit L_{max} increases as Planck's constant goes to 0, or equivalently, as the wavenumber k becomes large. When L in the calculation of the rigidity becomes large with respect to L_{max}, the relevant expression in terms of the coefficients A_j in the trace formula (16.16) is a convergent sum. As a function of L the rigidity for an integrable system is then seen to *saturate,* as was first noted in the numerical computations of Casati et al. (1985); cf. Section 16.6.

17.8 Homogeneous Hamiltonians

Many of the most popular Hamiltonian systems have a special property that simplifies some of the analytical and computational work: all their trajectories can be organized into one-parameter families, where the individual family members are transformed into one another by changing the scales on all the coordinates appropriately. The Anisotropic Kepler Problem (AKP) belongs here as shown at the beginning of Section 11.2; so does the free motion of a particle inside a region with hard boundaries, or in a Riemannian space, and also problems where the potential energy is a homogeneous polynomial of the position coordinates; but the Hénon-Heiles model and the Toda lattice are not of this type.

The trace formula (17.13) now becomes very simple. The numerical input is rather small and gives us hope of understanding more intimately the relation between periodic orbits and energy levels. As in (11.5), the action integral can be written as a product

$$\frac{1}{\hbar} \int p \, dq = \sigma \, \Phi \, , \tag{17.21}$$

where Φ is a purely geometric quantity belonging to a particular periodic orbit (like 2π for the circle), and all the physics is combined in σ. For example, one finds that

$$\sigma = \begin{cases} \sqrt{2mER^2/\hbar^2} & \text{for free motion,} \\ \sqrt{m_0 e^4/2\kappa^2 \,|\, E|\hbar^2} & \text{for the AKP,} \end{cases} \tag{17.22}$$

where R is a characteristic length for the region where the particle moves, such as the radius of curvature of a Riemannian space or the diameter of a cavity. The parameters for the AKP are explained in Section 11.1.

The energy E determines the scaling factor in all these cases; the classical trajectories are computed at some arbitrary reference energy. For a fixed value of $\hbar$, however, the eigenvalues of Schrödinger's equation are not subject to scaling. In quantum mechanics, the parameter σ can be interpreted as the wave vector. It has very simple values in certain easy cases: it is an integer for both the particle in a one-dimensional box and for the electron in the hydrogen atom; but in the last example, the simplicity is bought at the price of a very high degeneracy for the energy level.

Not only the action (17.21) simplifies, but also the other ingredients in the trace formula: both the stability exponent χ and the number of conjugate points ℓ are independent of E, and the period is given by $T = \sigma/2|E|$ in both examples (17.22). If a particular periodic orbit comes from repeating its primitive progenitor n times, we will write $\chi = n\alpha$ and $\ell = n\beta$, and end up with

$$\frac{\sigma}{2i|E|}\sum_{ppo}\Phi\sum_{n=1}^{\infty}\frac{1}{1-e^{-n\alpha}}\exp\left[n\left(i\sigma\Phi-\frac{\alpha+i\beta\pi}{2}\right)\right]\tag{17.23}$$

for the classical trace $g_c(E)$ where the outer summation goes over all primitive periodic orbits (ppo).

If α has values, say, larger than 1, it is tempting to drop the term $\exp(-n\alpha)$ from the denominator. Such a simplification would require some corrections in Selberg's trace formula (cf. Chapter 19). The term $\exp(-n\alpha)$ will definitely be dropped in Section 17.10, when the trace formula is applied to the AKP. Since the singularities in $g_c(E)$ come from the long periodic orbits whose instability is invariably large, there seems no great loss in simplifying (17.23) accordingly. If we do so, the sum over n becomes a geometric series and can be carried out.

The next step in the reduction concerns the exact values of α and β. There are no conjugate points on a surface of negative curvature so that β=0. In the AKP, however, $\beta \neq 0$; but its value depends in a very simple manner on the periodic orbit, and is easy to incorporate into the computations. On a surface of constant negative curvature $\alpha = \Phi$; in the AKP, the values of α vary quite a bit, although in the mean they must increase linearly with Φ. Indeed, if we write $\alpha \simeq \mu\Phi$, the constant of proportionality μ is the *metric entropy* of Section 10.3.

As a purely mathematical model of the trace formula, therefore, we can assume $\alpha = \mu\Phi$ and $\beta = 0$, so that

$$g_c(E) = \frac{\sigma}{i|E|} \sum_{ppo} \frac{\Phi}{\exp[(\frac{\mu}{2} - i\sigma)\Phi] - 1} . \tag{17.24}$$

The only ingredients are the values of Φ for the primitive periodic orbits, and the entropy constant μ; the energy dependence is contained in σ through (17.22). We can also study g_c directly as a function of σ whose physical meaning is the wave vector conjugate to Φ and normalized to Planck's constant $\hbar$.

The last expression for the classical trace can hardly be surpassed in simplicity, and we can, therefore, hope to get some simple answers concerning its validity, or usefulness in numerical computations. Can we expect to find simple poles as a function of σ, and are they located on the real axis? Or even the preliminary question: does the sum over the periodic orbits converge in the upper half-plane of σ? There are no good answers available at this time; this is not surprising since the mathematicians have not succeeded even in the special case to be discussed in the next section.

Some of these questions can be reduced to a comparison between the metric entropy μ and the *topological entropy* τ (cf. Sections 10.3 and 10.5): the number of terms grows as $\exp(\tau\Phi)$, whereas their absolute value decreases as $\exp(-\mu\Phi/2)$. Absolute convergence requires that $\text{Im}(\sigma) > \tau - \mu/2$. Conditional convergence can be obtained, provided the complex phase $\exp(i\sigma\Phi)$ causes the right amount of destructive interference among terms with similar values of Φ. If the phase angles are distributed at random, we can indeed expect a further reduction in absolute value by the square root of the number of terms, i.e., by a factor $\exp(-\tau\Phi/2)$. If Pesin's theorem holds (cf. Section 10.5), i.e., $\tau = \mu$, we find the right-hand side of (17.24) convergent for positive $\text{Im}(\sigma)$. The fluctuations of the action integral Φ for different periodic orbits enter as a crucial ingredient into the validity of the trace formula, as was pointed out by the author (Gutzwiller 1986a); one is tempted to speak of a *third entropy,* although nothing has been done as yet to develop this idea.

17.9 The Riemann Zeta-Function

There exists a special mathematical model for the set of action integrals Φ, which is almost too good to be true: $\Phi = \log p$ where p runs through the set of prime numbers. According to the Prime Number Theorem (cf. Section 16.1), the number $N(\Phi < \xi) \simeq e^{\xi}/\xi$ so that the topological entropy equals $\tau = 1$. Therefore, we set $\mu = 1$ in (17.24);

the sum over the primitive periodic orbits becomes a sum over the prime numbers, and one can write for the classical trace $g_c(E)$,

$$\frac{\sigma}{i|E|}\sum_{\text{primes}} \frac{\log p}{p^{(\frac{1}{2} - i\sigma)} - 1} = \frac{\sigma}{i|E|} \frac{\zeta'(\frac{1}{2} - i\sigma)}{\zeta(\frac{1}{2} - i\sigma)}\,; \tag{17.25}$$

the second equality introduces the logarithmic derivative of the Riemann zeta-function $\zeta(z)$ for the complex value $z = 1/2 - i\sigma$. What is that zeta-function?

The definition is straightforward: there are two possibilities,

$$\zeta(z) = \sum_{n=1}^{\infty} \frac{1}{n^z} = \prod_{\text{primes}} \frac{1}{1 - \frac{1}{p^z}}\,. \tag{17.26}$$

The first equality gives the usual definition, while the second represents the infinite sum as an infinite product. The proof is almost trivial, although it took Euler to think of it; the second expression is referred to as Euler's product for Riemann's zeta-function, a peculiar nomenclature since Euler lived a century before Riemann; formula (17.25) results from differentiating the logarithm of Euler's product. Euler used the divergence of the sum for $\mathrm{Re}(z) < 1$ to show the existence of infinitely many primes.

The zeta-function is probably the most challenging and mysterious object of modern mathematics, in spite of its utter simplicity. The reader will find a concise introduction in Chapter XIII of Whittaker and Watson's treatise (1927 and later), and a more exhaustive presentation in a book by Titchmarsh (1951); but the best source for the mathematical novice is a monograph by Edwards (1974), which includes a lot of the historical background as well as the recent numerical results.

The main interest comes from trying to improve the Prime Number Theorem, i.e., getting better estimates for the distribution of the prime numbers. The secret to the success is assumed to lie in proving a conjecture which Riemann stated in 1859 without much fanfare, and whose proof has since then become the single most desirable achievement for a mathematician.

The zeta-function $\zeta(z)$ is well defined and analytic in the whole complex plane with the exception of a simple pole at $z = 1$. It can rather easily be shown to vanish when z is a negative, even integer, $\zeta(-2m) = 0$ for m integer and positive. Otherwise $\zeta(z) \neq 0$ outside the critical strip $0 < \mathrm{Re}(z) < +1$. The *Riemann Hypothesis* states that all additional zeros are located on the critical line $\mathrm{Re}(z) = 1/2$, the median of the critical strip.

The number of zeros $N(Y)$ inside the critical strip between the real axis and an upper limit $y = \mathrm{Im}(z) < Y$ can be estimated for large Y,

$$N(Y) \simeq \frac{Y}{2\pi}\log\frac{Y}{2\pi} - \frac{Y}{2\pi} + \frac{7}{8}\ ; \qquad (17.27)$$

these 'non-trivial' zeros are at first rather sparse (at $y_1 = 14.135, y_2 = 21.022, y_3 = 25.011, \ldots$), but eventually manage to have a density $\rho_0(Y) = (1/2\pi)\log(Y/2\pi)$ going to ∞ very slowly.

The numerical check of Riemann's hypothesis gets very difficult as y increases, but the evidence for this conjecture is overwhelming since it has now been established for $y < 10^9$. Moreover, Odlyzko (1987) has computed the first 10^5 zeros above $10^{12} + 1$ with an accuracy of 10^{-8}. The latest work of Berry (1988) is based on this sample, but his motivation for interpreting these data in the light of the general trace formula (17.13) dates further back.

Already Selberg had hoped to find a connection between his trace formula (cf. Chapter 19) and the Riemann conjecture, but without avail. A classical mechanical system with a well-behaved Hamiltonian is needed, which also has a good quantum-mechanical relative, such that, on the one hand, the lengths of the periodic orbits are given by the logarithms of the prime numbers, and on the other hand, the energy levels are the zeros of the Riemann zeta-function. If we set $\sigma = E$, the comparison between (17.24) and (17.25) shows that our mathematical model for the trace formula has a simple pole with residue 1 at each one of the non-trivial zeros $z = 1/2 + y_j$ with $E = y_j$.

The existence of a corresponding mechanical analog is not necessarily 'an impossible dream'. Most remarkably, the asymptotic distribution (17.27) is realized for the eigenvalues of the Laplacian in the non-compact Euclidean domain that is bounded by the positive Cartesian axis and the hyperbola $xy = 1$. Steiner and Trillenberg (1990) have shown, however, that the further terms in (17.27) do not agree with the asymptotic distribution in the equivalent of Weyl's formula (16.3) for this particular domain. In any case, the closer study of Riemann's zeta-function will help us to understand the trace formula.

A first analogy comes from the distribution of the zero spacings; it conforms very closely with the *Gaussian Unitary Ensemble (GUE)*, which is similar to the GOE of Section 16.4; but its elements are the Hermitian matrices, whose spectrum is invariant under (complex, linear) unitary transformations. The GUE is natural in chaotic systems without time-reversal invariance; the analog of Wigner's surmise (16.10) for the spacing of neighboring eigenvalues becomes

$$P(x) = \frac{32}{\pi^2} x^2 \exp[-4x^2/\pi] \tag{17.28}$$

for the GUE, and is again very close to the exact result. Berry (1986) argues that such statistics for a set of numbers $\{y_j\}$ are not easy to justify, in contrast to a Poisson distribution, unless the eigenvalues of a complex Hermitian matrix are involved.

A more compelling argument goes back to the analysis at the end of Section 17.7. Instead of the rigidity, Berry (1988) computes the number variance $V(L, x)$, the mean square of the difference between the actual and the average number of zeros in an interval of length L near x; the distribution is normalized by studying the values $x_j = N(y_j)$ with the help of (17.27). Against all the rules of mathematical rigor, the poorly convergent series (17.24) is inserted, and $V(L, x)$ is expressed as a sum over the periodic orbits, i.e., the prime numbers.

The limit of very large prime numbers yields the number variance to be expected from the GUE for moderate values of L. As L becomes greater then 5, however, the short periodic orbits become dominant in the formal expression for the number variance; $V(L, x)$ deviates strongly from the slowly increasing function of L for the GUE. Such behavior was already noted at the end of Section 17.7.

The resulting highly unexpected behavior is plotted in Figure 50, which shows the practically perfect fit with the data of Odlyzko. Most remarkably, the mechanical interpretation has yielded a sharp mathematical result. At the same time a model has been found for investigating the level statistics in bona fide mechanical systems where the available data are not in general as extensive as for the Riemann zeta-function.

The short periodic orbits obviously play a significant role which will come out even more clearly in the next chapter. Formula (17.24) seems to suggest that singularities in $g_c(E)$ arise from the vanishing of a denominator which is associated with a particular periodic orbit. The author (Gutzwiller 1971) promptly succumbed to this temptation when he tried to apply the newly found trace formula (17.13) to a system with isolated unstable periodic orbits. He had computed only the simplest periodic orbit in the AKP so that (17.24) had only one term; the quantization condition reduces to $\sigma\Phi = 2m\pi$ with positive integer m. Such a simple-minded interpretation of the trace formula has received support from Ozorio de Almeida (1989) who studies the homoclinic neighborhood of a periodic orbit. Nevertheless, great care is required because the primitive periodic orbits proliferate exponentially.

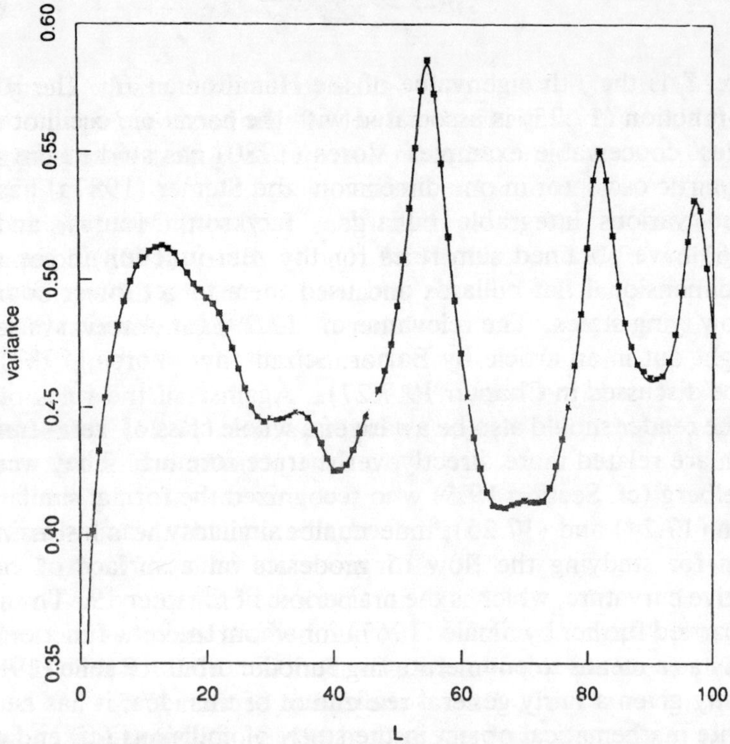

Figure 50 The number $V(L, x)$ for the zeros of the Riemann zeta-function where for $x \simeq 10^9$, as computed from Odlyzko's data (dots), and from Berry's semiclassical approximation (continuous line); the GUE result increases monotonically as a logarithm, and fits only up to $L = 3$ [from Berry (1988)].

The mathematical background for this difficulty has been exposed by Voros (1988). The denominators in (17.24) and (17.25) can be directly reduced to the individual factors in Euler's product (17.26). The Riemann zeta-function $\zeta(z)$ does not have poles at the zeros in the Euler product because the latter does not converge at the presumed location of these poles on the imaginary axis. It is the conspiracy of the periodic orbits that leads to the singularities in the trace, but as Berry (1985) and the author demonstrated, it takes relatively few primes in (17.24) to bring out the zeros of $\zeta(z)$. The sparseness of the spectrum greatly favors such a simple and mathematically dubious approach. For the AKP and the free motion on a surface of constant negative curvature (cf. Section 19.6), an enormous number of periodic orbits are required to isolate the energy levels.

Before ending this section, the reader has to be warned that the term 'zeta-function' is commonly applied to a different function of the spectrum,

$$\zeta_H(z) = \sum_j \frac{1}{E_j^z} , \tag{17.29}$$

where E_j is the j-th eigenvalue of the Hamiltonian H. The Riemann zeta-function (17.25) is associated with the harmonic oscillator as the simplest conceivable example. Voros (1980) has studied the case of the quartic oscillator in one dimension, and Steiner (1987a) has investigated various integrable billiards. Itzykson, Moussa, and Luck (1986) have obtained sum rules for the zeta-function in the case of two-dimensional flat billiards and used them to get lower bounds for the low lying states. The relevance of (17.29) for chaotic systems was brought out in an article by Balazs, Schmit, and Voros (1987) which will be discussed in Chapter 19.

The reader should also be aware of a whole class of 'zeta- functions' which are related more directly to the trace formula. They were used by Selberg (cf. Section 19.5) who recognized the formal similarity between (17.24) and (17.25); indeed, this similarity was his main inspiration for studying the flow of geodesics on a surface of constant negative curvature, which is the main topic of Chapter 19. The analogy was carried further by Smale (1967) for whom the zeta-function served mainly as a means to enumerate the periodic orbits. Ruelle (1986) has recently given a fairly general realization of this idea; it has become a favorite mathematical object in the study of collisions (cf. end of Section 20.1, in particular Gaspard and Rice 1989). One of its basic properties is inherent in the formulas (17.37) and (17.38): the classical approximation for the energy levels is given by the zeros of a zeta-function that can also be defined as an infinite product over all periodic orbits.

17.10 Discrete Symmetries and the Anisotropic Kepler Problem

Even a chaotic dynamical system may have discrete symmetries; e.g., the Hamiltonian (11.6) of the AKP remains the same if we change the signs in both components of the conjugate pairs (u, x) or (v, y). (As explained at the end of Section 11.2, the radial coordinate ρ is called y when the angular momentum $M = 0$, in order to stress the planar nature of the trajectories.) Most trajectories, and in particular most periodic orbits, do not have any of the symmetries of their Hamitonian; and yet, the spectrum is well known to split into sets of eigenstates characterized by these discrete symmetries. How can the trace formula accommodate this contrast between classical and quantal behavior? The answer was given for the first time by the author (Gutzwiller 1980 and 1982 with the details provided in the later paper); the same ideas were formulated more generally by Robbins (1989).

Rather than to set up general rules, we will discuss the AKP where all the typical features of the problem are present. There are two reflection symmetries: the *X-symmetry* transforms a particular trajectory, i.e., a solution of the equations of motion (11.7), into another trajectory by sending $u \to Xu = -u$, $v \to Xv = v$, $x \to Xx = -x$, and $y \to Xy = y$; the *Y-symmetry* transforms a trajectory according as $u \to Yu = u$, $v \to Yv = -v$, $x \to Yx = x$, and $y \to Yy = -y$.

In quantum mechanics, these two reflection symmetries allow us to split the eigenstates of the Hamiltonian into four distinct, and mutually orthogonal sets, each characterized by the obvious names:

$$\begin{aligned} \text{X-even: } \phi(Xq) &= \phi(q)\,, \\ \text{X-odd: } \phi(Xq) &= -\phi(q)\,, \\ \text{Y-even: } \phi(Yq) &= \phi(q)\,, \\ \text{Y-odd: } \phi(Yq) &= -\phi(q)\,. \end{aligned} \tag{17.30}$$

We will discuss the effect of the X-symmetry on the trace formula in some detail, and only quote the results for the Y-symmetry.

Let us give the names ϕ to the X-even states, and ψ to the X-odd ones; the expansion (13.10) for Green's function then becomes,

$$G(q''\,q'\,E) = \sum_{j=0}^{\infty} \frac{\phi_j(q'')\,\phi_j^+(q')}{E - E_j} + \sum_{k=0}^{\infty} \frac{\psi_k(q'')\,\psi_k^+(q')}{E - E_k} \tag{17.31}$$

The X-even and the X-odd parts can be separated from each other so as to have a Green's function G_e for the X-even eigenstates, and G_o for the X-odd ones:

$$\begin{aligned} G_e(q''\,q'\,E) &= [G(q''\,q'\,E) + G(Xq''\,q'\,E)]/2\,, \\ G_o(q''\,q'\,E) &= [G(q''\,q'\,E) - G(Xq''\,q'\,E)]/2\,. \end{aligned} \tag{17.32}$$

The definitions (17.4) for the trace $g(E)$ and (17.6) for the density of states $\rho(E)$ are now applied to the X-even and the X-odd Green's functions to yield the corresponding traces and densities of states; they are distinguished by carrying the appropriate indices.

The derivation of the trace formula in Section 17.4 has to be repeated for G_e and for G_o. The new feature is the second term on the right-hand side of (17.32); if $q'' = q' = q$, then we have to deal with an excitation of the system in q and its response in Xq, rather than with the response in the same place q as the excitation. When G on the right-hand side of (17.32) is replaced by G_c, we have to consider trajectories that start in q and end in Xq ; they are not closed any longer. As the starting point q ranges over the whole position space that is classically accessible at the stipulated energy E, the endpoint Xq runs over the mirror images with respect to the y-axis, while the trajectory connecting them deforms smoothly.

In the spirit of the stationary-phase method, we ask as in (17.7) for the rate of change of the exponent $S(Xq\,q\,E)/\hbar$ with respect to q; it now becomes $Xp'' - p'$. Constructive interference in the integral over q requires that the trajectory from q to Xq end up with a momentum that is X-symmetric with respect to the starting momentum. Such a trajectory will be called an X-periodic orbit; it actually closes smoothly like an ordinary periodic orbit, if it is continued to twice its running time.

The trace formulas for $g_e(E)$ and for $g_o(E)$ look just like (17.13); but apart from an overall factor 1/2, the sum over periodic orbits is now extended to include the X-periodic orbits. Notice that these new orbits are affected with different signs, + for X-even trace $g_{ec}(E)$ and − for X-odd trace $g_{oc}(E)$.

The same reasoning applies to the mirror symmetry with respect to the x-axis, called the Y-symmetry. At this point, however, the basic three-dimensional nature of the problem makes itself felt. With the angular momentum around the x-axis $M = 0$, the eigenstates can depend only on the absolute value of the radial distance which is y in our nomenclature. Therefore, we will consider only the Y-even states, and their Green's function.

The traces g_e and g_o will from now on be understood as belonging to the even Y-symmetry. Nevertheless, the trace formula (17.13) gets further amplified by the inclusion of the Y-periodic as well as the XY-periodic orbits; but the signs are determined by the X-symmetry. Notice that the collection of periodic orbits gets larger as the set of corresponding eigenstates becomes smaller because of their symmetry.

17.11 From Periodic Orbits to Code Words

The most remarkable feature about the AKP was the possibility to code all the trajectories with infinite binary sequences, in a one-to-one and continuous manner (cf. Section 11.3); all the periodic orbits can be enumerated by the binary sequences of even length (cf. Section 11.5). Before trying to calculate the trace formula (17.13), this scheme has to be enlarged so as to include the three new kinds of periodic orbits.

An X-periodic orbit ends up on the same side of the x-axis, whose intersections define the binary code; there is an even number $2k$ of intersections between q and Xq, say the sequence $(a_1, ..., a_{2k})$. When the trajectory is continued, it will close itself smoothly after $2k$ more intersections with the x-axis; but the intersections take place at the X-symmetric locations on the x-axis, so that corresponding binaries are $(\bar{a}_1, ..., \bar{a}_{2k})$ where $\bar{a}_i = - a_i$. As one keeps on running around the X-periodic orbit and its extension, the binary sequence consists of alternating segments with opposite signs.

This feature is important when the length of the X-periodic orbit is approximated by the formula (11.15), which uses the binaries directly. The second summation in (11.15), for j from $-\infty$ to $+\infty$ was defined by stipulating that $a_{j+2k} = a_j$, but this condition for extending the summation now gets changed into $a_{j+2k} = -a_j$. This change in sign will allow us in the next section to see how the quantum-mechanical spectrum actually divides into completely separate sets.

The Y-periodic orbits have to be treated in the same manner. Their endpoint has the opposite y-coordinate from the starting point; therefore, an odd number $2k + 1$ of intersections occurs in between; let the corresponding binary sequence be $(a_1, ..., a_{2k+1})$. As the Y-periodic orbit is continued until it closes smoothly, the same sequence of intersections with the x-axis takes place. This sequence of length $2k + 1$ repeats itself indefinitely; the formula (11.15) stays again the same, except that the first summation, over i, goes from 1 to $2k + 1$, and the second summation requires that $a_{j+2k+1} = a_j$.

The trace formula was originally derived in order to find an approximate spectrum for a classically chaotic system on the basis of the classical trajectories. Einstein (1917) was the first to recognize that a real problem lay ahead when the classical system was no longer integrable. If the summation over all the periodic orbits can be carried out, the singularities in the E-dependence can be identified as energy levels. The results of Berry (1985) as reported in Section 17.7 apply only to the correlations in the spectrum, but not to the calculation of individual energies. This challenge will be taken up in this section, based upon the author's work (Gutzwiller 1980 and 1982) on the AKP.

The starting point is the expression (17.23) where we have to insert the approximation (11.15) for the normalized action integral in terms of the binary sequences that characterize the periodic orbits, including the modifications of the preceding section due to the discrete symmetries. The trace formula now becomes a sum over binary sequences; two comments of a general nature are called for.

The binary sequences constitute a *code*. Such a code describes the dynamical system almost as well as the original equations of motion, provided it is supplemented with a formula like (11.15) for the action integral. Hard chaos, in contrast to soft chaos, is expected to offer a good code for its systems; *the problem is no longer how to solve the equations of motion with general initial conditions, and for arbitrarily long times, but to find the appropriate code.* The AKP must be an exceptionally simple case in this respect, but general procedures to establish appropriate codes have been in use for some time to describe the geodesics on a surface of negative curvature. Some special cases will be discussed in Chapter 20.

The general requirements for the use of a code in connection with the trace formula can be gathered from the way it works out in the AKP. The connection with classical statistical mechanics is hard to avoid. A binary sequence, in conjunction with a formula like (11.15), is the most elementary model of a spin system. The imaginary factor $i\sigma$ is replaced by the real negative quantity $-1/k_B T$, with the Boltzmann constant k_B and the absolute temperature T. Notice that the signs are correct: if the energy E in the formulas (17.22) is replaced by $E + i\varepsilon$ with a positive ε, we go smoothly into statistical mechanics. The method of transfer matrices in the next section is taken from there, although much more is demanded of it in our context than in its field of origin.

Before applying this method to the trace formula (17.23), the stability exponent α and the phase loss β in the primitive orbit have to be determined. The rule on β is simple: each crossing of the x-axis, or equivalently each binary, leads to a loss of π. The corresponding two conjugate points are found, the first somewhere between two consecutive crossings, and the second at the crossing itself. The latter arises only in three dimensions where the whole plane of the trajectory is rotated around the x-axis. The restriction to Y-even orbits and the extra conjugate point are the main differences between two and three dimensions, exactly as in the ordinary Kepler problem.

The stability exponent α cannot be handled so convincingly. The numerical data from all the periodic orbits up to 10 binaries (cf. Gutzwiller 1981) show that they cannot be approximated by as simple a formula as (11.15) for the action Φ. If the mean for each length of orbit (in terms of binaries) is taken, a fairly good proportionality with the number of binaries is seen. Without thinking of the entropy, the author originally settled for the value $\alpha = .75$, and noticed only many years later that this value is tantalizingly close to log(2). The discrepancy, if it is real, could probably be found in the wide scatter of the Φ-values for a given binary length; it is such as to cover the average of the actions for both shorter and longer orbits. For the numerical calculations, $\alpha = 2k\alpha_0$ was assumed, with $\alpha_0 = .75$, for an orbit of $2k$ binary length.

If the binaries are used as independent parameters for the sum over the periodic orbits, each orbit is overcounted by the number of binaries defining the period. This difficulty is actually more subtle, because if the periodic orbit is the repetition of a shorter one, the overcounting factor is only as large as the length of the primitive periodic orbit; moreover, some orbits have symmetries with respect to the two coordinate-axes and/or time-reversal. These are reflected in their binary representation and lead again to lesser overcounting. All these features, however, tie in perfectly with the primitive period T_0 which multiplies each term in the trace formula. Everything is accounted for if T_0 is replaced by T/N, where T is the ordinary period, and N is the number of binaries.

A further step in the reduction of (17.23) uses $T = dS/dE$, and the fact that neither α nor β depend on E. Both sides can be integrated with respect to E, leaving the lower limit open. Thus, we arrive at

$$\int^{E} g_c(E)\, dE = -\sum_{N} \frac{(-1)^N \exp(-N\alpha_0/2)}{N} \sum_{\text{bin.seq.}} \exp(i\sigma\Phi)\ , \quad (17.33)$$

where the inner sum goes over the binary sequences of length N. The Y-periodic orbits are already included in the terms with odd N. Their phase loss of π yields the factor $(-1)^N$. The uniform binary sequences (+++ ...) or (--- ...) are included in the addition, although they have no correspondent among the periodic orbits in the AKP, but their contribution can be calculated in closed form, and subtracted. Similarly, we keep the term $N = 1$, although there are no such Y-periodic orbits. The separation into an X-even and an X-odd part according as (17.32) has not been carried out as yet at this point.

17.12 Transfer Matrices

In order to explain the idea of the transfer matrix, the expression (11.15) is simplified for the time being. Instead of two binaries a_i and a_j contributing according as $\exp(-\gamma\,|j-i|)$, however far apart, the range of interaction is limited to the ℓ nearest neighbors; therefore,

$$\Phi = \sum_{i=1}^{N} [A(1 - a_i a_{i-1}) + B(1 - a_i a_{i-2}) + ... + L(1 - a_i a_{i-\ell})]; \quad (17.34)$$

it is always assumed that $a_{i-N} = a_i$. The coefficients $A, B, ..., L$ are positive and decreasing. The peculiar combinations $(1 - a_i a_{i-1})$ have been chosen to make sure that Φ=0 when $a_1 = a_2 = ... = a_N$.

A 2^ℓ by 2^ℓ matrix $\mathbf{T}_\ell$ is now defined, where the rows are labeled by the binary sequences $(a_i, a_{i-1}, ..., a_{i-\ell+1})$, and the columns by the binary sequences $(a_{i-1}, a_{i-2}, ..., a_{i-\ell})$; the matrix element is $\exp(i\sigma\Phi)$ as given by (17.34). As an example,

$$T_2 = \begin{pmatrix} 1 & b & 0 & 0 \\ 0 & 0 & a & ab \\ ab & a & 0 & 0 \\ 0 & 0 & b & 1 \end{pmatrix}, \tag{17.35}$$

where $a = \exp(2i\sigma A)$ and $b = \exp(2i\sigma B)$.

The inner sum of (17.33), over the binary sequences of length N, can now be expressed as the matrix multiplication of $(\mathbf{T}_\ell)^N$; the reader will need a little patience to see this fact clearly. $\mathbf{T}_\ell$ effects the transfer from the subsequence $(a_{i-1}, a_{i-2}, \ldots, a_{i-\ell})$ to the subsequence $(a_i, a_{i-1}, \ldots, a_{i-\ell+1})$, hence the name *transfer matrix.*

The invariance of Φ with respect to the transformation $a_j \to -a_j$ for $j = 1, \ldots, N$ allows us to split $\mathbf{T}_2$ into two square blocks of equal size along the diagonal, by a similarity transformation $\mathbf{S}_2$,

$$\mathbf{T}_2 = \frac{1}{2}\mathbf{S}_2 \begin{pmatrix} 1 & b & 0 & 0 \\ ab & a & 0 & 0 \\ 0 & 0 & 1 & b \\ 0 & 0 & -ab & -a \end{pmatrix} \mathbf{S}_2^{-1}, \mathbf{S}_2 = \begin{pmatrix} 1 & 0 & 1 & 0 \\ 0 & 1 & 0 & 1 \\ 0 & 1 & -1 & 0 \\ 1 & 0 & 0 & -1 \end{pmatrix} \tag{17.36}$$

If $\ell = 1$, each block has only one row and one column; the upper block is $1 + a$, and the lower block $1 - a$. For $\ell > 2$, these blocks have a simple structure, provided Φ is given by the expression (17.34); cf. Gutzwiller (1988a).

Notice that this method could equally well take care of an expression for Φ that includes terms such as $Ma_i a_{i-1} a_{i-2} a_{i-3}$, in addition to the ones in (17.34). The splitting into two diagonal blocks is still feasible, as long as there are only terms of even order in the binaries; that requirement goes directly back to the time-reversal invariance of the AKP.

When taking the trace of $(\mathbf{T}_\ell)^N$, the chain of binaries gets wrapped around according as $a_{i-N} = a_i$, corresponding to the ordinary periodic orbits. The closure condition $a_{i-N} = -a_i$ for the X-periodic orbits is obtained, if all the binaries are inverted before taking the trace. This inversion J is accomplished by inserting an 2^ℓ by 2^ℓ matrix which has 1's in the diagonal from the upper right to the lower left, and 0's elsewhere. After the similarity transformation (17.36), the inversion J takes the 2-block diagonal form, with $+1$ in the diagonal of the upper block, and -1 in the diagonal of the lower block, with 0's elsewhere. Again, the reader has to check these elementary facts.

At last, the X-even and the X-odd traces can be worked out by combining (17.32) with (17.33), and summing over the binary sequences of length N by taking the trace of $(\mathbf{T}_\ell)^N$. Since the trace of a matrix does not change under similarity transformations, the block-diagonal form (17.36) of $\mathbf{T}_\ell$ is applied. The X-even trace $g_{ec}(E)$ uses only the upper block, while the X-odd trace $g_{oc}(E)$ depends only on the lower block.

If $\ell = 1$, the upper trace reduces to $(1 + a)^N$ and the lower trace to $(1 - a)^N$. If $\ell > 1$, however, each block has to be diagonalized. The diagonal elements of T_ℓ are the eigenvalues $\Lambda_m(\sigma)$ with $m = 1, 2, ..., 2^{\ell-1}$; their dependence on the physical parameter σ is kept explicitly in view. For $\ell = 2$, the two eigenvalues in each block can be obtained from solving a simple quadratic equation in accordance with (17.36), but one can hardly avoid numerical calculations for $\ell > 2$. The dependence of the eigenvalues $\Lambda_m(\sigma)$ on the parameters $A, B, ..., L$ is not easy to visualize.

The trace of the transfer matrix in terms of the eigenvalues Λ is now inserted into (17.33),

$$\int^E g_c(E)\, dE = -\sum_N \frac{(-1)^N \exp(-N\alpha_0/2)}{N} \sum_{m=1}^{2^{\ell-1}} [\Lambda_m(\sigma)]^N , \quad (17.37)$$

which is valid both for the X-even and the X-odd spectrum, provided the appropriate eigenvalues Λ are used. Miraculously, the sum over N is the expansion of $\log(1 + x)$. If both sides are exponentiated, and we recall the original expansion (17.4) for the quantum-mechanical trace $g(E)$, we get the final formula,

$$\prod_{j=1}^{\infty} (E - E_j) \simeq \prod_{m=1}^{2^{\ell-1}} [1 + \Lambda_m(\sigma) \exp(-\alpha_0/2)] . \quad (17.38)$$

The $\simeq$ sign refers mainly to the classical approximation that justifies the right-hand side, but also to a number of minor details such as a constant factor, which were neglected in the derivation.

The classical approximation to the spectrum is now obtained by finding the values of σ where one of the factors on the right vanishes. The explicit computation of these zeros is based on inserting $\Lambda(\sigma) = -\exp(\alpha_0/2)$ into the characteristic polynomial of the transfer matrix, so that

$$\det |\mathbf{T}_\ell(\sigma) + \exp(\alpha_0/2)| = 0. \quad (17.39)$$

If $\ell = 1$, this condition becomes trivial, because $\mathbf{T}_1$ is already diagonal after the similarity transform (17.36). The solution is multiple-valued because σ occurs in the exponents of the matrix elements $a, b, ...$ Thus,

$$\begin{aligned} \text{X-even}: \quad 2\sigma A &= (2n+1)\pi - i\log(1 + \exp(\alpha_0/2)) , \\ \text{X-odd}: \quad 2\sigma A &= 2n\pi - i\log(1 + \exp(\alpha_0/2)) . \end{aligned} \quad (17.40)$$

When we go back all the way to (17.22) and insert these values for σ, while ignoring the imaginary part in (17.40), a Balmer-type formula for

1S	1.56751	1.568	1.457	2P	0.57620	0.577	0.526
2S	0.44413	0.443	0.419	3P	0.27506	0.275	0.256
3S	.23955	.238	.232	4P	.16595	.167	.156
3D	.18808	.188	.183	4F	.11729	.117	.115
4S	.14607	.143	.141	5P	.11207	.112	.108
4D	.10743	.106	.106	5F	.08179	.081	.082
5S	.09677	.094	.094	6P	.07571	.076	.075
5D	.07762	.076	.077	6F	.06233	.060	.066
5G	.07320	.069	.072	6H	.05526	.055	.055
6S	.06422		.063	7P	.05032		.050
6D	.05837		.057	7F	.04921		
6G	.05245		.052	7H	.04223		

Table The energy-levels of the electron in the donor impurity of Silicon according to the Hamiltonian (11.2) with the angular momentum around the heavy axis = 0, normalized to $\sqrt{m_1 m_2}\, e^4/2\kappa^2\hbar^2$; the left-hand four columns for even, and the right-hand four columns for odd eigenfunctions with respect to reflexion on the light plane. Columns 1 and 5 give the notation of Faulkner (1969); columns 2 and 6 are the results of Wintgen, Marxer, and Briggs (1987) using a very large basis of trial functions; columns 3 and 7 give Faulkner's results with a basis of 9 functions for each symmetry class; columns 4 and 8 are obtained from a formula like (17.38) based on the sum (17.23) over the periodic orbits, cf. Gutzwiller (1980 and 1982).

the energy levels is obtained, with alternating X-even and X-odd states; the lowest state is X-even.

This lengthy development was explained to the reader because it gives a systematic method for finding approximate energy levels from the trace formula (17.13), assuming that the periodic orbits have been coded. In exactly this way, the author (Gutzwiller 1980 and 1982) was able to get a spectrum for the AKP, the first time such a calculation for a chaotic system had ever been completed on the basis of the classical periodic orbits.

The relation between the binary code and the action integral is given by (11.15) in the AKP. The coefficients decrease exponentially and do not have a finite range ℓ as in (17.34). The transfer matrix is infinite; but its simple structure allows an approximate treatment (cf. the 1982 paper), and leads to rather good results as shown in Table I which gives the comparison with the recent extensive calculations of Wintgen, Marxer and Briggs (1987).

Nevertheless, (17.40) shows that we have to expect to find complex quantum numbers σ, if the right-hand side of the trace formula (17.14) is evaluated. In other words, the density of states $\rho_c(E)$ does not consist of δ-functions, but rather of some broadened peaks. In the case of (17.40), the imaginary part, $\log(1 + \exp(\alpha_0/2)) = .88$ for $\alpha_0 = \log 2$, has to be compared with the separation 2π between the levels of the same symmetry; the result is satisfactory. There is no general argument available at this time, however, to suggest that the trace formula always performs at least that well. Our confidence still rests mainly on the Selberg trace formula which gives perfect results as will be explained in Chapter 19.

The locations of the singularities in the complex plane were calculated by the author (Gutzwiller 1988a) for a finite range of the interactions A , ... , L, rather than for the infinite range that leads to Table I. Even for ℓ up to 4, the zeros keep drifting away from the real E-axis. This phenomenon may be related to the correlations found by Pandey, Bohigas, and Giannoni (1989) when the number $m + \alpha n + \beta k$ are mapped onto the unit-circle; m, n, k are integers (cf. Section 10.2).

Cvitanovic (1988) together with Eckhardt (1989) has proposed a different method for numerically summing the right-hand side of the trace formula (17.13). They generate all the periodic orbits in terms of a few elementary 'cycles'; those are the simplest periodic orbits. If appropriately chosen, all the terms in (17.13) can be very nearly compensated with the right combination of cycles, at least when the reciprocal of the zeta-function is calculated; this object is the exponential of the left-hand side in (17.33) or (17.4) and becomes the product $\Pi(E - E_j)$, exactly what was calculated in (17.38).

This new scheme has been worked out for the scattering from the circular disks in the open Euclidean plane, and gives excellent results for the poles in the scattering amplitudes; it has not been tried as yet for bound states as in the AKP. The trajectories or periodic orbits are essentially reduced in a symbolic manner to the combination of the simplest of their kind; this construction of the symbolic sequences appears somewhat different from the binary sequences for the AKP in Chapter 11, or similar ones to be discussed in Chapter 20. It may possibly yield a more direct access to the singularities of Green's function than the transfer matrices of this section, or it may ultimately turn out to be quite similar.

CHAPTER 18

The Diamagnetic Kepler Problem

The effect of a magnetic field on an atom has been studied for a long time and has played a crucial role in the understanding of atomic physics. The magnetic fields available in the laboratory are small, however; their action in the atom used to be well explained by perturbation theory. The evidence for strong magnetic fields was first found by the astronomers in white dwarfs some twenty years ago, and extremely strong fields are assumed to exist in neutron stars. This discovery led to laboratory experiments that somehow manage to imitate the extraordinary conditions in these exceptional stars.

The main idea is to prepare the atom in a very highly excited, but still bound state, so that the orbit of the electron encloses a large area, and thus a large magnetic flux even in a relatively small field. Perturbation theory is then no longer applicable; the dominant term in the Hamiltonian is no longer the linear one (in the field strength) which leads to the ordinary *Zeeman effect*.

By contrast, the quadratic term in the Hamiltonian is ordinarily not as interesting, because it comes into play only when the atom or molecule does not have a magnetic moment to start with, and it leads to a small negative magnetic susceptibility, called *diamagnetic* rather than the much larger positive *paramagnetic* one. The quadratic, or diamagnetic term, however, eventually leads to chaotic classical motion, whereas the linear, or paramagnetic term can be effectively eliminated by using a rotating frame of reference as in Hill's theory of the Moon (cf. Chapter 5).

Experiments on atoms near the ionization threshold have become possible in the last decade because of the great strides in optical technology, as well as in the preparation of atomic beams and vapors. Whereas the pioneering work was done on alkali metals and alkaline earths, the last few years have produced detailed spectra for hydrogen atoms. They seemed at first utterly confusing, but they now appear understandable in terms of classical periodic orbits, isolated and unstable like the ones in the preceding chapter. The underlying classical behavior is probably a case of soft chaos since the system is integrable in both limits, very weak and very strong magnetic fields.

This chapter does not try to discuss all the recent work in this active area nor is it possible to offer a definitive picture since not all the evidence is in as yet. We will, therefore, concentrate on the aspects most directly related to our main concern, the symptoms of chaos in quantum mechanics and their explanation in terms of classical orbits. Atomic physics in high magnetic fields has been surveyed by Gay (1984 and 1985); extensive reviews on the Diamagnetic Kepler Problem have recently been written by Hasegawa, Robnik, and Wunner (1989) as well as by Friedrich and Wintgen (1989).

18.1 The Hamiltonian in the Magnetic Field

The magnetic force on an electric charge e has its origin in the theory of relativity. It arises only when the charge is moving, and its strength is proportional to the ratio of its velocity to the velocity of light. The equations of motion in the combined electric field $\mathbf{F}$ and magnetic field $\mathbf{B}$ are given by

$$m \frac{d^2\mathbf{q}}{dt^2} = e\,(\mathbf{F} + \frac{1}{c}\frac{d\mathbf{q}}{dt} \wedge \mathbf{B}) \,, \tag{18.1}$$

where the symbol $\wedge$ indicates the vector product. The second term on the right is called the *Lorentz force,* in honor of H.A. Lorentz who was one of the first to reduce atomic physics to the interplay of electric charges around 1900. Both the electric and the magnetic fields are assumed to be known functions of the position $\mathbf{q} = (q_1, q_2, q_3) = (x_1, x_2, x_3) = (x, y, z)$ and the time t.

Maxwell's equations, which govern all of electrodynamics, allow us to introduce the *four-component vector potential* (A_0, A_1, A_2, A_3), which determines the electric and magnetic fields through the relations

$$F_i = -\frac{\partial A_0}{c\,\partial t} - \frac{\partial A_i}{\partial x_i} \,, \quad B_i = \frac{\partial A_k}{\partial x_j} - \frac{\partial A_j}{\partial x_k} \,, \tag{18.2}$$

where the indices i, j, k are a cyclic permutation of (1, 2, 3). The 0-component A_0 is the ordinary electrostatic potential, usually called $V = e A_0$. The vector potential is not uniquely determined by the fields **F** and **B**, however, because the fields on the left in (18.2) remain the same if we replace A by A', where

$$A'_0 = A_0 - \frac{\partial \chi}{c\, \partial t} , \quad A'_i = A_i + \frac{\partial \chi}{\partial x_i} , \tag{18.3}$$

and χ is an arbitrary function of (x, y, z) and t.

The equations of motion (18.1) can be derived from the Hamiltonian

$$H(p\,q\,t) = \frac{1}{2m} \sum_1^3 \left(p_i - \frac{e}{c} A_i\right)^2 + eA_0 , \tag{18.4}$$

i.e., the equations (2.2) yield (18.1), if applied to (18.4). This Hamiltonian can be obtained from the expression (7.11), if we set $M = m^2c^2$, and solve for p_0; if the resulting square root is expanded in powers of the three spatial components of the momentum, the lowest term is indeed (18.4). The intrinsic angular momentum of the electron, the electron spin, is completely left out of the discussion in this chapter.

Most remarkably, the Hamiltonian involves the vector potential, while the equations of motion contain only the fields. The Hamiltonian is sensitive to the transformation (18.3), but the equations of motion (18.1) are not; this property is called *gauge invariance,* because the vector potential, in particular the electrostatic potential $A_0 = V/e$, defines a gauge (voltage) for any measurements.

Since Schrödinger's equation is taken directly from the Hamiltonian, it is sensitive to the choice of the gauge as well. A change of the gauge (18.3) has to be compensated by a change in the phase of the wave function $\psi(x, y, z, t) \rightarrow \psi'(x, y, z, t)$, where

$$\psi' = \psi \exp(ie\chi/\hbar c) . \tag{18.5}$$

This explicit dependence on the gauge leads to measurable consequences, whenever the spatial region for the wave function is multiply connected. The interference between the waves from two topologically inequivalent paths is known as the *Aharonov-Bohm effect.*

A uniform magnetic field B in the z-direction can be obtained from $\mathbf{A} = \mathbf{x} \wedge \mathbf{B}/2$, i.e., $A_x = -By/2$, $A_y = Bx/2$, and $A_z = 0$, also called the *symmetric gauge,* because x and y occur in a somewhat symmetric fashion. The Hamiltonian now becomes

$$H = \frac{p^2}{2m} - \frac{e^2}{r} - \frac{eB}{2mc} L_z + \frac{e^2B^2}{8mc^2} (x^2 + y^2) , \tag{18.6}$$

where $L_z = x\,p_y - y\,p_x$ is the angular momentum in the z-direction; the right-hand side contains the kinetic energy, the Coulomb potential, the (paramagnetic) Zeeman energy, and the (diamagnetic) quadratic term. We will now switch over to quantum mechanics without much ado, and assume that the reader is familiar with elementary atomic physics.

18.2 Weak Magnetic Fields and the Third Integral

Until recently, the last term in (18.6) was always assumed to be much smaller than the third in the usual applications, unless the angular momentum L_z happened to vanish. To a first approximation, the last term was neglected. Since L_z commutes with the Hamiltonian, every level in the hydrogen atom is split by the Zeeman energy $e\hbar B/2mc$ according as the z-component of the angular momentum. That leaves a large degeneracy, because for each principal quantum number n, there are n different values of the total angular momentum ℓ ; a particular eigenvalue ℓ_z of L_z can belong to different values of ℓ . The further splitting according as ℓ is due to the diamagnetic term, but its calculation by perturbation theory is not easy and cannot be pushed very far.

A deeper understanding of the Kepler problem involves the *Runge-Lenz vector*

$$\Lambda = \mathbf{p} \wedge \mathbf{L} - me^2\mathbf{r}/r\,. \tag{18.7}$$

Since the scalar product $(\mathbf{L}, \Lambda) = 0$, the Runge-Lenz vector lies in the plane of the Kepler orbit. It points toward the perinucleus (point of closest approach to the nucleus), and its length, $\Lambda^2 = 2mEL^2 + m^2e^4$, differs by a factor m^2e^4 from the square of the eccentricity.

In quantum mechanics, the operators associated with $\mathbf{L}$ and Λ generate a group that is isomorphic with the rotations in a four-dimensional space, rather than only the three-dimensional rotations from $\mathbf{L}$ alone. Λ commutes with the Kepler Hamiltonian just as $\mathbf{L}$ does; the spectrum of the Kepler problem can be treated, therefore, exactly as the spectrum of a particle on a four-dimensional sphere. Pauli (1926a) was the first to exploit this fact to obtain the levels in the hydrogen atom by purely algebraic manipulations. Fock (1935) then made the connection explicit by going into momentum space; the discussion in Section 12.4 and 12.8 is based on this idea.

How much of this high degeneracy survives in a magnetic field? The first indications came from calculations by Zimmerman, Kash, and Kleppner (1980) who obtained the energies as a function of the field strength for even parity, $L_z = 0$ states with principal quantum numbers n from 1 to 22. When plotted against B^2, the individual levels run along roughly straight lines that converge to their points of degeneracy at $B = 0$. Since there is no symmetry left in the problem, these lines are not expected to cross, or even get very close. They should repel and avoid one another like the two branches of a hyperbola. Such is indeed the case for $n < 8$. For $10 \leq n \leq 16$, however, the 'anti-crossings' become ever tighter as n increases, and they do so with an exponential dependence on n.

Robnik (1981) found evidence in the classical surfaces of section that the diamagnetic Kepler problem is less chaotic than our inability to solve the equations of motion seemed to suggest. Reinhardt and Farrelly (1982) used the Birkhoff-Gustavson normalization (cf. Sections 8.4 and 14.3) to match the results of Zimmerman et al.. Delos, Knudson, and Noid (1983) obtained the energy as a function of three actions I by perturbation theory, and then calculated the spectrum as in an integrable system, again with good results beyond expectations.

The explanation involves a 'third integral' like the one invoked by the astronomers to explain the motion of stars in the galactic gravitational potential (cf. Section 8.1). The first concrete proposal came from Solovev (1981 and 1982), and Herrick (1982) who had discovered an approximate third integral for the He-atom (Herrick and Sinanoglu 1975). Hasegawa with Harada and Lakshmanan (1983, 1984, and later) as well as Delande and Gay (1984) worked out many more of the geometric details; Deprit together with Coffey, Ferrer, Miller, and Williams (1986 and 1988) applied the perspective of celestial mechanics and the mathematics of dynamical systems.

A linear combination of $\Lambda_x^2 + \Lambda_y^2$ and Λ_z^2, whcse coefficients are independent of the field B, is the best thing one can hope for as a third integral. When the time derivatives of these two quantities are calculated, the terms proportional to B^2 can be canceled out in

$$\mathrm{K} = 4\,(\Lambda_x^2 + \Lambda_y^2) - \Lambda_z^2 \,, \tag{18.8}$$

provided the time derivative of K is averaged over a periodic orbit. The explicit construction of such an approximate constant of motion, and the particular criterion for justifying it, seems to be new in classical mechanics. There may be other problems where this method leads to a better understanding of an almost integrable system.

18.3 Strong Fields and Landau Levels

The terms in the Hamiltonian (18.6) are ordered in decreasing strength for the case of a weak magnetic field; the diamagnetic term was treated as a perturbation to the Coulomb potential. In very strong magnetic fields, the ordering is opposite; the Coulomb attraction to the nucleus becomes a small disturbance. This situation was first studied by Yafet, Keyes, and Adams (1956). It arises near the impurity in a semiconductor, both because the effective mass of the electron is small, and because the dielectric constant of the crystal is large (cf. Section 11.1), which does not affect the two magnetic terms in (18.6), however. The whole question was reviewed by Hasegawa (1969).

The lowest order approximation ignores the Coulomb potential; the electrons are free except for the presence of the magnetic field, a situation encountered in metals, and that was first investigated by Landau. Classically, the electron moves along a helix whose axis is parallel to the magnetic field. The motion is uniform in the z- direction, whereas it is a uniform rotation in the xy-directions, on a circular track of radius $\sqrt{2mE - p_z^2}\ /m\omega_c$. The rotational speed is the *cyclotron frequency* $\omega_c = eB/mc$. Notice that the time for the electron to go around its circular track is independent of its kinetic energy $E - p_z^2/2m$ in the xy-plane; that is the basic idea behind the original cyclotron accelerators.

The quantum description is more complicated and somewhat confusing because various gauges for the magnetic field can be used. Contrary to the treatment in the books on solid-state physics, we will stay with the symmetric gauge; the location of the atom and its Coulomb potential will be the origin for the coordinate system. The wave function becomes a product of a plane wave $\exp(ikz)$ in the direction of the magnetic field, and a two-dimensional harmonic oscillator function in the xy-plane. The paramagnetic term in (18.6) makes us organize the oscillator states according as their angular momentum, but the energy of the electron,

$$E = (N + 1/2)\,\hbar\omega_c + \hbar^2 k^2/2m\,, \tag{18.9}$$

does not depend on the value $M\hbar$ of L_z. For fixed k, the spectrum looks like a harmonic oscillator with steps $\hbar\omega_c$, although each level is highly degenerate; these are the *Landau levels*.

Garton and Tomkins (1969) were the first to see a Landau-like spectrum near the ionization limit of Ba in a field of 2.4 Tesla. Very surprisingly, however, the spacing was in steps of about 1.5 $\hbar\omega_c$, a result that was consistently confirmed in later measurements on other hydrogen-like atoms in very high magnetic fields. An explanation was given by Edmonds (1970) using the Bohr-Sommerfeld (or EBK) quantization condition (14.4) for the motion at right angles to the z-axis. Although such an argument seems quite arbitrary, it does give an answer in accord with the experiments, and has to be considered seriously.

The momentum p_z is set to 0 for this calculation; polar coordinates are used in the xy-plane, with the radial distance ρ and the conjugate momentum p_ρ. The quantization condition is $\int p_\rho \, d\rho = (\lambda + 1/2)\pi\hbar$, with

$$p_\rho^2 = 2mE + \frac{2me^2}{\sqrt{\rho^2 + z^2}} - \left(\frac{L_z}{\rho} - \frac{eB}{2c}\rho\right)^2 \; ; \qquad (18.10)$$

the integration is between the limits ρ_1 and ρ_2 where $p_\rho = 0$, all at a fixed value of z. If z is very large, the Coulomb potential becomes small for all ρ so that we are left with the pure magnetic field, and the Landau spectrum is obtained. As z decreases, however, the energy differences increase, while remaining roughly constant as a function of E. When $z = 0$, the steps are $1.58\ \hbar\omega_c$ around $E = 0$, as in the experiment.

Fonck, Tracy, Wright, and Tomkins (1980) went up to 4 Tesla in both barium and strontium and got almost perfect agreement with the formula (18.10) for the spectrum near the ionization threshold. More comprehensive data of the same kind are reported by Gay and coworkers (1980). The individual lines in the quasi-Landau spectrum have a very complicated shape with many subsidiary peaks that still have to be explained; on this finer scale the spectrum looks quite chaotic.

A remarkable interpretation of these details was given by Castro, Zimmerman, Hulet, Kleppner, and Freeman (1980). The spectra for sodium in magnetic fields from 1 to 5.5 Teslas, in roughly even steps for B^2, show that after a very complicated intermediate region, the oscillator strengths tend to group themselves preferentially around the lines of the quasi-Landau spectrum. Some of the details can be seen in the results of a straightforward diagonalization of (18.6) in a basis of zero-field eigenfunctions for the Coulomb potential.

More sophisticated calculations were carried out by Clark and Taylor (1980 and 1982) in a *Sturmian basis,* i.e., using again the eigenfunctions of the Kepler problem, but with the scale in the Laguerre functions independent of the principal quantum number. This basis is complete without requiring the positive-energy eigenfunctions and still preserves some of the advantages of the hydrogen eigenfunctions. Delande and Gay (1986) provide the arguments for the best choice of the Sturmian functions on the basis of the underlying group theory.

The resulting spectra are quite complicated. They arise from the mixing of states that were either degenerate or well separated in a zero

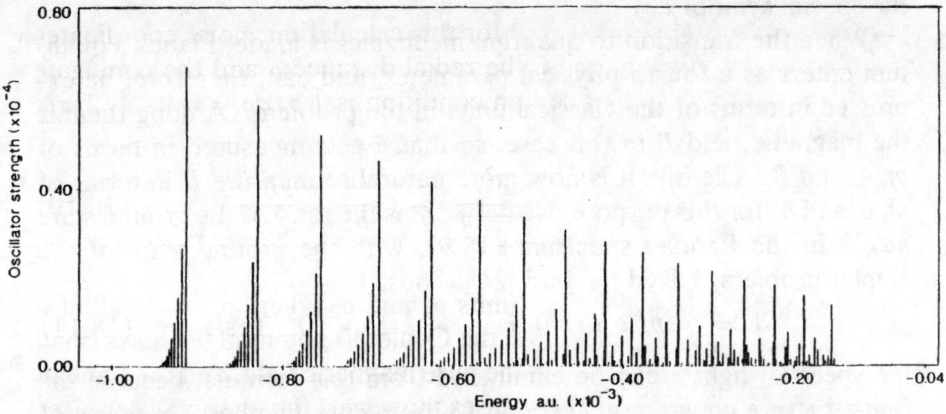

Figure 51 Computed oscillator strengths for the transitions from the ground state (Lyman series) to high Rydberg levels with $|M| = 1$ in a field of 4.7 T as a function of the energy in atomic units [from Clark and Taylor (1980)].

magnetic field. Thus, there are two regimes as the field increases: in the ℓ-mixing, states with different values of the total angular momentum $\ell\hbar$, but belonging to the same principal quantum number n, get scrambled; in the n-mixing regime, states from different levels in the field-free atom get mixed. Of course, the spectrum observed in the laboratory consists of transitions between these energy levels; in the absence of selection rules, only a plot of the computed oscillator strengths gives a realistic picture of the difficulties in the physical interpretation. Figure 51 shows such a plot; the impression of chaos is hard to avoid.

18.4 Scaling the Energy and the Magnetic Field

The structure of the classical trajectories in phase space has to be freed of all incidental features, in order to establish a connection between the classical and the quantal regime. Three physical parameters occur in (18.6): the mass m, the squared charge e^2, and the magnetic field B in the combination of the cyclotron frequency $\omega_c = eB/mc$. As in any good non-linear classical system, like the Toda lattice (cf. Section 3.6) or the Hénon-Heiles model (cf. Chapter 8), all dynamical quantities can now be scaled in a natural manner; the reference distance a is given by Kepler's third law, $m\,\omega_c^2 a^3 = e^2$, and so on. We will call these the *classical units*. The nature of the trajectories in phase space depends on the value of the classically normalized energy, which will be given the special symbol E_c.

When the transition to quantum mechanics is made, Planck's quantum enters as a fourth physical parameter, and can, therefore, be expressed in terms of the classical units in the problem. Among them is the magnetic field B in this case, so that $\hbar$ gets measured in terms of m, e, and B. Clearly, it is now more natural to measure B in terms of m, e, and $\hbar$; for this purpose, we define γ as the ratio of the ground state $\hbar\omega_c/2$ in the Landau spectrum (18.9), with the ground state of the Kepler problem, 1 Ryd $= me^4/2\hbar^2$. Thus,

$$B = \gamma\,(m^2e^3c/\hbar^3) = \gamma \bullet 2.35 \bullet 10^5 \text{ Tesla} ; \qquad (18.11)$$

the speed of light c can be eliminated from this formula, because the fine-structure constant $e^2/\hbar c \simeq 1/137$ is a pure number. In terms of these *atomic units,* the various quantities will be designated by their usual symbols. In particular, if the energy E is measured in Rydbergs, then the classically normalized energy $E_c = E/\gamma^{2/3}$.

This scaling argument has been made by a number of authors, e.g., Robnik (1982), Harada (1983), and Hasegawa (1983); but its impact on the interpretation of experimental or computational spectra was first realized by Wintgen and his coworkers (1987a and b), and we will follow his line of reasoning. As a first step which has become the tradition in this trade, we use Hill's trick in the treatment of the lunar orbit and give our coordinate system a uniform rotation of angular speed $n' = -\omega_c/2$ around the z-axis. A check with (5.11) shows that the Zeeman term, $\omega_L L_z$, is thereby eliminated from the Hamiltonian (18.6). This transformation to rotating coordinates is an application of Larmor's theorem in atomic physics and the frequency $\omega_L = \omega_c/2$ is known as the *Larmor frequency* $\omega_L = eB/2mc$.

Cylindrical coordinates (ρ, ϕ, z) are natural; the coordinates are expressed in the atomic units, as stated above. (The classically scaled distances are then obtained by multiplying with $\gamma^{2/3}$, momenta with $\gamma^{-1/3}$, time with γ, angular momentum with $\gamma^{1/3}$; and the value of the classically scaled Hamiltonian E_c follows by dividing with $\gamma^{2/3}$.) The Hamiltonian (18.6) becomes

$$H = p_\rho^2 + p_z^2 + \frac{\ell_z^2}{\rho^2} - \frac{2}{r} + \frac{\gamma^2}{4}\rho^2 , \qquad (18.12)$$

which is measured in Rydbergs. In order to explain Wintgen's ingenious use of scaling, we continue his arguments in a somewhat simplified form while giving concrete details about the computations.

Semiparabolic coordinates (μ, ν) are used: $\rho = (\mu^2 - \nu^2)/2$, $z = \mu\nu$, and $r = (\mu^2 + \nu^2)/2$. Schrödinger's equation becomes

$$[-\Delta_\mu - \Delta_\nu - E(\mu^2 + \nu^2) + (\gamma^2/4)\mu^2\nu^2(\mu^2 + \nu^2) - 4]\psi = 0\,, \quad (18.13)$$

where the two Laplacians have the same form

$$\Delta_\mu = \frac{1}{\mu}\frac{\partial}{\partial\mu}\mu\frac{\partial}{\partial\mu} - \frac{\ell_z^2}{\mu^2}\,.$$

If we assume that $E < 0$ and fixed, the two first terms in (18.13) represent two decoupled two-dimensional harmonic oscillators of angular momentum ℓ_z. The third term is a sixth-order perturbation, whereas the last term, coming from the Coulomb potential, is simply a numerical constant.

Harmonic-oscillator wave functions are called for; they are used in some standard form, $w_m(\mu)$ and $w_n(\nu)$. Their matrix elements, for the kinetic energy $T = -\Delta_\mu - \Delta_\nu$, for the harmonic potential $V = (\mu^2 + \nu^2)$, and for the anharmonic potential $W = \mu^2\nu^2(\mu^2 + \nu^2)$ are simple numbers. The corresponding matrices are very sparse, with non-vanishing elements close to the diagonal. We are still able to pick a scale b, which means that $\psi(\mu, \nu)$ is expanded in terms of the functions $w_m(\mu/b)w_n(\nu/b)$. The eigenvalue problem is now written as

$$[b^{-2}T - E\,b^2V + (\gamma^2b^6/4)W - 4]\,\psi = 0\,,$$

in terms of the standardized matrices T, V, and W.

There are different ways of choosing the scale b ; but if we want to get back to the classically scaled energy E_c, we have to use $b = \gamma^{-1/3}$. The condition for ψ now becomes

$$[E_c V - (1/4)W + 4]\,\psi = \gamma^{2/3}T\psi\,, \quad (18.14)$$

where E_c is given some fixed value, while we try to find γ such that (18.14) is satisfied.

The problem has been turned around: rather than to find the energy levels for a given magnetic field B, we now determine the special values of $\gamma^{2/3}$ where an eigenstate exists for a given classically normalized energy E_c . These special values $\gamma_j^{2/3}$ are found again by solving an eigenvalue problem, and they yield the energy levels in the atomic units through $E_j = \gamma_j^{2/3} E_c$. Another way of stating this result: in a plot of E versus γ, we get the intersections γ_j of the curves $E_j(\gamma)$ with the curve $E = \gamma^{2/3} E_c$.

The trace $g(E)$ as defined in (17.4) becomes a function of γ; the typical denominator $E - E_j$ is replaced by $E_c(\gamma^{2/3} - \gamma_j^{2/3})$, where E_c has a fixed value. If we use the trace formula (17.13) to evaluate the classical approximation $g_c(E)$, the main object is to write the exponents as functions of γ.

The classical action $S(E)$ around a particular periodic orbit depends on the magnetic field in a very simple manner, once E_c is given. $S(E)$

in (17.13) is $\int p\, dq$ in atomic units; but since the classically normalized energy E_c is held fixed while the magnetic field varies, it is natural to evaluate $S(E)$ in classical units, indicated by a bar, and then revert to the atomic units.

In the discussion preceding (18.12), we stated that $\bar{q} = \gamma^{2/3} q$ and $\bar{p} = \gamma^{-1/3} p$; therefore,

$$S_k(E) = \gamma^{-1/3}\, \bar{S}_k(E_c) = 2\pi(n + \beta_k)\,, \qquad (18.15)$$

where $\bar{S}_k(E_c)$ depends on the periodic orbit (indexed by k), but is independent of γ, because the magnetic field does not appear in the Hamiltonian if classical units are used. The first equality simply relates the action integral in the atomic and in the classical scale. The second equality goes further, however, and applies the quantization rule for integrable systems (14.4). The Maslov index β_k arises from the count of conjugate points as usual. The conditions under which this last equality is valid will be discussed at some length in Section 18.5.

To complete the discussion of the trace $g_c(E)$ according as (17.13), the amplitudes A also can be expressed in both systems of units. Thus, the period $T_0 = \gamma^{-1} \bar{T}_0$, while the the stability exponent χ and the number of conjugate points do not depend on γ at all. The remarkable feature in this interpretation of (18.15) is that the action is a linear function of the variable $\gamma^{-1/3}$.

Therefore, the trace formula gives the Fourier analysis of $g_c(E)$ with respect to the variable $\gamma^{-1/3}$. Such an analysis can be carried out directly on the experimental or computational density of states, $\rho(E)$ at constant E_c and varying magnetic field $\gamma^{-1/3}$. For the lack of a better name, the variable in the Fourier transform ρ_F is called $\gamma^{1/3}$; a plot of ρ_F versus $\gamma^{1/3}$ should then have strong spikes at $\gamma^{1/3} = \bar{S}(E_c)$ for any periodic orbit at the clasically normalized energy E_c . The evidence for this striking result will be discussed in the next section; in essence, the rather chaotic looking spectrum of the diamagnetic Kepler problem can then be understood in terms of the periodic orbits at the classically normalized energy E_c .

18.5 Calculation of the Oscillator Strengths

The standard experiment for studying the energy levels in an atom or a molecule is optical absorption. The frequency of the light can nowadays be very sharply controlled with the help of tunable lasers. If the atom is initially in the state designated by ϕ_i, the probability for ending up in the state designated by ϕ_f rather than in some other state, is given by the *oscillator strength,*

$$f = \frac{2m(E_f - E_i)}{\hbar^2} \mid < \phi_f | D_{op} | \phi_i > \mid^2 ; \qquad (18.16)$$

where D_{op} is the component of the dipole operator in the direction of polarization of the absorbed light; $< \phi_f | D_{op} | \phi_i >$ is the matrix element between the initial and final state. The reader will find this formula in any textbook on modern optics; in the further development, we follow the paper of Du and Delos (1988).

The interesting final states for the hydrogen atom in a magnetic field are near the ionization threshold; the level density is very high, and it is often not possible to separate out completely one particular state. In order to define the proper average, we start with the formula (13.10) for Green's function. Then we take the difference between its values for $E \rightarrow E + i\varepsilon$ and $E \rightarrow E - i\varepsilon$, exactly as in (17.6), but without the integration over $q'' = q' = q$; thus,

$$G_\varepsilon(q''q' E) = (i/2\pi) \left[G(q''q' E + i\varepsilon) - G(q''q' E - i\varepsilon) \right] . \qquad (18.17)$$

This function of the energy E is now smoothed with the help of a function $\theta(E)$ which has a somewhat broadened peak near 0, and yields $\overline{G}_\varepsilon(q''q' E) = \int dE' \, \theta(E' - E) \, G_\varepsilon(q''q' E')$. The smoothed oscillator strength is defined as

$$f(E) = \frac{2m(E - E_i)}{\hbar^2} < D_{op}\phi_i | \overline{G}_\varepsilon | D_{op}\phi_i > , \qquad (18.18)$$

where the variables q'' and q' in $\overline{G}_\varepsilon$ have been integrated over when the matrix elements were calculated. The factor $(E - E_i)$ should be included into the smoothing operation (18.17); but it can be taken outside, if the smoothing function is narrow compared with the energy difference $E - E_i$.

Green's function in (18.18) is now replaced by its classical approximation (12.28), to yield the classical oscillator strength $f_c(E)$, exactly as in the derivation of the trace formula (cf. Section 17.4). There is an important modification, however, which was first discussed in detail by Du and Delos (1988). The initial state ϕ_i is close to the ground state of the hydrogen atom. It extends over a few atomic units at best, while the final states may have a radius 1000 times as large. The magnetic field has very little effect on the initial state; the electron leaves the neighborhood of the proton as in a pure Coulomb field. On the other hand, the classical approximation to the hydrogen wave-functions near the nucleus is poor, and it is better to use the exact Green's function out to a distance of 50 atomic units. The computational labor is manageable, because the electron leaving the neighborhood of the nucleus has an energy $E \simeq 0$; the asymptotic form for the wave functions can be used.

The electron is first represented by an outgoing wave for a small distance. Then it follows a classical trajectory for the main part of its journey. Finally, it reaches again the sphere of transition near the nucleus, and becomes an incoming wave. The integration over the initial and the final coordinate with the dipole operator D_{op} takes place in the wave region. The amplitude A for a particular trajectory consists, therefore, of several factors: the dipole operator D_{op} at the departure and at the arrival provides some dependence on the polarization of the absorbed light, on the angular momentum of the trajectory, as well as on the initial and final angle with respect to the z-axis. The trajectory itself contributes a factor related to the stability exponent, just as in the trace formula.

The critical energy dependence is in the phase factor, however, which has the usual form: the action-integral $S(E) = \int p\, dq$ for the whole trajectory, minus $\pi/2$ for each conjugate point along the way. Planck's constant $\hbar$ does not appear explicitly because we are using atomic units. These trajectories are periodic only in the sense that they start at the nucleus and end up there again, but the initial and the final momenta may have different directions.

Delos and coworkers (1988) have investigated the structure of trajectories at low magnetic fields, in an effort to understand the transition from the elliptical, Kepler-type regime to the helical, Landau-type behavior; bifurcations play a critical role. More generally, but in less detail, Delande and Gay (1986a) have calculated surfaces of section for different values of E_c. The first signs of chaos appear when $E_c \simeq -.5$; the radial trajectory in $z = 0$ that was used in connection with (18.10) is seemingly the last to become unstable, at $E_c = -.127$.

Around the ionization threshold, the orbits out of the nucleus and back into it are isolated, i.e., they have only the energy as a continuous parameter; moreover, they maintain their basic shape over a relatively wide range of the energy, including $E = 0$. The magnetic field tends to confine them, at least in the x, y - directions because of the diamagnetic term in (18.12) and forces them to turn around rather than to escape to infinity.

The phase $S(E)$ is, therefore, written as an expansion in powers of the energy, $\Phi + E\,T$. The elementary relation (2.4) has been used to get $T = dS/dE$; both the phase Φ and the period T are evaluated at $E = 0$. The typical period T is very large in atomic units; if it is expressed in classical units, i.e., in terms of the cyclotron period $2\pi/\omega_c$, it will be called T_c which is smaller by a factor γ. Du and Delos give a list of 65 periodic orbits, probably all with $T_c < 10$; the list includes the various factors that make up the amplitude A, as well as the period T_c and the phase Φ. They also provide a picture for each; since the classical angular momentum $\gamma^{1/3}\ell_z$ is negligible, the orbits are confined to a plane through the z-axis. A coding scheme has been found only recently by Eckhardt and Wintgen (1990), a major achievement that will be mentioned in Chapter 20.

The classical oscillator strength $f_c(E)$ is now given by a formal expression very much like the trace formula (17.13), but we still have to carry out the smoothing operation which got us from (18.17) to (18.18). Very reasonably, Du and Delos argue that *the averaging over an interval of width ΔE is achieved, if the summation over the periodic orbits is restricted to $T < 2\pi/\Delta E$.* This idea will be given a mathematically clean expression in the next chapter on the basis of Selberg's trace formula; but in the absence of a rigorous formula, we continue to depend on physical arguments.

The various ingredients from the last few paragraphs now yield

$$f_c(E) = \sum_{T_k}^{2\pi/\Delta E} A_k \sin(T_k E + \Phi_k) \; ; \tag{18.19}$$

the individual orbits are distinguished by the index k, and are assumed to be ordered according as their period T_k; the sine rather than the exponential appears because of (18.17). The finite resolution ΔE is normal in a Fourier series such as (18.18), if the sum is limited to periods smaller than $2\pi/\Delta E$.

The discussion in the preceding section was phrased in terms of the usual trace $g(E)$, whereas in the present section we have concentrated on the oscillator strength $f(E)$. In both cases, we derived a classical expression for a quantum-mechanical function of the energy, with the help of the same reasoning which went into the trace formula (17.13). The trace $g_c(E)$ in the preceding section was treated as the Fourier expansion of $g(E)$ in the variable $\gamma^{-1/3}$ for fixed E_c, whereas (18.19) is the Fourier expansion of $f(E)$ in the variable E. The scaling with the help of $\gamma^{-1/3}$ applies just as well to the oscillator strength $f_c(E)$.

The expansion parameter $\gamma^{-1/3}$ has been interpreted so far as relating mostly to the magnetic field B through (18.11). Another view has a wider range of application, although it seems rather artificial: at a fixed value of B, the parameter $\gamma^{-1/3}$ varies as $1/\hbar$. Wintgen's use of the trace formula is tantamount to a Fourier series in $1/\hbar$, where each term $\exp(iS/\hbar)$ corresponds to the frequency S, and $1/\hbar$ plays the role of time. The classical limit appears in the same light as a Fourier series in the limit of large time.

18.6 The Chaotic Spectrum in Terms of Closed Orbits

The Hamiltonian for the hydrogen atom in a magnetic field has inversion symmetry through the origin, and is, therefore, expected to obey Gaussian Orthogonal Ensemble (GOE) statistics in spite of the lacking time reversal, because the Hamiltonian (18.6) is symmetric with respect to inversion at the origin. This fact has been confirmed through numerical calculations of the spectrum in strong fields by Wintgen and Friedrich (1986) as well as Delande and Gay (1986a), although there are deviations in the higher-order correlations such as the rigidity. Poisson statistics applies in the low-field region as expected, so that the Diamagnetic Kepler Problem provides a fine example of a transition in the spectrum from integrable to chaotic.

Optical absorption in a beam of hydrogen atoms going through a strong magnetic field can be measured with very high resolution. The reader has to study the experimental details in the papers of the group working in Bielefeld comprising Welge and his coworkers, in particular Holle et al. (1986) and Main et al. (1986). The spectrum can be sorted out experimentally into a fixed angular momentum and parity both in the initial and in the final state. Many of the lines are as narrow as .1 cm^{-1} which is the resolution of the UV-laser, and are spread over a typical range of 60 cm^{-1} A layperson looking at a plot of the output signal such as Figure 52, cannot help but call this kind of data chaotic, although it is quite reproducible.

The reliability of these results comes out very clearly, when they are compared with the spectrum which is obtained from diagonalizing a large matrix such as (18.14). The agreement is excellent, as shown in the work of Wintgen et al.(1986), covering dozens of lines, including their oscillator strengths; more such comparisons are given in Holle et al. (1987). Still, these results only show that Schrödinger's equation gives the correct foundation, but they do not really help us gain any insight into the physical processes involved.

The argument of Edmonds (1970), leading to the formula (18.10), is certainly a step in the right direction; it was put into a more convincing form by Reinhardt (1983). He started from a formula for the dipole photo-absorption cross-section quoted by Heller (1978),

$$\sigma(\omega) = 2\pi\omega \int_{-\infty}^{+\infty} d\tau \exp(i\omega\tau) < i|D_{op}(\tau)\, D_{op}(0)\, |i> \quad ,(18.20)$$

written in atomic units; the dipole operator is now taken in the Heisenberg representation as explicitly time-dependent; the system's

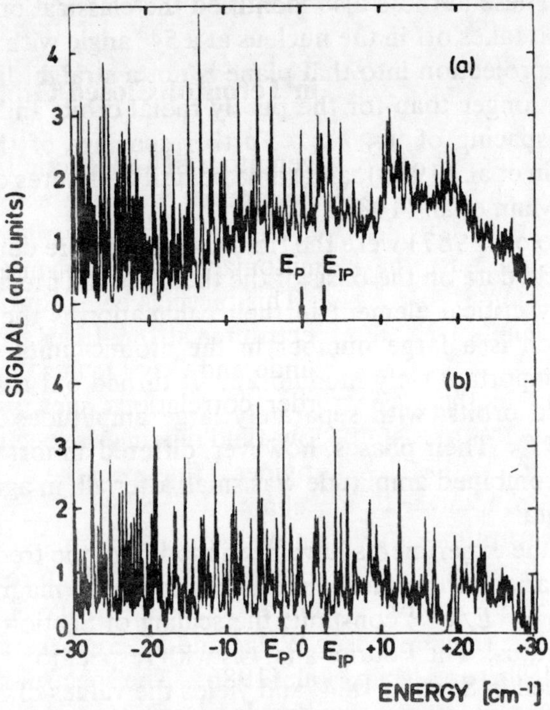

Figure 52 Excitation ionization spectrum of H-atom Balmer series around the ionization limit in a static homogeneous magnetic field of 5.96 T with spectral resolution of .07/cm; E_P = linear Zeeman shift, E_{IP} = Landau zero-point energy, E_0 = field-free ionization limit: (a) transition from ($2p, m = 0$) to $m = 0$ even parity; (b) from ($2p, m = -1$) to ($m = -1$) even parity [from Main, Wiebusch, Holle, and Welge (1986)].

initial state is again ϕ_i . This expression is a different version of (18.18); it uses the propagator K rather than Green's function G, and emphasizes the time development of the dipole moment.

The initial state is taken to be a wave packet near the nucleus; it moves away, say along the ρ-axis, and returns after a well-defined time τ_{recur}. At that time the matrix element in (18.20) becomes large, and the integral over τ gets a maximum at the frequency $2\pi/\tau_{recur}$. Conversely, if $\sigma(\omega)$ has been measured, its Fourier transform should yield the matrix element $< i|D_{op}(\tau) \bullet D_{op}(0)|i >$. Reinhardt remarks correctly that Edmond's reasoning is just the WKB version of this general argument; any reliance on classical trajectories leads to the same results as in the past two sections.

This kind of analysis got a big boost when Holle et al. (1986) made a complete Fourier analysis of their data and found a second peak corresponding to a level spacing of .64 $\hbar\omega_c$, in addition to the spacing of 1.5 $\hbar\omega_c$. These authors also identified the classical orbit leading to this spacing: it takes off in the nucleus at a 54° angle with respect to the xy-plane; its projection into that plane is not a straight line; its period is, of course, longer than for the purely radial orbit, and accounts for the reduced spacing of .64 $\hbar\omega_c$. Further spacings of this kind were found by Main et al. (1986), and some beautiful pictures of these orbits are given in Main et al. (1987).

Du and Delos (1987) were the first to make a more detailed analysis of the Bielefeld data on the basis of the theory in the preceding section. A particularly critical element in their calculation is the phase Φ_k in (18.19), which is a large number in the atomic units whose value, however, is important only modulo 2π. It turned out that two otherwise unrelated orbits, with separately large amplitudes A_k, had very close periods T_k. Their phases, however, differed almost exactly by π, so that their combined amplitude was small after all, in agreement with the optical data!

Recently, the experiments have been modified; the frequency of the UV-laser is varied at the same time as changing the magnetic field, so as to keep $E_c = E/\gamma^{2/3}$ constant; the scaling of Section 18.4 is built into the apparatus! The Fourier analysis with respect to $\gamma^{-1/3}$ is carried out as explained in Section 18.4, and yields the values of $\overline{S}_k(E_c)$.

Holle et al. (1988) have varied E_c in small steps from -.5 to +.2 ; the values of $\gamma^{1/3}$ where the Fourier amplitude is large, lead to a set of curves in the E_c versus $\gamma^{1/3}$ plane which gives a good picture of the classical orbit structure. The authors have identified a number of classes of orbits, called primary vibrators, primary rotators, and "exotics," whose representatives can be followed from the regular into the chaotic region, including various bifurcations. The origin of the chaotic features in the experimental spectrum is, therefore, directly related to the classical orbits. Figure 53 shows both the experimental spectrum and its Fourier transform with respect to $\gamma^{-1/3}$; the most prominent peaks are identified with closed orbits.

In contrast to the original analysis which led to the quasi-Landau spectrum by Garton and Tomkins (1969), the regularities associated with the classical orbits are not easily recognized in the output signals of these more sophisticated experiments. The classical periodic orbits manifest themselves as correlations that are found when the spectrum is subjected to a Fourier analysis.

The analysis has not been pushed to the point where the confusing set of sharp peaks can be shown to result directly from a superposition of terms as in (18.19) that would require very many terms with long periods, i.e., a much smaller energy interval ΔE. The second equality in (18.15) gives a qualitative idea of the physical process leading to the

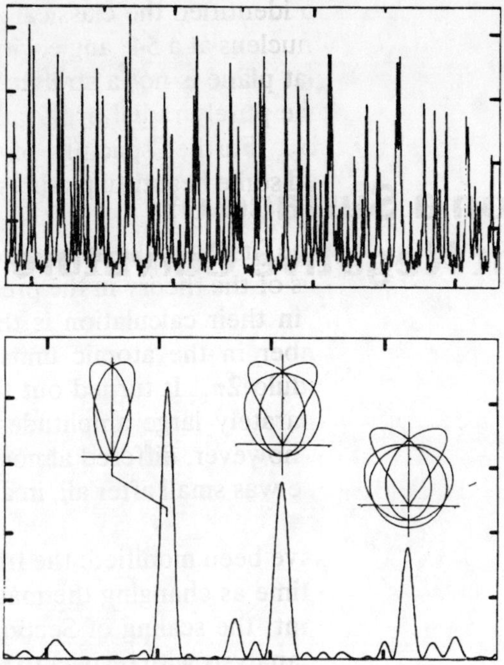

Figure 53 (a) Scaled-energy spectrum at E_c = -.45 as a function of $\gamma^{-1/3}$ in the range of excitation energy -77.7/cm $\leq E \leq$ -54.4/cm and the the range of magnetic field 5.19 $\geq B \geq$ 3.03 T; (b) Fourier-transform with resonances correlated to closed orbits as projected onto the (ρ, z) plane [from Holle, Main, Wiebusch, Rottke, and Welge (1988)].

formation of regular features in the spectrum, such as the bunching of energy levels at the intervals specified in (18.19); but this relation does not give us the exact location of approximate eigenvalues, as one expects from the WKB quantization condition.

The apparent fallacy of quantizing a single isolated periodic orbit was mentioned at the end of Section 17.9, and appears now in a new light. The condition (18.15) gives the energies in whose neighborhood the level density is higher than average. The shortest periodic orbits have the strongest effect. In intervals of low average level density, as at the bottom of the spectrum, they may even yield the individual energy levels, as happened in the author's earliest work on the Anisotropic Kepler Problem (Gutzwiller 1971). It may not be possible to recognize such short periodic orbits in the lowest eigenstates, however, because the corresponding wave functions have long wave lengths and very little structure.

CHAPTER 19

Motion on a Surface of Constant Negative Curvature

The first concrete example of a dynamical system with hard chaos was given by *Jacques Hadamard* in 1898. Although he was primarily a geometer, to use the traditional name for the members of the mathematics section in the French Academy of Sciences, he saw the physical application immediately. A short account of his work in the same year carries the title, *On the billiard game on a surface of negative curvature.*

His arguments are all given in the language of differential geometry; this field started as the study of arbitrary two-dimensional curved surfaces imbedded in three-dimensional Euclidean space. Hadamard chose to express himself in this manner and make his ideas easily accessible and intuitively appealing. As a bonus, he shows that all the important features in this model depend on the curvature being negative, but not necessarily constant, and that most of the reasoning can be phrased in terms of the Gauss-Bonnet theorem (19.10).

While this approach is very helpful to the student who likes to have a picture to fix her ideas, it does not embrace some interesting examples of hard chaos; they can only be reached in the more abstract setting of Riemannian geometry. Indeed, Hilbert (1901) proved that it is not possible to imbed a smooth, compact, and complete surface of strictly negative curvature into three-dimensional Euclidean space $\mathbf{R}^3$.

Poincaré and Klein, on the other hand, gave very simple and effective constructions of such two-dimensional Riemannian spaces of constant negative curvature as polygons in the hyperbolic geometry of

Lobatchevsky. If the reader gets terrified by the idea of *non-Euclidean geometry,* he should devote some time to get acquainted with Poincaré's model. The basic concepts of Euclidean geometry, such as straight lines, angles, translations, and rotations around a point, are all represented by traditional objects; in particular, nothing more complicated than ordinary circles are ever used and all the classical problems can be solved by elementary means.

The calculations in this model are also reduced to what must be a minimum of algebraic complication, namely dealing with groups of 2 by 2 real matrices, normalized to have a determinant = 1. Most remarkably, an amazing variety of different possibilities and phenomena is contained in this seemingly tight, mathematical framework. The author (Gutzwiller 1980) was the first to emphasize that the many results of hyperbolic geometry, in particular Selberg's trace formula (19.20), are of central importance for the understanding of quantum chaos. Although a systematic account of this rich field cannot be condensed into one chapter, the essential ingredients will be presented to make some of the recent work by physicists accessible to the reader.

The whole discussion will be limited to two dimensions. Some of the methods can be generalized to three dimensions, while keeping the elementary character of the Poincaré model. A first step in this direction has been taken by Tomaschitz (1989), but there are significant differences, both geometric and algebraic, that would require a whole new chapter to explain (cf. Gutzwiller 1985b).

The goal is twofold: On the one hand, Selberg's trace formula is explained in some detail, because it is the testing ground for the more general form (17.13) of the trace formula; the relation between classical and quantum mechanics in a chaotic system has to be understood in this special context. On the other hand, the possibly chaotic features of quantum mechanics itself, whatever they turn out to be, have a better chance to manifest themselves on surfaces of negative curvature than anywhere else; also, they may be more easily computed thanks to Poincaré's model, as will be demonstrated explicitly in a scattering problem.

19.1 Mechanics in a Riemannian Space

A particle of mass m which moves with velocity v on the surface of a sphere of radius R, experiences a force directed toward the center of the sphere, and equal to mv^2/R. If the surface is a triaxial ellipsoid, the force keeping the particle there is given by (3.9) with (3.10). This view of a particle, as being forced to stay on the given surface S by applying the appropriate force along the normal, is unnecessarily complicated; the equations of motion for the particle can be written directly in terms of the *two intrinsic coordinates* on the surface, without using *the third, extrinsic coordinate.*

The extrinsic coordinate is ambiguous in most cases; not only can the location of S in $\mathbf{R}^3$ be shifted, and its orientation turned; but it can be deformed without in any way changing the trajectories with respect to the intrinsic coordinates. It takes a rather difficult theorem to show that a complete surface of strictly positive curvature, like the sphere or the ellipsoid, can actually not be deformed; but an incomplete one, like one-half of a grapefruit shell, or any piece of negative curvature, can be deformed 'isometrically', a term to be explained shortly along with the concept of completeness for a surface S.

Before going on, however, the notion of curvature has to be defined. It starts again from imbedding the surface S in $\mathbf{R}^3$; at the point P we erect the normal, and provide it with a direction called 'inward'. A plane in $\mathbf{R}^3$ containing the normal, intersects S in a curve C which has a well-defined curvature κ in P, positive if C is turned inward, and negative if C is turned outward. As the plane through the normal is allowed to rotate around the normal, κ has a maximum value κ_1 and a minimum value κ_2. The *Gaussian curvature* K at P is defined as $K = \kappa_1 \kappa_2$.

If $K > 0$, the surface looks locally like a sphere, although it would be more exact to compare it locally to an ellipsoid; similarly, if $K < 0$, the surface looks locally like a piece of the one-sheeted hyperboloid, i.e., like the top of a mountain pass. An approximate description of S in $\mathbf{R}^3$ near P would have to specify not only K, but also the mean curvature $(\kappa_1 + \kappa_2)/2$; the neighborhood of a point on an ellipsoid (where $\kappa_1 \neq \kappa_2$) may look quite different from a point on the sphere (where $\kappa_1 = \kappa_2$), since the mean curvatures are different, even though $K = \kappa_1 \kappa_2$ is the same.

Gauss discovered that K is all that matters when only the intrinsic properties of a surface are of interest, i.e., those that are independent of the way S is imbedded in $\mathbf{R}^3$; it is called his *theorema egregium* where the word egregious has its original meaning as 'outstanding'. In particular, the geodesics on S, the shortest connections between two, not too distant, points do not depend on the mean curvature; they are also the trajectories of a particle which moves freely on S.

The intrinsic description of an f-dimensional Riemannian space is made in terms of the position coordinates $q = (q_1, ..., q_f)$, and the local metric ds. In an obvious generalization of Pythagoras's theorem, the distance of two neighboring points is given by a quadratic function in the differentials,

$$ds^2 = \sum_{j,k} g_{jk}\, dq_j\, dq_k \ , \tag{19.1}$$

where the *metric tensor* g_{jk} is a sufficiently smooth function of the coordinates q.

Two Riemannian spaces, the first with coordinates q' and metric tensor g', the second with coordinates q'' and metric tensor g'', are called *isometric* if there exists a transformation $q' \to q'' = h(q')$ such that the distance elements ds' and ds'' become equal.

The *volume-element* is $dV = \sqrt{g}\ dq_1 ... dq_f$ with $g = \det(g_{jk})$. A direction at a point q is given by a unit vector $u = (u_1, ..., u_f)$ where $\Sigma g_{jk} u_j u_k = 1$. The *angle* $0 \leq \alpha \leq \pi$ between two such directions u' and u'' is given by $\cos\alpha = \Sigma g_{jk} u'_j u''_k$.

The velocity of a particle is ds/dt ; only the kinetic energy enters into the *Lagrangian* for a freely moving particle, so that

$$L(\dot{q}, q) = \frac{m}{2} \sum_{jk} g_{jk}\, \dot{q}_j\, \dot{q}_k \ , \tag{19.2}$$

where $\dot{q}_j = dq_j/dt$ as in Chapter 1. The transition from Lagrangian to Hamiltonian mechanics is made according as the rules in Chapter 2 with the help of (2.1), so that the *Hamiltonian* becomes

$$H(p, q) = \frac{1}{2m} \sum_{jk} g^{jk}\, p_j\, p_k \ , \ \text{with}\ p_j = \sum_{\ell} g_{j\ell}\, \dot{q}_\ell \ . \tag{19.3}$$

The tensor g^{jk} is the inverse of g_{jk}, i.e., $\Sigma g_{j\ell}\, g^{\ell k} = \delta_{jk}$.

The metric tensor is symmetric, and positive definite in Riemannian geometry; in the better known application of Riemannian geometry to general relativity, the metric tensor is semidefinite, with one positive and three negative eigenvalues, as in the relativistic Hamiltonian (7.11) of the electron. Except for the formal similarities, such as for a Hamiltonian like (7.11), and the resulting equations of motion, the two situations are entirely different; we will strictly adhere to the assumption of a positive definite metric. Also, except where stated explicitly, we will stay with two dimensions, $f = 2$.

The curve Γ in the Riemannian space which is defined by a solution of the equations of motion (1.3) resulting from the Lagrangian (19.2), or (2.2) from the Hamiltonian (19.3), is called a *geodesic*. The integral $\int ds$ along a geodesic, with ds given by (19.1), defines a length; distances in a Riemannian space will always be measured along a geodesic. Since the velocity for the particle is defined as $v = ds/dt$, its energy becomes simply $E = mv^2/2$ as in Euclidean space. A Riemannian space is called *complete* if every geodesic can be continued indefinitely, as on a sphere, but not on a half-sphere.

If s is the distance from q' to q'', as measured along a geodesic which joins these two points, we find for Hamilton's principal function R and for the action integral S,

$$R(q''\, q'\, t) = m\, s^2/2t \;, \quad S(q''\, q'\, E) = \sqrt{2mE}\;\, s \;, \qquad (19.4)$$

in exact agreement with (1.5) and (2.5) in Euclidean space.

The equations of motion for a geodesic through an arbitrary point P_0 can be solved in powers of the time t, if we expand the metric tensor g_{jk} in powers of the coordinate difference $q_j - q_{0j}$. In this way, we obtain *geodesic circles* centered in P_0 as the locus of points at a fixed distance r from P_0. The length L of the circumference, and the area A of such a circle, become

$$L = 2\pi r - \pi K r^3/3 + \ldots \,, A = \pi r^2 - \pi K r^4/12 + \ldots \,, \qquad (19.5)$$

where the omitted terms in the expansions have higher powers in r. The coefficient K combines various derivatives (up to the second) of the metric tensor, all evaluated at P_0; it is the *Gaussian curvature* at P_0, which is now seen to have a simple, intrinsic, geometric meaning. Obviously, such a definition makes no longer any reference to an imbedding in Euclidean space $\mathbf{R}^3$.

Let now Γ_0 be some fixed geodesic, and write its coordinates as a function of the distance s measured along it; then consider a geodesic Γ nearby. Through the point P on Γ_0 with the distance parameter s, draw the geodesic perpendicular to Γ_0, until it intersects Γ; the distance $\zeta(s)$ between the two geodesics at P is defined thereby. The differential equation for Γ, as obtained from the equations of motion for the particle, can be reduced to a linear second-order equation for $\zeta(s)$, in lowest order with respect to the distance between Γ and Γ_0,

$$d^2\zeta(s)/ds^2 = -\; K(s)\,\zeta(s)\;. \qquad (19.6)$$

The coefficient $K(s)$ is again the Gaussian curvature in P.

This relation is sometimes called *Jacobi's equation;* it may be used to define the Gaussian curvature K at P. It tells us to what extent the neighboring geodesic Γ gets pulled back toward Γ_0 if $K > 0$, or pushed away from Γ_0 if $K < 0$. Thus, the Gaussian curvature K becomes a local measure of the stability of the geodesic; $K > 0$ as on the sphere means stability, whereas $K < 0$ implies instability; $\log\sqrt{K}$ plays the role of the Lyapounoff exponent in (1.6) or again in (10.7).

19.2 Poincaré's Model of Hyperbolic Geometry

The hyperbolic world as first imagined and investigated by Lobatchevski and Bolyay in the first half of the nineteenth century is not so different from our own Euclidean world. It has all the familiar objects, like straight lines and planes, and most importantly, it has the same basic symmetries which lead to a six-parameter invariance group of translations and rotations; distances and angles are defined in a natural manner. Can we draw pictures of it, even if they are distorted?

Among several equivalent models, we will use only Poincaré's *upper half-plane U* for the two-dimensional hyperbolic geometry because it is technically the easiest to handle. There is a three-dimensional generalization that is even more amazing in its hidden richesse; but much more work on the classical as well as on the quantal aspects has been done in two dimensions. There is obviously still a large field waiting to be explored.

Poincaré's model can be approached from several directions; e.g., Balazs and Voros (1986) start with one sheet of a two-sheeted hyperboloid in three dimensions, which they call the *pseudosphere*. We will get straight to the final product: a Euclidean plane with a Cartesian coordinate system (x, y), where we consider only the upper half $y > 0$.

There is a slight semantic problem with the names of the geometric objects: since we are dealing with a model, one familiar object like a circle in Euclidean geometry may represent a completely different object like a straight line in hyperbolic geometry. We will, therefore, attach the adjective 'Euclidean' or 'hyperbolic' whenever necessary, so that the reader knows which one of the two interpretations is called for.

The points in U are best designated by a complex number $z = x + iy$ with $y > 0$; but we will not limit our attention to analytic functions, also called harmonic functions $f(x, y)$ satisfying Laplace's equation $\Delta f = 0$, as is usual in the theory of complex variables. The complex notation is useful, however, when the translations and rotations of the hyperbolic plane are investigated.

Hyperbolic geometry is invariant with respect to the *Möbius transformations*

$$z \to z' = \frac{a z + b}{c z + d} \quad \text{with } a d - b c = 1 , \tag{19.7}$$

where a, b, c, d are real numbers. It is easy to check that these transformations form a group under the usual multiplication of 2 by 2 matrices; we will write those sometimes in the *flattened notation* $[\, a, b; c, d\,]$.

There are three continuous parameters in this group of *Special* (because their determinant = 1) *Linear* (if applied to vectors rather than complex numbers) transformations of *2-component Real* vectors $SL(2,R)$. When the quotient with $\{\pm I\}$ is formed in (19.7), this group is reduced to its *Projective* normal subgroup $PSL(2,R) = SL(2,R)/\{\pm I\}$, also called the *Möbius group.* This group plays the same role in hyperbolic geometry as the group of rotations in spherical geometry, and the translation-rotation group in Euclidean geometry.

The distance s between two points, z_1 and z_2, is given by the formula

$$\cosh s = 1 + \frac{(x_2 - x_1)^2 + (y_2 - y_1)^2}{2\, y_1 y_2}\,,\ ds^2 = \frac{dx^2 + dy^2}{y^2}\,. \tag{19.8}$$

The second equation is obtained, if the left-hand side of the first equation is expanded in s, and the right-hand side is expanded in powers of the coordinate differences; this metric is easily checked to keep the same functional form under the transformation (19.7).

The angle α between two directions in hyperbolic geometry is the same as the Euclidean angle of the corresponding two directions in U; the model is, therefore, called *conformal.* The lengths, however, are badly distorted. Notice that all the points on the x-axis are infinitely far away, and so is the *point at infinity,* i.e., $y \to \infty$ while x = constant. The integral $\int dy/y$ for the length along the y-axis does not converge as y goes either to 0, or to ∞.

The geodesics are represented by Euclidean circles whose center is on the real axis; they play the role of the straight lines. They are infinitely long; but in this model, they can be given two *end-points,* namely the coordinates on the real axis, ξ and η, where the corresponding Euclidean circle intersects the real axis. If these geodesics become the trajectory for a particle, then there is a direction of motion. We will, therefore, adopt the convention that the particle comes from ξ and moves toward η, as in Figure 54a.

Instead of Poincaré's upper half-plane U, it is sometimes convenient to use a closely related model, also due to Poincaré; U with coordinates $z = x + iy$ is mapped along with all its geometric objects into the *unit circle* $u^2 + v^2 < 1$ with the complex coordinates $w = u + iv$ through

$$w = \frac{-z + i}{z + i}\,,\ z = i\,\frac{1 - w}{1 + w}\,. \tag{19.9}$$

Since circles go into circles in such bilinear maps of the complex plane, and angles are preserved in any analytic map, the geodesics are now represented by Euclidean circles which intersect the unit circle at a right angle, as shown in Figure 54b. The metric in the unit circle is given be $ds^2 = 4(du^2 + dv^2)/(1 - u^2 - v^2)^2$.

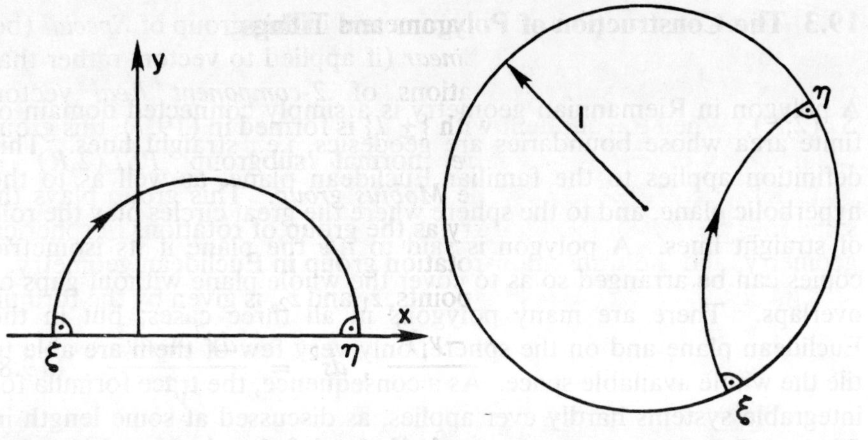

Figure 54 A straight line in Poincaré's model of hyperbolic geometry is a Euclidean circle that intersects the boundary at a right angle, (a) in the upper half-plane U as well as (b) in the unit circle; the free motion of a particle goes from the point ξ on the boundary to the point η.

The typical reasoning in hyperbolic geometry may be illustrated by the following example: Let us try to calculate the curvature at some arbitrary point z in U! As a first step, z is moved into the point i ; according as (19.7), this requires finding $[a, b; c, d]$ such that $z = x + iy = (ai + b)/(ci + d)$ with $ad - bc = 1$. There is more than one solution; this ambiguity simply means that we are still free to rotate around the point i. Now we go into the unit circle with (19.9) so that i goes into $w = 0$. The geodesics through 0 are represented by the Euclidean straight lines, and there is circular symmetry around the origin; the hyperbolic distance of a point $w = u + i\,0$ from the origin is $r = 2\int du/(1 - u^2)$, so that $u = \tanh(r/2)$. The hyperbolic circumference of a geodesic circle with radius r is $L = 2\pi\, 2u/(1 - u^2)$; if we insert for u its value in terms of r, it follows that $L = 2\pi \operatorname{Sinh}(r)$. Finally, we apply (19.5) to find the Gaussian curvature $K = -1$.

One could have used a general formula for K in terms of the metric tensor g_{jk} and its derivatives; this might have entailed some tedious, but not difficult calculations. Instead, we have used first the translational invariance to move z into i, and then the invariance with respect to rotations around the origin in the unit-circle model. The calculations become almost trivial, and they allow us to understand the overall picture much better.

19.3 The Construction of Polygons and Tilings

A polygon in Riemannian geometry is a simply connected domain of finite area whose boundaries are geodesics, i.e., straight lines. This definition applies to the familiar Euclidean plane, as well as to the hyperbolic plane, and to the sphere where the great circles play the role of straight lines. A polygon is said to *tile* the plane if its isometric copies can be arranged so as to cover the whole plane without gaps or overlaps. There are many polygons in all three cases; but in the Euclidean plane and on the sphere, only very few of them are able to tile the whole available space. As a consequence, the trace formula for integrable systems hardly ever applies, as discussed at some length in Section 17.3.

There is a very general formula for the area of polygons in Riemannian spaces, which is the global version of Jacobi's equation (19.6). Consider an n-sided polygon with the vertices $P_1, ..., P_n$, enclosing the simply connected domain D; it is assumed to be on the left as one goes counterclockwise around the polygon, as shown in Figure 55. The tangent vector has to rotate counterclockwise through the *outer angle* α_j at the vertex P_j ; we will restrict this angle to $-\pi < \alpha_j \leq +\pi$, although one could allow larger or smaller angles at the price of considering multiple coverings of D.

The *Gauss-Bonnet theorem* states that

$$\int_D K\, dS = 2\pi - \sum_j \alpha_j ; \tag{19.10}$$

in words, the sum of the outer angles fails to add up to 2π by the amount $\int K\, dS$, i.e., the total Gaussian curvature inside D.

This proposition contains as a special case the statement from elementary geometry ($K = 0$) that the sum of the inner angles equals π in a triangle, whereas in spherical geometry ($K = 1/R^2$), the area equals $(2\pi - \Sigma\alpha_j)\, R^2$. In hyperbolic geometry, this last relation is reversed since we get the area of a polygon as $(\Sigma\alpha_j - 2\pi)/|K|$. We will pay special attention to *singular polygons* in the last sections of this chapter, where $\alpha_j = \pi$; the vertex P_j is then located on the real axis of U or at ∞.

The simplest kind of tiling is obtained as follows: copies of the original polygon are translated and rotated so as to join along boundaries, until a simply connected domain is filled without gaps or overlaps. In the case of the sphere, a finite number of polygons fills out the full area of 4π; the most famous tilings are the projections onto a concentric sphere of the five regular polyhedra , tetrahedron, cube, etc.; but there

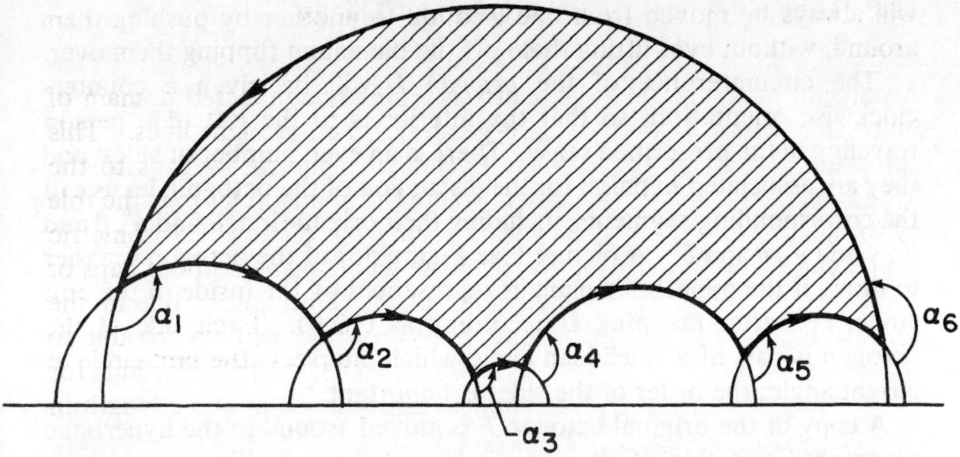

Figure 55 Boundary of an arbitrary polygon consisting of straight line segments; the inside of the polygon is on the left when going counterclockwise around the perimeter; the outer angle α_j at the vertex P_j is counted counterclockwise.

are others which can be derived from them. If the radius of the sphere is fixed at 1, none of these tilings has a further parameter to vary.

The Euclidean plane is already much trickier. If we insist on the required translations and rotations of the original polygon forming a group, there are very few groups, although the shapes of the polygons can vary within certain limits. The group property is the crucial ingredient in finding an integrable system and being able to apply the method of images. Some tilings have a continuous parameter besides a scale, e.g., parallelograms with different angles and side ratios. Quite recently, tilings of the Euclidean plane without the group property have been discovered and are studied extensively (cf. Grünbaum and Shephard 1987); some of these are projections of regular tilings in spaces of higher dimensions. Maybe some method of images still applies, and the trace formula takes on a pseudointegrable appearance.

There is a large variety of tilings in the hyperbolic plane, all of them with continuous parameters besides the Gaussian curvature; we will set $K = -1$ henceforth. The construction of a tiling polygon will be explained in this section; but we will not attempt to show how to obtain all possible tilings. The octogon F will be our principal example for a compact tile; the simplest non-compact tile of finite area will be disussed in Section 19.7. A slight knowledge of topology will be helpful; but every reader will be able to follow the simple arguments.

The transformations (19.7) preserve the orientation; we will not consider reflexion symmetries in this simple discussion; the polygons will always be moved from one location to another by pushing them around, without ever lifting them off the plane and flipping them over.

The circumference of the *octogon* F will be given a counter-clockwise orientation, so that the interior is to the left of a person traveling in the prescribed sense. There is an even number of sides, and they are associated in pairs; for the discussion of the octogon, let us call the corresponding segments, including their orientation, α and α', β and β', γ and γ', δ and δ'. A picture (see Figure 56) of the octogon is easier to draw, if the hyperbolic plane is represented by the inside of the unit circle, i.e., after mapping U according as (19.9). Each side of the octogon is part of a Euclidean circle which intersects the unit circle at a right angle; the order of the sides is important.

A copy of the original octogon F is moved around in the hyperbolic plane until one side of the copy is brought into congruence with its partner in the original octogon; e.g., α in the copy is brought into congruence with α' in the original. Notice that when the two octogons are put next to each other, the arrow on α in the copy points in the direction opposite to the arrow on α' in the original. If a curve is trying to leave the original octogon through the side α', it reenters the copy through the side α. If we want to continue showing the curve in the original octogon, we have to transform the neighborhood of α' back into α; the name for this operation is A. Similarly, we define the operations B: $\beta \to \beta'$, C: $\gamma' \to \gamma$, and D: $\delta \to \delta'$; notice the slight differences in these definitions.

Thus, four hyperbolic transformations, A, B, C, and D, each given by a 2 by 2 real matrix of determinant 1, have to be found to define the four congruence operations for the octogon. Among the many possibilities, we will discuss only the assignment in Figure 56, where the counterclockwise sequence of the sides is $\alpha\,\beta\,\alpha'\,\beta'\,\gamma\,\delta\,\gamma'\,\delta'$. Will this assignment give a tiling of the hyperbolic plane? This question can be answered if we know how the various copies of the original fill out the angle 2π around a particular vertex. Again without going into the deeper reasons for this crucial step in the argument, let us just follow the recipe. The reader will find an exhaustive, yet readable account of this combination of geometry and algebra in the textbook of Beardon (1983).

Suppose that we are trying to go counterclockwise around the vertex numbered 1 in Figure 56, where the side δ' joins the side α, starting in P_0 on α and moving as indicated by the little arrow. When our little circuit hits δ', we have to move over to δ using the operation D^{-1}, which brings us near the vertex called 2 in Figure 56. Now we continue to go counterclockwise, but this time around the vertex 2, until we hit the side γ, and we have to move over to γ' with the help of C^{-1} so as

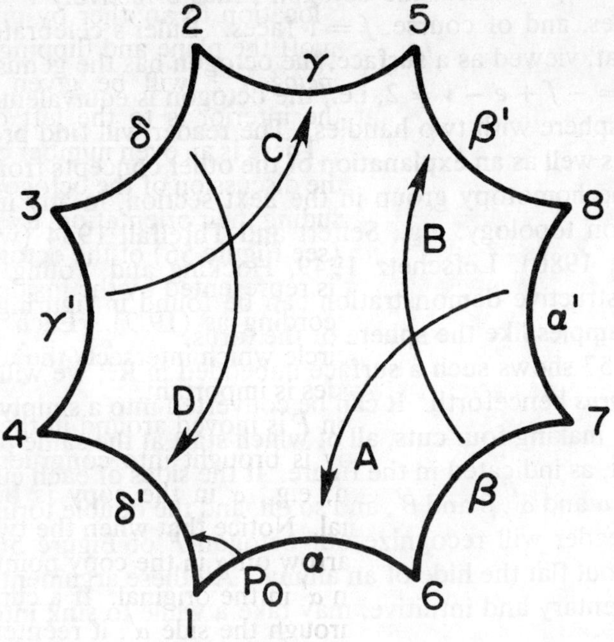

Figure 56 The octogon in the unit circle with the identification of different sides, α' being mapped into α by A, etc.; a small circuit starting in P_0 to go around the vertex 1 ends up going around all the other vertices as well, and finishes in P_1 near P_0 ; the octogon in this figure has all inner angles equal to $\pi/4$, corresponding to the 'regular octogon', but the designations and identifications are valid for any octogon.

to continue our trip around the vertices. We are now near the vertex 3, and will hit the side δ before long, as we go counter-clockwise; the operation D then gets us to the vertex 4 on δ'. As we continue counterclockwise around 4, we hit γ' and need C to get to vertex 5 on γ, and so on.

Eventually, after the counterclockwise journey around vertex 8 has been completed, and the operation A has brought us back to α near vertex 1, we find ourselves in a point P_1, which is a copy of the starting point P_0. The neighborhood of P_0 has undergone a succession of transformations, first D^{-1}, followed by C^{-1}, then D, etc., and ultimately A. By the vicissitudes of the customary notation, this whole sequence gets its name in the reverse order, $ABA^{-1}B^{-1}CDC^{-1}D^{-1}$.

The eight vertices of the octogon are now recognized as representing only one single point. Similarly, each side and its partner can be viewed as one segment, if we agree to deal with the octogon in the hyperbolic plane in the same manner as with a parallelogram in the Euclidean plane, and identify opposite sides. In a somewhat strained analogy with polyhedra, the octogon F has effectively $v = 1$ vertices, $e = 4$ edges, and of course, $f = 1$ faces. Euler's celebrated theorem tells us that, viewed as a surface, the octogon has the genus g given by $2(g-1) = -f + e - v = 2$, i.e., the octogon is equivalent to the surface of a sphere with two handles. The reader will find proofs of this formula, as well as an explanation of the other concepts from topology, such as the homotopy group in the next section, in any introductory textbook on topology, e.g., Seifert and Threlfall 1934 (with English translation 1980), Lefschetz 1949, Hocking and Young 1961. An equally instructive demonstration can be found in figuring out a few special examples like the sphere or the torus.

Figure 57 shows such a surface imbedded in $\mathbf{R}^3$; we will also call it a *double torus* henceforth. It can be converted into a simply connected domain by making four cuts, all of which start at the same point P, and return to it, as indicated in the figure. If the sides of each cut are given the names α and α', β and β', and so on, and the double torus is laid out flat, the reader will recognize our octogon F of Figure 56; it is like stretching out flat the hide of an animal. All these arguments, although quite elementary and intuitive, may take a while to sink into the reader's mind.

To what extent can F in the hyperbolic plane be compared to the double torus of Figure 57 ? It is natural to associate the single vertex of the octogon with the point P on the double torus from which the four cuts were made. P does not have any distinguishing features; it belongs to a neighborhood where the double torus is smooth. In particular the eight corners, which come from the four cuts, fit together so as to fill out the full angle of 2π around P. The same situation has to obtain for the octogon.

The condition to be imposed on the octogon requires, therefore, that upon completion of the circuit around the vertex, the neighborhood of the starting point P_0 in the hyperbolic plane coincides with the neighborhood of the endpoint P_1; or equivalently,

$$A\,B\,A^{-1}\,B^{-1}\,C\,D\,C^{-1}\,D^{-1} \;=\; I\,(\text{Identiy})\;, \qquad (19.11)$$

in terms of the 2 by 2 matrices. This condition also insures that the eight angles on the vertices of F add up to 2π; we shall not attempt to show that this condition is indeed sufficient for the tiling.

Since the sum of the inner angles in the octogon equal 2π, the outer angles add up to 6π; the Gauss-Bonnet theorem yields, therefore, the area 4π for the double torus, the same as for the sphere. As a consequence, Weyl's formula tells us that the average density of states for the Laplacian on the double torus is 1. The simplest realization of the octogon makes all the inner eight angles equal to $\pi/4$; the octogon then

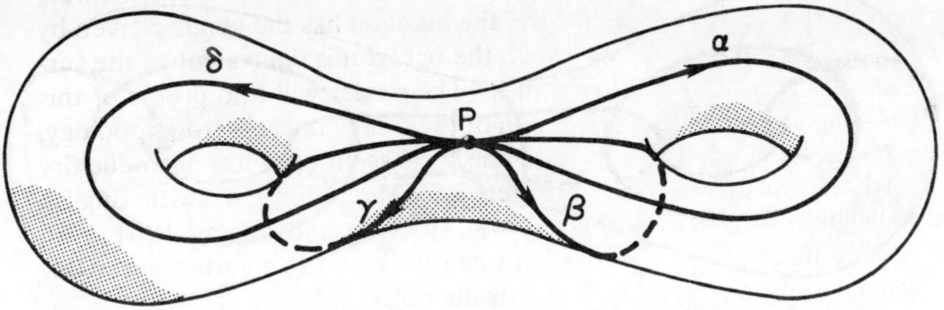

Figure 57 Surface of genus 2, alias Double Torus (or sphere with two handles attached), embedded in three-dimensional Euclidean space; four contours starting and ending the arbitrary point P are used both to define the homotopy group G, and to show the relation with the octogon of the Figure 56.

takes the regular shape shown in Figure 56; we will refer to this special case as the *regular octogon.*

It is important to count the number of parameters which this construction allows us to choose freely. Each real 2 by 2 matrix with determinant $=1$ has three parameters; since we have to find four such matrices, there are twelve unknowns to be determined. The equation (19.11) spells out the equality of two matrices, where the condition on the determinant $=1$ is already satisfied on both sides; thus, (19.11) represents three independent algebraic equations for the matrix elements of A, B, C, and D. The location of the octogon in the hyperbolic plane is not relevant; two solutions of (19.11), A, ... and A', ..., describe the same octogon, if they differ only by a change of coordinates, $A' = T^{-1} A T$, where T is an arbitrary regular matrix. Therefore, with twelve unknowns, three algebraic equations, and three parameters to locate and orient the octogon in the hyperbolic plane, we are left with six essential parameters.

A systematic procedure to define these six parameters in terms of certain natural quantities on the double torus has been discussed by Keen (1966a and b). One particular choice is shown in Figure 58: the six closed loops, C_1 , ..., C_6, are topologically different; as will be discussed in the next section, there is exactly one closed geodesic (periodic orbit) for each such topologically independent loop. The lengths of these six closed geodesics, Γ_1 , ..., Γ_6, determine the shape of the octogon. Two octogons with different sets of these lengths are not isometric, i.e., they cannot be mapped into each other so that the length of all corresponding curves is preserved.

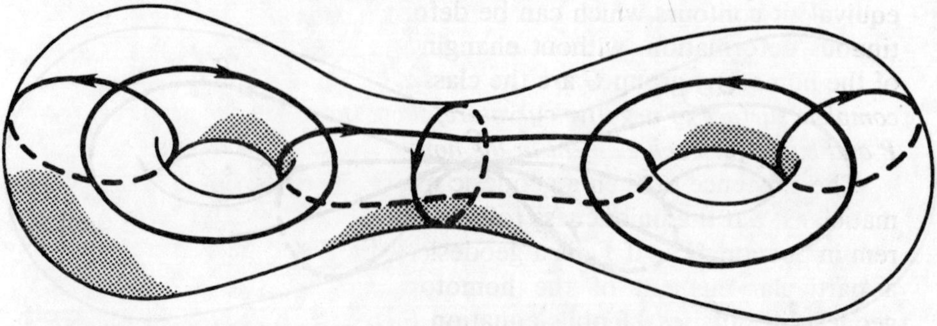

Figure 58 Six closed loops whose minimal length uniquely characterize a double torus; they can be interpreted as giving the physical parameters for a particle that is trapped in a two-dimensional double well.

Although this six-parameter family of double tori was constructed in a rather abstract manner, it has an appealing physical interpretation. Each torus represents a box in which the particle moves; these two boxes are coupled by the pipe connecting them. The shape of each box is specified by two parameters, the lengths of Γ_1 and Γ_2 for the first box, and the lengths of Γ_5 and Γ_6 for the second box; notice that a Euclidean box, i.e., a parallelogram, is also specified by two parameters in addition to a scale. Finally, the pipe between them is specified by its length as given by the length of Γ_3, and by its circumference as given by the length of Γ_4. The octogon is seen as a purely geometric model for a particle in a two-dimensional double well, with all the freedom to imitate the various features of a realistic physical situation, such as the *hindered rotation in a molecule.*

19.4 The Geodesics on a Double Torus

The construction of the geodesis on a surface of negative curvature is closely related to its topology, as Hadamard (1898) was the first to recognize; the essential tool is the *homotopy group G* which was invented by Poincaré. A point like P in Figure 57 is chosen; the four oriented loops α, β, γ, and δ are defined so as to start in P and return to it; every other contour Γ on the double torus, from P and back, can be obtained by deforming continuously a particular sequence of α, β, γ, δ, and their inverses.

These oriented contours form a group; the product of two contours, $\Gamma_1\Gamma_2$, is the contour which first follows Γ_1, and then Γ_2; the inverse,

Γ^{-1}, is the contour Γ with its orientation reversed. There are many equivalent contours which can be deformed into one another by continuous deformation, without changing the orientation; the elements of the homotopy group G are the classes of equivalent contours. *On a complete surface of negative curvature, there is exactly one geodesic, from P and back, for each element in the homotopy group.*

The existence of such a geodesic may be a problem for the mathematicians; but the uniqueness follows immediately from Morse's theorem in Section 1.5. If Γ_0 is a geodesic from P and back, belonging to a particular element of the homotopy group, then a neighboring geodesic Γ satisfies Jacobi's equation (19.5). If both start in P, they will not intersect anymore; there are no conjugate points when the Gaussian curvature $K < 0$. Therefore, the second variation of the length of Γ_0 is positive; the geodesic realizes a minimum in its equivalence class of contours. There is no second minimum of this kind, because if the first geodesic could be deformed continuously into the second, they would be separated by a contour with vanishing first variation, i.e., a geodesic, but not with a positive definite second variation.

The trace formula requires a knowledge of the periodic orbits, or equivalently, of the geodesics which close smoothly after a finite length. If a particular equivalence class of contours is chosen, and the point P is varied continuously, the length of the geodesic in this class has a first-order variation with respect to the displacement of P. If the space of negative curvature is not only complete, however, but also compact, there is a minimum with respect to the variation of P, yielding a unique closed (periodic) geodesic.

There is no maximum of length in the particular equivalence class, however, because there is no upper limit to lengths of the geodesics, as P is allowed to roam freely. As an example, we might start with a very short geodesic loop like α in Figure 59. Then P is taken on a trip around the double torus. When it passes by its original place, it can be said to have followed along the contours of some equivalence class, say τ. Thus, by letting P run around without restrictions, the class α has become the class $\tau^{-1}\alpha\tau$, where τ can be any element of the homotopy group G. Quite generally in group theory, two elements β and $\tau^{-1}\alpha\tau$ are called *conjugate*. *A periodic orbit is, therefore, defined by a conjugacy class of the homotopy group G.*

The actual construction of a particular periodic geodesic in the octogon of Figure 56 is not difficult; let us take the conjugacy class of β as an example. The side β can be viewed as going from α to α'; but it is not the shortest connection between these two sides, because otherwise it would cut both of them at a right angle. The problem, therefore, is to find a Euclidean circle that intersects the boundary of the

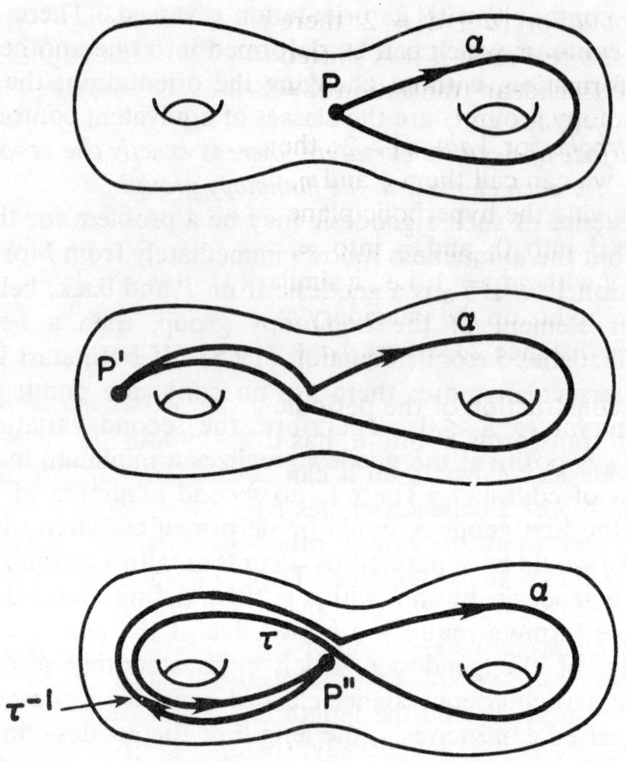

Figure 59 The original contour α on the double torus is deformed by moving the point P around the contour τ; the resulting contour, starting and ending again in P, belongs to the element $\tau^{-1}\alpha\tau$ of the homotopy group G, and to the same conjugacy class as α.

hyperbolic plane (unit circle or real axis) at right angles, and does so also for α and α'. The solution, either geometrically or algebraically, is straightforward; it can also be interpreted as finding the shortest loop around the right-hand torus in Figure 57.

Before getting the same information directly out of the 2 by 2 matrices A, B, C, and D, the elements of the group $SL(2,R)$ have to be put into equivalence classes. For this purpose, we look at the fixed points; they result from (19.7) by setting $z' = z$, which yields a quadratic equation in z; its discriminant is $D = (a + d)^2 - 4$. Three cases arise:

i) *elliptic* for $|a + d| < 2$: there is a single fix-point in the upper half-plane U; if it is shifted into i, and thence into the origin of the unit circle, the transformation becomes a rotation around the origin by an angle ϕ where $\cos(\phi) = (a + d)/2$;

ii) *parabolic* for $|a + d| = 2$: there is one fix-point which is located on the real axis; if it is shifted to ∞, the transformation becomes the Euclidean translation $z' = z + b$;

iii) *hyperbolic* for $|a + d| > 2$: there are two fix-points on the real axis; we can call them ξ and η, and interpret the transformation as moving the hyperbolic plane away from ξ and toward η; if ξ is shifted into 0, and η into ∞, the transformation becomes $z' = az/d$ with $ad = 1$, i.e., a similarity of the whole upper half-plane; the points on the imaginary axis are moved upwards by the hyperbolic distance 2σ, where $|a + d| = 2\cosh(\sigma)$.

For our construction of the periodic geodesic β to succeed, the matrix A which transforms α' into α, has to be hyperbolic. Indeed, if the periodic geodesic β exists, then it can be transformed into the imaginary axis of U; the Euclidean circles corresponding to α and α' both become circles around the origin; otherwise they don't intersect β as well as the real axis at right angles. The hyperbolic distance σ from α' to α is given by $2\cosh(\sigma) = \text{trace } A$.

By the same arguments, the other 2 by 2 matrices, B, C, and D, are also hyperbolic; and so are $ABA^{-1}B^{-1}$ and $A^{-1}B^{-1}CD$, which correspond to the thickness and the length of the pipe connecting the two tori; see Figure 58. Quite generally, there is a 1-to-1 correspondence between the homotopy group and the group of 2 by 2 matrices that is generated by A, B, C, and D, subject to the condition (19.11). This restriction means that the contour $\alpha\beta\alpha'\beta'\gamma\delta\gamma'\delta'$ can be contracted to the base point P, as can be seen either from Figure 56 or from Figure 57.

The exact shape of the octogon F depends on the matrices A, B, ...; and conversely, the octogon with the identification of sides in Figure 56, as well as the sum 2π for the angles on the vertices, defines the group G of matrices which is generated by A, B, ... The octogon is, therefore, called the *fundamental domain* of G. The hyperbolic plane U is divided up into fundamental domains by G, an idea that leads to the expression $F \simeq U / G$.

The hyperbolic geometry of the double torus F, the combinatorial properties of the homotopy group G as defined by the generators α, β, γ, δ, and their inverses, and the subgroup of PSL(2,R) generated by the corresponding matrices, and also called G for simplicity's sake, are obviously different interpretations of the same underlying mathematical structure; but lest the reader thinks that all our questions can, therefore, be answered, a look at the literature on the subject will tell otherwise; cf. Magnus, Karass, and Solitar (1966), Beardon (1983), Balazs and Voros (1986).

The effective enumeration of all the periodic geodesics remains an elusive goal; all the words in the alphabet of $\alpha, \beta, \gamma, \delta$, and their inverses can be written down, and any occurrences like $\alpha\, \alpha^{-1}$ or $\alpha^{-1}\alpha$ can be eliminated. The resulting group is called a *free group* on four generators; but the condition (19.11) is essential, and the resulting group on *four generators with one relation* is different when long words in the eight-letter alphabet are considered. The problem is further complicated by the requirement of finding the conjugacy classes; an algorithm was found by Dehn (1911) and worked out by Greendlinger (1961), to decide whether two words are equivalent, or even conjugate; but the number of steps in the algorithm increases quickly with the length of the word. We will discuss a very different method for finding a code at the end of Section 20.2.

19.5 Selberg's Trace Formula

Schrödinger's equation in a Riemannian space can be derived from the classical Hamiltonian by applying the usual procedures for going from the classical dynamical variables to the corresponding linear operators. The ordering of these operators is determined by the rule that the probability density $|\psi|^2$ has to be conserved; also the expectation value of the kinetic energy has to be positive definite, and its expression in terms of the metric tensor has to remain the same in all coordinate systems. Thus, one finds that

$$i\hbar \frac{\partial \psi}{\partial t} = -\frac{\hbar^2}{2m} \Delta\psi + V(q)\psi\,, \quad \Delta = \frac{1}{\sqrt{g}} \frac{\partial}{\partial q_k} (\sqrt{g}\, g^{kj} \frac{\partial}{\partial q_j}) \qquad (19.12)$$

The potential energy $V(q)$ will be a multiple of the Gaussian curvature $K(q)$ in two dimensions. In higher dimensions, the curvature becomes a rather complicated tensor; but its trace K over the components is not trivial and can be used for V. The curvature term is included in Schrödinger's equation, because Van Vleck's approximation then becomes exact in a three-dimensional space of constant curvature, provided $V(q) = \hbar^2 K/2m$. In two dimensions, Van Vleck's formula becomes exact only in the limit of large distances $|q'' - q'|^2 >> 1/K$, provided $V(q) = \hbar^2 K/8m$. When applied to the sphere, this extra term changes the eigenvalues from the usual $\ell(\ell+1)\hbar^2$ to the more intuitive $(\ell + 1/2)^2\hbar^2$.

The stationary Schrödinger equation is, therefore, written as

$$(\Delta + \frac{1}{4} + \kappa^2)\, \phi = 0\,, \text{ where } \kappa = \sqrt{2mE/\hbar^2}\ ; \qquad (19.13)$$

κ can be interpreted as the wave number with respect to the radius of curvature $\sqrt{-1/K}$; occasionally, we will use the customary $\lambda = \kappa^2 + 1/4$ for the eigenvalue of the Laplacian. When the coordinates in U are written as $z = x + iy$, the Laplacian becomes $\Delta = y^2 \, (\partial^2/\partial x^2 + \partial^2/\partial y^2)$. We will be interested in the spectrum of the Laplacian for various domains; if it is a polygon that does not tile U, then the boundary conditions have to be specified, e.g., the Dirichlet condition $\phi = 0$.

If the polygon does tile U, however, we naturally choose periodic boundary conditions, $\phi(z') = \phi(z)$, where $z' = Tz$, and T is one of the transformations (19.7) which define the polygon. Functions of this type are called *automorphic,* without necessarily being eigenfunctions of the Laplacian. Their automorphism refers to the specific tiling polygon like F, or equivalently, to the corresponding homotopy group G. Again, the reader has to be warned about overlapping nomenclature: a lot of mathematical results have been accumulated about automorphic, analytic functions; they can be regarded as the kind of functions we are looking for, but belonging to the eigenvalue 0 of the Laplacian; obviously, there are not many of them for any fixed polygon.

The essential steps in the derivation of the Selberg trace formula (STF) will now be recounted, without any of the detailed mathematical arguments, but with some hints concerning the purely algorithmic or computational aspects. The purpose of this short exposition is to convey an idea of the reasoning behind the STF, and to show how it differs from the justification of the trace formula in Section 17.4. The details can be found most completely in the first chapter of Hejhal's treatise (1976a); the relations with the Riemann zeta-function are emphasized by Hejhal (1976b). A somewhat abbreviated version was given by McKean (1972) and Randol (1984); the book of Terras (1985) presents an appealing mixture of ideas and insights; Balazs and Voros (1987) as well as Steiner (1987b) discuss the fine points in the derivation of the STF.

The first ingredient is a quite arbitrary function $k(z_2, z_1)$, which is defined for any pair of points, z_1 and z_2, in U ; but which depends only on their hyperbolic distance s, so that we can write it as $Q(|z_2 - z_1|^2/y_2 y_1)$ in view of (19.8). Such a function is used as a kernel to define a linear operator M on the functions in U, and gives us the first unexpected, although not difficult result: If $f(z)$ is any function in U which satisfies $(\Delta + \lambda)f = 0$, then it is also an eigenfunction of the operator M,

$$M f = \int_U k(z_2, z_1)\, f(z_1) \frac{dx_1 dy_1}{y_1^2} = \Lambda(\kappa)\, f(z_2) \;, \quad (19.14)$$

where the function $\Lambda(\kappa)$ depends only on the function Q, but not on the eigenfunctions $f(z)$; equivalently, the linear operator M and the Laplacian commute.

The function Λ can now be computed explicitly in terms of Q by inserting the particularly simple eigenfunction $f(z) = y^{i\kappa + 1/2}$. If the integral (19.14) is calculated for $z_2 = i$, one finds

$$\Lambda = \int_0^\infty \frac{dy}{y^2} \int_{-\infty}^{+\infty} dx\, Q\left(\frac{x^2 + (y-1)^2}{y}\right) y^{i\kappa + 1/2} \;. (19.15)$$

Since the hyperbolic distance along the imaginary axis is $u = \log(y)$, the function $f(z) = y^{i\kappa + 1/2}$ represents a plane wave; its amplitude is $\sqrt{y}$ because of the distortion in the x-direction by the factor $1/y$ in the metric. With a minor rearrangement, one can write,

$$\Lambda(\kappa) = \int_{-\infty}^{+\infty} du\, e^{i\kappa u} P(u) \;, \quad P(u) = \int_w^{+\infty} dv \frac{Q(v)}{\sqrt{v - w}} \;, (19.16)$$

where $w = (2 \sinh(u/2))^2$.

Instead of functions which spread indifferently all over U, let us now specialize $f(z)$ to be an automorphic function for our octogon F. Since $f(z)$ repeats itself in every tile gF of U, where g is an element of the homotopy group G, the integral (19.14) can be divided into a sum over all the group elements $g \in G$, and an integral over each tile. The linear operator M on the automorphic functions in F is now

$$M f(z_2) = \int_F \frac{dx_1 dy_1}{y_1^2} K(z_2, z_1) f(z_1) \;, \quad K(z_2, z_1) = \sum_{g \in G} k(g z_2, z_1) \quad (19.17)$$

Both points, z_1 and z_2, belong to F, but z_2 is spread over all of U by the action of the group elements $g \in G$. The convergence of the sum over G in the definition of the kernel $K(z_1, z_2)$ requires that the function Q in (19.15) go to zero sufficiently quickly for large hyperbolic distances.

The operator M can be expanded in its eigenfunctions $\phi_j(z)$; but according to (19.14), they are also eigenfunctions of the Laplacian, and since F is compact, the spectrum is discrete;

$$\sum_{j=0}^{\infty} \Lambda(\kappa_j)\, \phi_j(z_2)\, \phi_j^+(z_1) = K(z_2, z_1) \;. \quad (19.18)$$

Now the trace is taken, yielding $\Sigma\, \Lambda(\kappa_j)$ on the left-hand side, and that is already one-half of Selberg's trace formula. On the right-hand side, we go back to the definition (19.17) of K as a sum over all elements in the group G.

This sum is now broken into conjugacy classes, exactly as it came up in the discussion of the periodic geodesics in the preceding section; but this break-up requires some care, and the reader has to be convinced that there is no double counting or leaving out of certain group elements. The conjugacy class C_q of $q \in G$ has a primitive element $p \in C_q$, such that all the other elements of C_q can be written as $g^{-1}\, p^n\, g$ with positive integer n. Every element in C_q is generated exactly once when n runs through the positive integers, and g runs through G/G_p ; these are the cosets in G with respect to the *centralizer* G_p ; the latter is the subgroup of all the elements of G which commute with p. The proof of these statements depends on all of the 2 by 2 matrices representing G being hyperbolic.

The integrand of any particular term in the trace on the right of (19.17) now becomes $k(g^{-1}p^n gz\,,z) = k(p^n gz\,, gz)$, since k depends only on the geodesic distance. The integration can then be made over the tile gF with the integrand $k(p^n z\,, z)$. The sum over these tiles, where $g \in G/G_p$, is a domain F_p that is determined by p alone; it can be constructed quite easily as soon as p has been given the simple form $(\mu, 0; 0, 1/\mu)$ by an appropriate choice of the coordinates. F_p is then bounded by the horizontal lines at $y = 1$ and $y = \mu^2$ in U. The primitive element p gives rise to a primitive periodic orbit of length $\ell(p) = 2\log\mu$.

If we now go back to the function $Q(|z_2 - z_1|^2/y_2 y_1)$ as in (19.13) and (19.14), and set $z_2 = p^n z = \mu^{2n} z$ as well as $z_1 = z$, the trace for (19.17) becomes

$$\sum_{j=0}^{\infty} \Lambda(\kappa_j) = \sum_{\text{pcc}} \sum_{n=1}^{\infty} \int_1^{\mu^2} \frac{dy}{y^2} \int_{-\infty}^{+\infty} dx\, Q\Big(\frac{(\mu^{2n}-1)^2}{\mu^{2n}} \frac{x^2+y^2}{y^2} \Big) \tag{19.19}$$

where 'pcc' stands for 'primitive conjugacy classes'. The integrations present no difficulty, and the result can be expressed in terms of the function $P(u)$ in (19.16), where the lower limit of integration is $w = (\mu^{2n}-1)^2/\mu^{2n} = 4\,\mathrm{Sinh}^2[\ell(p^n)/2]$; more directly, the argument of $P(u)$ is simply $u = \ell(p^n) = n\,\ell(p)$, where ℓ is the length of the periodic geodesic belonging to $g \in G$.

Before writing the final formula, however, we must treat the exceptional term in (19.19) that belongs to the conjugacy class of the unit element in G. It requires the integration over F of $k(z, z) = Q(0)$, according as (19.17) and (19.14); thus, we get simply $4\pi\, Q(0)$, since the area of the octogon F is 4π. The function $\Lambda(\kappa)$ on the left-hand side, and the contributions from all the other conjugacy classes on the right-hand side, are directly expressed in terms of the function $P(u)$; therefore, the definition of $P(u)$ in terms of $Q(v)$ according as (19.16) is now inverted. This inversion is known as *Abel's integral equation* (cf. Courant-Hilbert vol. 1, a particularly simple treatment is given by Hejhal 1976a, p.37), and requires no more than elementary manipulations.

As a last step, the exceptional term is expressed as a function of $\Lambda(\kappa)$, because it will ultimately be shifted to the left-hand side, so that the right-hand side depends only on the fluctuating part of the spectrum. Selberg's trace formula now becomes

$$\sum_{j=0}^{\infty} \Lambda(\kappa_j) = \int_{-\infty}^{+\infty} \nu\, \Lambda(\nu)\, tanh(\pi\nu)\, d\nu \qquad (19.20)$$

$$+ \sum_{\text{pcc}} \sum_{n=1}^{\infty} \frac{\ell(p)}{2\,\text{Sinh}[n\,\ell(p)/2]}\, P(n\,\ell(p)) \;.$$

The summation over the primitive conjugacy classes (pcc) is, of course, equivalent to the summation over the primitive periodic orbits (ppo). Notice that $\Lambda(\kappa)$ and $P(u)$ are Fourier transforms of each other; the derivation of (19.20) can, therefore, be carried out for any pair of functions Λ, P with good convergence properties, such as the example given with (16.13).

In order to make contact with the more general, but only asymptotic trace formula (17.13), we notice from the definition (19.16) that $P(u)$ is symmetric in u. Therefore, we can set $P(u) = \exp(i\,|\,u\,|\sqrt{2mE/\hbar^2}\,)$, and give E a small positive imaginary part so as to make the exponential vanish for large $|\,u\,|$. The Fourier integral (19.16) is elementary, and yields $\Lambda(\kappa_j) = i\hbar\sqrt{2E/m}\,/(E - E_j)$, where we have set $E_j = \hbar^2\kappa_j^2/2m$. To recover (17.13), the expression (19.4) for the action integral is used, and the primitive period T_0 is obtained as the primitive length $\ell(p)$ divided by the the velocity $\sqrt{2E/m}$; finally the instablility exponent χ is equal to the length $n\,\ell(p)$ because of Jacobi's equation (19.6), and there are no conjugate points for the same reason.

Sieber and Steiner (1990) have generalized the trace formula (17.13) for billiard systems, and probably homogeneous Hamiltonians in general (cf. Section 17.8). An equation like (19.20) with an arbitrary pair of functions $\Lambda(\kappa)$ and $P(u)$ is obtained, which leads to absolutely convergent series on both sides of (19.20). Although this version of (17.13) is correct only in the limit of small $\hbar$, it could be used for numerical calculations similar to the ones in the next section.

Many authors have pointed out that Selberg's trace formula is a generalization of the Poisson formula (16.13), because it equates two summations, the first with the function $\Lambda(\kappa)$, and the second involving its Fourier transform $P(u)$. Whereas Poisson's formula is based on the group of discrete translations in one dimension, the STF comes directly out of the discrete group G whose fundamental domain is F. Such an interpretation relies on the tiling of the hyperbolic plane, and does not really account for the more general, but asymptotic trace formula (17.13). Conversely, if we apply (17.13) to an arbitrary polygon in the hyperbolic plane, we get (19.20) with a $\simeq$ sign, and it is not clear why there is an equality when the polygon tiles the plane. Nevertheless, the STF gives a large class of examples where the general trace formula (17.13) can be greatly strengthened.

19.6 Computations on the Double Torus

Charles Schmit of the Nuclear Physics Institute in Orsay (France) was the first to carry out large-scale computations in order to obtain the spectrum of the Laplacian on a surface of constant negative curvature. Unfortunately, his results have not been published systematically; some of them are appended to the report by Balazs and Voros (1987), and others are in a paper by Balazs, Schmit, and Voros (1987); but further work that investigates the relation between the spectrum and the periodic orbits remains in the realm of private communications. This work is restricted to triangles in the hyperbolic plane with Dirichlet boundary conditions; this limitation is both a matter of choosing the most practical shape and of being able to apply a stable computing algorithm. A discussion of Schmit's work must start with explaining in more detail these two aspects.

Reflection symmetries were explicitly excluded from the construction of the tiling polygons in Section 19.3. Most members of the six-parameter family of double tori do not have such symmetries, and their spectra cannot be obtained by a method which relies on them. Aurich, Sieber, and Steiner (1988) have worked out a scheme which, at least in principle, is free of this restriction; most of the eigenvalues for the regular octogon have to be computed by this method. Meanwhile, they have confirmed some of Schmit's results; their numerical check of Selberg's trace formula will be discussed at the end of this section.

Just as with the *14 Bravais lattices* in three-dimensional Euclidean space, the double tori in the hyperbolic plane and the paving they generate can be distinguished by the additional symmetries in the group G of 2 by 2 matrices which is generated by A, B, etc. The largest number of such symmetries is undoubtedly found in the tiling of the regular octogon (cf. Figure 60); when placed at the center of the unit circle, the symmetry lines divide the octogon into 96 congruent triangles. The angles in one of them are $\pi/2$, $\pi/3$, and $\pi/8$; their sum misses π by $\pi/24 = 4\pi/96$, and that is its surface according to the Gauss-Bonnet theorem (19.10). By way of comparison, the most symmetric Bravais lattice in the Euclidean plane is the square lattice; its reflection symmetries divide the square into only eight congruent triangles.

The spectrum for the regular octogon divides itself essentially into 96 subspectra, because the various reflection symmetries all commute with the Laplacian. Each subspectrum is characterized by the way the eigenfunction in one of the small triangles repeats itself in any of the other 95. The most obvious of these transformations is a reflection at the boundary with a change in sign of the eigenfunction. The eigenfunction then vanishes at the boundary of the triangle, yielding the Dirichlet condition. The corresponding eigenvalues are large even for the lowest state, compared with the eigenfunctions with vanishing derivatives, or other conditions, across the sides of the triangle.

The eigenfunctions are constructed for the triangle OEF in Figure 60 as follows: Schrödinger's equation is solved in Euclidean polar coordinates (ρ, ϕ) around the center of the unit circle; the variables separate, and the angular dependence is the usual $\exp(i\ell\phi)$ with integer ℓ; the radial function is a Legendre function with the indices $(i\kappa+1/2)$ and ℓ, where κ can be any real number at this point; ℓ, however, is limited to multiples of 8 by the Dirichlet boundary condition along the sides OF and OE. The eigenfunction then becomes

$$\psi_\kappa(\rho, \phi) = \sum_{\ell=1}^{\infty} C_\ell \, F_{\kappa\, 8\ell}(\rho) \sin(8\ell\phi) , \tag{19.21}$$

where $F_{\kappa\, 8\ell}(\rho)$ is the appropriate, real Legendre function.

The boundary condition on the side EF is now enforced: let σ be the hyperbolic length along EF, and call $\rho(\sigma)$ the polar coordinate of EF as function σ; it is then required that

$$\int_0^L d\sigma \, F_{\kappa\, 8\ell}(\rho(\sigma)) \sin(n\pi\sigma/L) = 0 \, , \text{for } n = 1, \ldots, N \, ; \tag{19.22}$$

the hyperbolic length of EF is called L. If the sum (19.21) is truncated at $\ell = N$, a set of N linear conditions for the N coefficients C_ℓ is obtained; its determinant is set to zero, and then yields a discrete set of values for κ. The integral condition turns out to be crucial for the stability of the computation. Although the general idea of this calculation is easy to explain, its execution is a delicate task.

The resulting 369 eigenvalues λ are listed by Balazs and Voros (1986); they fit Weyl's asymptotic distribution very well, but it is necessary to include the boundary and curvature terms as in (16.3). There

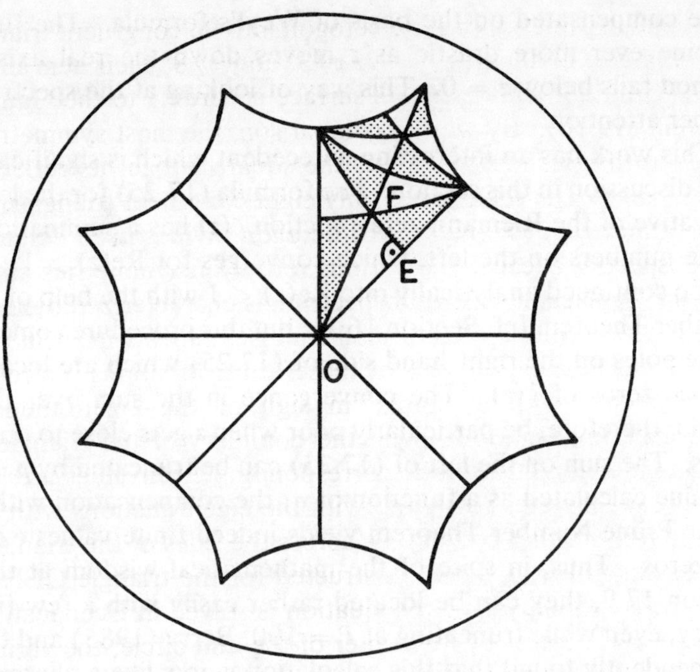

Figure 60 The regular octogon is divided into 96 congruent triangles with the inner angles $\pi/2$, $\pi/3$, and $\pi/8$; Schmit calculated the spectrum of the Laplacian with Dirichlet boundary conditions for this triangle.

are a few near-degeneracies, however, which would be very unlikely in a sample of this size, if the GOE were applicable. The reasons are not clear at this time; but the regular octogon where this particular triangle originates has very high symmetry contrary to most of the other tiling octogons. The generic octogons conform to the GOE statistics according to the recent calculations of Aurich and Steiner (1990). Also, Schmit has calculated the eigenvalues of the Laplacian with Dirichlet boundary conditions for the hyperbolic triangle with the inner angles $\pi/8$, $\pi/2$, and $67\pi/200$. Although it differs minimally from the previous triangle, but does not tile the hyperbolic plane, it has a spectrum with very good GOE statistics, and no near degeneracies.

The fluctuations in the spectrum for the triangle *OEF* were investigated by Balazs, Schmit, and Voros (1987) with the help of a larger sample of 1500 eigenvalues. The zeta-function (17.29) is examined in the complex z-plane; its values for $\text{Re}(z) > 1$ can be calculated from the known sample, because the series is convergent; the authors ask whether the function can be continued analytically to the left of the line $\text{Re}(z)=1$. The known poles are directly related to Weyl's asymptotic distribution; convergence for $\text{Re}(z) < 1$ requires each term in (17.29) to be compensated on the basis of Weyl's formula. The fluctuations become ever more drastic as z moves down the real axis, and the method fails below $z = 0$. This way of looking at the spectrum merits further attention.

This work has an interesting antecedent which is significant for the later discussion in this section. The formula (17.25) for the logarithmic derivative of the Riemann zeta-function $\zeta(z)$ has a summation over all prime numbers on the left, which converges for $\mathrm{Re}(z) > 1$. This sum can be continued analytically into $\mathrm{Re}(z) < 1$ with the help of the Prime Number Theorem (cf. Section 16.1); but this procedure comes to a halt at the poles on the right-hand side of (17.25) which are located at the famous zeros of $\zeta(z)$. The convergence in the sum over the primes should, therefore, be particularly poor when z gets close to one of these zeros. The sum on the left of (17.25) can be truncated by $p < P$, and its value calculated as a function of σ; the compensation with the help of the Prime Number Theorem yields indeed finite values except near the zeros. Thus, in spite of the mathematical wisdom at the end of Section 17.9, they can be located rather easily with a few figures accuracy, even while truncating at $P = 100$; Berry (1985) and the author independently found that this calculation works like a charm. It is, of course, directly related to the remarks at the end of the preceding chapter.

Aurich and Steiner (1989) have recently developed a method for finding the spectrum on a double torus without Dirichlet boundary conditions along the symmetry lines. They have used the 200 lowest energy levels to make a numerical check on Selberg's trace formula. Such a task requires that the periodic orbits be enumerated and calculated; the mathematicians have worked on this problem for some time, and have established, in the special case of constant negative curvature, a number a theorems for what they call the *length spectrum* of a Riemannian surface. In particular, Huber (1959) showed that the number of periodic geodesics of length up to ℓ is given by $\exp(\ell)/\ell$ in the limit of large ℓ; this estimate confirms the identity of the topological and the metric entropy which was discussed in Section 10.5; indeed, the Lyapounoff number is 1 on a surface of constant negative curvature, as shown by (19.6).

Again, Aurich and Steiner (1988) have calculated the length spectrum for the octogon F to some upper limit. The author had done the same thing in 1979; but he only discussed the results with a few colleagues; his failure to publish anything had the following reason. The analogy with Riemann's zeta-function has just been recalled; the length spectrum is given by the logarithm of the prime numbers, and the eigenvalues are the zeros of $\zeta(z)$; even a short list of prime numbers

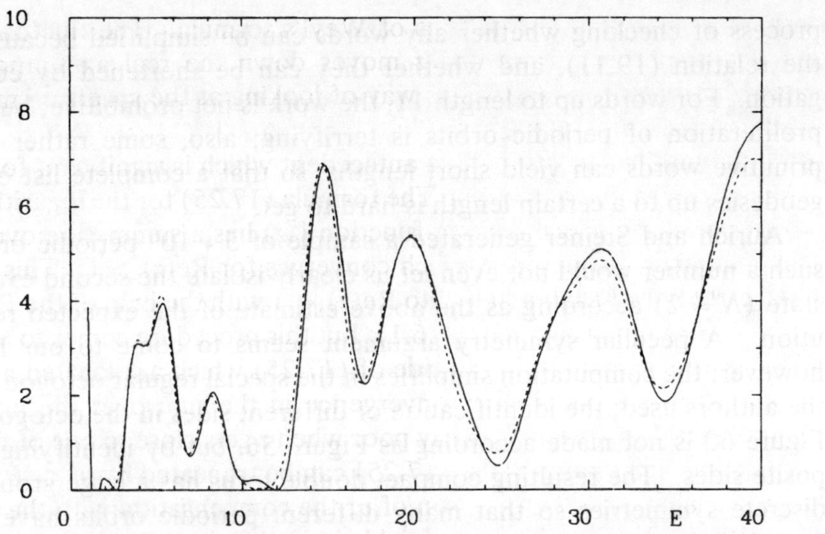

Figure 61 Computation of the STF for the pair (19.23), using the lowest 100 eigenvalues on the left (dashed line), and 10,000 primitive periodic orbits on the right (continuous line) [from Aurich, Sieber, and Steiner (1988)].

yields remarkably good values for the zeros. Our main goal, to get the quantum-mechanical energy spectrum from the classical length spectrum, as we did in the AKP, seemed to be within easy reach.

But Weyl's formula (16.2) shows that the mean spacing of the eigenvalues $\lambda = \kappa^2 + 1/4$ on the double torus is 1 because the area $A = 4\pi$. If we want to resolve the N-th excited level against the $(N+1)$-st with the help of the STF (19.20), we have $\Delta\lambda = 2\,\kappa\,\Delta\kappa \simeq 1$, or $\Delta\kappa \simeq 1/2(N+1/2)^{1/2}$. For a resolution of $\Delta\kappa$ in the Fourier transform (19.16), however, the integral over u has to be extended to $2\pi/\Delta\kappa$, as evidenced by the example given together with Poisson's formula (16.13). In the context of (19.20), the sum over the periodic orbits has to be extended to $\ell(p) \simeq 4\pi(N+1/2)^{1/2}$. There are $\simeq \exp(\ell)/\ell$ periodic orbits of length $\leq \ell$, so that the sum over the periodic orbits requires $\simeq \exp(4\pi(N+1/2)^{1/2})/4\pi(N+1/2)^{1/2}$ terms to yield even the N lowest excited states; thus, the computation seemed impossible.

In order to get the length spectrum for the double torus, the conjugacy classes of the homotopy group G have to be enumerated, and the lengths of the periodic geodesics are then obtained from the trace of the corresponding 2 by 2 matrices. The enumeration can be done as mentioned at the end of Section 3; first, all the words in the alphabet are written down systematically, and all adjacent pairs of letters like $\alpha^{-1}\alpha$ are eliminated right away; then follows the difficult process of checking whether any words can be simplified because of the relation (19.11), and whether they can be shortened by conjugation. For words up to length 11, the work is not prohibitive; but the proliferation of periodic orbits is terrifying; also, some rather long primitive words can yield short lengths, so that a complete list of all geodesics up to a certain length is hard to get.

Aurich and Steiner generated a sample of $5 \cdot 10^6$ periodic orbits; such a number would not even let us clearly isolate the second excited state ($N = 2$) according as the above estimate of the expected resolution. A peculiar symmetry argument seems to come to our help, however; the computation simplifies in the special regular octogon that the authors used: the identification of different sides in the octogon of Figure 60 is not made according as Figure 56, but by identifying opposite sides. The resulting compact double torus has a large group of discrete symmetries so that many different periodic orbits have the same length. According to Aurich and Steiner (1988), the traces of the corresponding 2 by 2 matrices have the value $m + n\sqrt{2}$, where n runs through all the positive integers, and m is the positive odd integer that minimizes $|\sqrt{2} - m/n|$; this remarkable result is complicated by the degeneracies which Bogomolnyi and Steiner (1990) are now able to calculate.

The pair $\Lambda(\kappa)$, $P(u)$ in formula (19.20) is first chosen to be

$$e^{-(\kappa-\bar{\kappa})^2/\varepsilon^2} + e^{-(\kappa+\bar{\kappa})^2/\varepsilon^2}, \quad (\varepsilon/\sqrt{\pi}) \cos(\bar{\kappa}u)\, e^{-\varepsilon^2 u^2/4}, \tag{19.23}$$

where $\bar{\kappa}$ is a parameter that varies from 0 to ∞. As a function of $\bar{\kappa}$, the left-hand side (lhs) of STF becomes a series of peaks, located near κ_j, and of root-mean-square width $\Delta\bar{\kappa} = \sqrt{2}\,\varepsilon$. The right-hand side (rhs) of STF becomes a Fourier series in $\cos(n\,\bar{\kappa}\,\ell(p))$; the resolution is $\Delta\bar{\kappa} \simeq 2\pi/\ell(p)$ by the same argument as above. With $\varepsilon = .2$, the periodic geodesics up to $\ell \simeq \sqrt{2}\,\pi/\varepsilon \simeq 22.2$ are required, or about the first 10,000 primitive ones; Figure 61 shows both sides of the STF as a function of λ near the origin. The agreement is very impressive, and seems to contradict our pessimistic estimate, at least for the highly symmetric octogon.

A similar result with the two sides of STF playing the reverse role, is obtained with the pair

$$\cos(\kappa L)\, e^{-\lambda_j t}, \quad (16\pi t)^{-1/2}\, e^{-t/4}\, [e^{-(u-L)^2/4t} + e^{-(u+L)^2/4t}] \tag{19.24}$$

the function $P(u)$ is the heat diffusion in one dimension, from the sources at $u = L$ and at $u = -L$, over the time t. For sufficiently short times these sources at the locations $n\ell(p)$ are separated because of the Gaussian nature of the heat kernel. The lhs yields these locations, provided the summation is carried over a sufficiently large set of energy

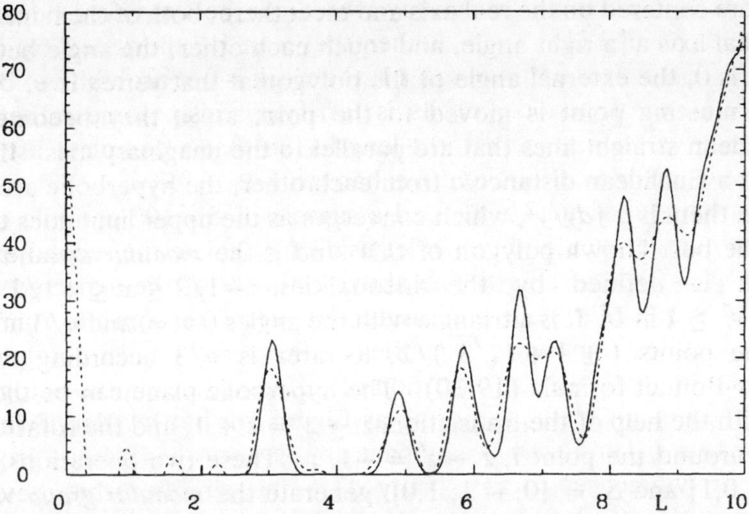

Figure 62 Computation of the STF in function of the length of the orbits, using the pair (19.24), with the same set of eigenvalues and periodic orbits as in Figure 61; dashed line for lhs, and continuous for rhs [from Aurich, Sieber, and Steiner (1988)].

levels. The continuous curve in Figure 62 shows the rhs for $t = .01$ with the same set of 10,000 periodic prime geodesics, while the lhs is represented by the dotted curve using the 75 lowest eigenvalues. This time, the lengths of the periodic orbits are obtained from energy levels, and the agreement is again remarkable.

19.7 Surfaces in Contact with the Outside World

A polygon was defined at the beginning of Section 19.3 as a simply connected domain of finite area, which is bounded by straight lines. The restriction to a finite area is artificial in Euclidean geometry, and may just as well be replaced right away by the stronger assumption of compactness. In hyperbolic geometry, however, non-compact polygons of finite area are easy to construct; they can also be regarded as mathematical models for many physical situations where a particle or a wave enters a container from the outside, coming from infinitely far away as in a scattering problem.

Non-compact polygons of finite area, also called *singular polygons,* have vertices that are infinitely far away, as measured in the hyperbolic metric. In Poincaré's upper half-plane U, such a vertex is located on the real line; the two adjacent sides of the polygon are Euclidean circles that are centered on the real axis and meet there; both of them intersect the real axis at a right angle, and touch each other; the angle between them is 0, the external angle of the polygon at that vertex is π. When their meeting point is moved to the point at ∞, they become two Euclidean straight lines that are parallel to the imaginary axis. If they are at a Euclidean distance a from each other, the hyperbolic area between them is $a \int dy/y^2$, which converges as the upper limit goes to ∞.

The best known polygon of this kind is the *modular domain* D_1, which is defined by the inequalities, $-1/2 \leq x \leq +1/2$ and $x^2 + y^2 \geq 1$ in U. It is a triangle with the angles 0 at ∞, and $\pi/3$ at each of the points $(\pm 1 + i\sqrt{3})/2$; its area is $\pi/3$ according to the Gauss-Bonnet formula (19.10). The hyperbolic plane can be tiled by D_1 with the help of the translation, $z \rightarrow z' = z + 1$, and the rotation by 180° around the point i, $z \rightarrow z' = -1/z$. These two operations, $T = [1,1;\ 0,1]$ and $S = [0, -1;\ 1,0]$, generate the *modular group* whose elements are the 2 by 2 matrices $[\, a, b\, ;\, c, d\,]$ with integer entries a, b, c, d and $ad - bc = 1$. The pieces numbered 2 and 3 in Figure 65 make up the modular domain.

The modular domain D_1 becomes a Riemannian surface, if the points on its boundary are identified pairwise with the help of T and S. There are two singularities on this surface besides the point at ∞; the neighborhood of the point i looks like a cone with a 60° opening, and the neighborhood of the point $(1 + i\sqrt{3})/2$ looks like a cone with an opening of $\arcsin(4\sqrt{2}/9)$. Topologically, this surface is a sphere; it has played a major role in number theory and in complex variable theory; but as model for some simple physical system, it does not appear to be very useful, because the second conical point has no good physical interpretation. With a little imagination, the first conical point can be understood as a localized electric charge, because a geodesic near it makes a sharp U-turn, exactly as the scattering in a Coulomb potential.

Three adjacent copies of D_1 can be put together so as to form a *singular triangle* , i.e., a hyperbolic triangle whose vertices are all infinitely far away. The three vertices can always be brought into the points $x = 0, 1, \infty$ on the real axis; this region D_2 has as its sides the two vertical Euclidean lines $x = 0$ and $x = 1$, and the Euclidean circle $(x - 1/2)^2 + y^2 = 1/4$; it is the gray region in Figure 65, above the interval (0,1) on the real axis. The relation between D_1 and D_2 can be seen in Figure 65. The group belonging to D_2 is obtained from the modular group by dividing out the threefold rotation around the point $(1 + i\sqrt{3})/2$, i.e., the elements TS and $(TS)^2$.

The singular triangle paves the hyperbolic plane just as the modular domain does; its group tells us how to continue a geodesic inside D_2.

Clearly, the two vertical sides are mapped into each other by a translation parallel to the x-axis, while the bounding circle is mapped into itself by a rotation of 180° around $(1 + i)/2$, i.e., TST^2S. The resulting surface is topologically again a sphere; it has a point at ∞, and a conical point with a 60° opening angle. With some poetic license, this geometry can be interpreted as relating to the motion of a particle that comes from far away to settle on a two-dimensional sphere, with an attractive Coulomb potential centered in a point opposite to the entry-exit.

The singular triangle D_2 can be doubled, to yield a *singular square;* the corresponding group is obtained by dividing out the 180° rotation, thereby getting rid of the last conical point. The new fundamental domain D_3 is bounded by the two vertical Euclidean lines $x = \pm 1$, and the two circles $(x \pm 1/2)^2 + y^2 = 1/4$, as shown in Figure 63. Opposite sides are identified as in the ordinary torus, with the help of the maps $A = [1,1;1, 2]$ and $B = [1, -1; -1, 2]$. The group which is generated by A, B, and their inverses, can be characterized by the arithmetical properties of the integers $(a, b; c, d)$, though they are not all that simple to state; it is also known as the commutator subgroup of the modular group, which is again not very helpful. The reader will find many fascinating details about this group in the work of Harvey Cohen (1972).

The singular square D_3 has a rather convincing physical interpretation and can even be compared directly to a simple surface in three-dimensional Euclidean space. Start with the square $-s \leq x, y \leq +s$ with $s \geq 1$, where points on opposite sides are identified by translation; then, cut the circular disk of radius 1 from the center of the square. Into the gaping hole, attach a horn with the shape of a trombone, as shown in Figure 64; it is centered on the positive z-axis, and its cross-section in the (x, z) plane is the so-called *tractrix*. Its equation in terms of a parameter $\rho \leq 0$ is

$$x = 1/\cosh \rho \ , z = \rho - \tanh \rho \ . \qquad (19.25)$$

This surface was called the *pseudosphere* by Liouville; it has the Gaussian curvature $K = -1$ as shown in the marvelous *Introduction to Geometry* by Coxeter (1961).

The tangent to the curve (19.25) at $z = 0$ is horizontal; but its curvature is ∞. Thus, the connection with the punched torus whose Gaussian curvature is 0, of course, is not smooth; nevertheless , one can imagine a particle or a wave swooping down the horn, and then spreading inside the torus as inside a closed box in a highly chaotic manner. This picture of something coming in from the outside, and then bouncing around inside a container, becomes more obvious if the domain D_3 is rearranged in the hyperbolic plane, as shown in Figure

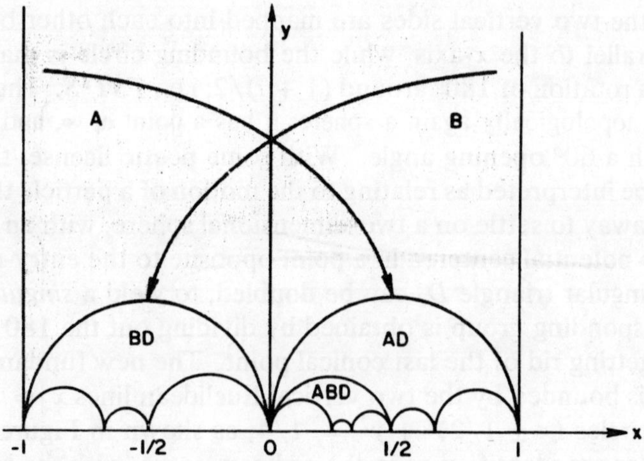

Figure 63 The singular square $D_3 = D$ with the maps A and B to identify opposite sides; copies of D, such as AD and ABD, are shown to demonstrate the trisection of the real axis.

65; it looks like six copies of the modular domain set down next to one another; the six circular arcs are identified in pairs by the maps A, B, and $C = (AB)^{-1}$, while the two outside vertical lines are mapped into each other by the translation R along the x-axis over six Euclidean units.

The construction of the singular square can be generalized to a *singular quadrangle* D_4. A closer look at Figure 63 shows that no more is required than the existence of two hyperbolic transformations A and B whose commutator is a parabolic transformation R,

$$ABA^{-1}B^{-1} = R \,. \tag{19.26}$$

As in the case of the double torus and the relation (19.11), we can count the essential parameters left to choose: R is parabolic iff $\mathrm{trace}(R) = 2$, so that (19.26) represents one algebraic condition on A and B; moreover, an overall coordinate transformation T allows for three inessential parameters; e.g., three of the four vertices are still located in the points 0, 1, and ∞ as in Figure 63, while the location of the fourth is somewhere on the negative x-axis, not necessarily in -1. Figure 65 gets modified into Figure 66; the six circular arcs are equal only in pairs as indicated by the transformations A, B and C.

The two essential parameters determine different singular quadrangles which cannot be mapped into one another isometrically. Since our Riemannian surface is topologically a torus, we end up with the same number of parameters for different hyperbolic tori, in addition to the scale factor K = Gaussian curvature, as in the Euclidean plane; there, the shape of the torus is determined by the ratio of its two

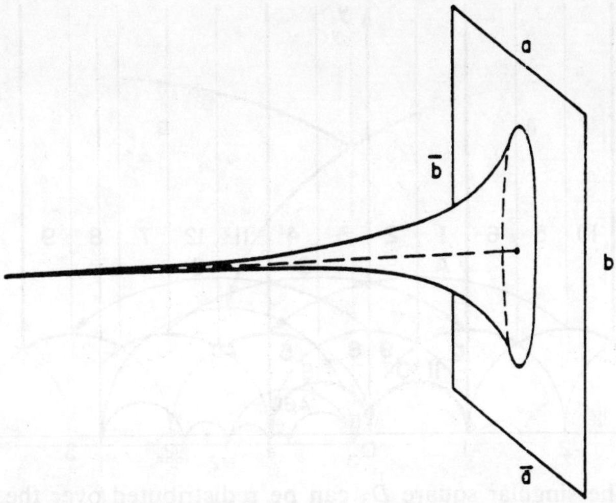

Figure 64 Euclidean model of the torus with the exponential horn attached, corresponding to Sinai's billiard, where the circular hole in the torus now serves as an entry-exit for a particle.

sides, α and β, taken to be complex numbers. Equivalently, we could replace the punched square with the attached horn by a punched parallelogram; the analogy breaks down, however, because the absolute size of the parallelogram now enters into the picture.

There is one essential parameter left, even after requiring the fourth vertex to be located in $x = -1$, so that D_4 still looks exactly as Figure 63. In terms of $e > 0$, we have $A = [\, e, e\, ; e, (1+e^2)/e\,]$ and $B = [\, e, -e\, ; -e, (1+e^2)/e\,]$; when e changes, the fit between opposite sides in D_4 gets modified; they slide internally with respect to each other, while keeping the endpoints on the real axis fixed.

Other singular polygons which tile the hyperbolic plane, and represent non-compact, complete Riemannian spaces of finite area, can be constructed in this manner. Their homotopy group is represented in terms of 2 by 2 real matrices. Instead of, or in addition to, relations like (19.11), each point at ∞ contributes a relation like (19.26) where the right-hand side is only required to be parabolic. As an exercise, the reader may try to write down the relations for a torus with two points at ∞, and count the number of independent parameters. Such a surface is a model for a metal ring with two leads attached, to allow a current to enter in one lead, and to exit in the other lead; a magnetic flux can traverse the ring, without generating a magnetic field in the space itself; the current in function of the magnetic flux shows the Aharonov-Bohm effect (cf. Gutzwiller 1986).

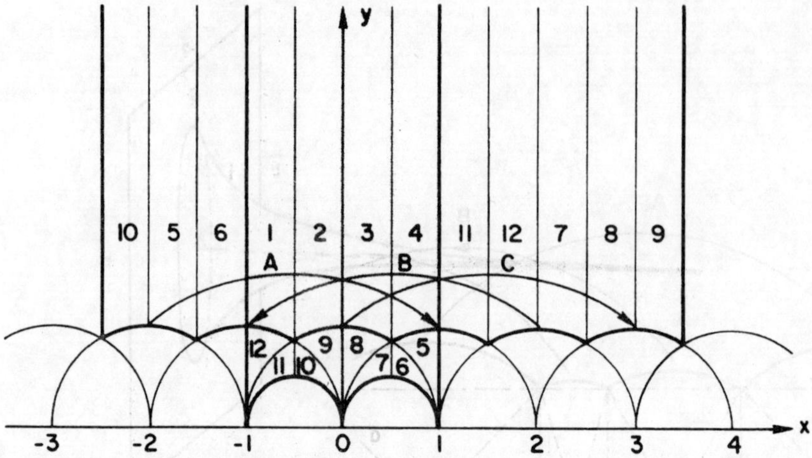

Figure 65 The singular square D_3 can be redistributed over the upper half-plane with the help of the transformations A and B, and becomes six adjacent copies of the modular domain D_1.

19.8 Scattering on a Surface of Constant Negative Curvature

The singular square allows us to give a complete solution for a problem in the scattering of waves where the behavior of the classical trajectories is known to be chaotic. The situation is most easily understood with the help of Figure 65: the solution $y^{1/2 - i\kappa}$ of Schrödinger's equation (19.13) represents an incoming wave which comes swooping down along the imaginary axis; as we argued in (19.15), its wave-vector is κ, and the amplitude $\sqrt{y}$ is necessary to give a flux independent of y in the metric (19.8). The scattering process will be explained in a language which implies a well-defined sequence of events; but as in the discussion of Section 17.5, we look at solutions of the wave equation for a constant frequency, and the scattered waves are present at all times. Only the more complicated ones have a smaller amplitudes (for more details, see Gutzwiller 1983).

The incoming wave has to be continued upon reaching the lower boundary, i.e., the six circular arcs in Figure 65. The procedure is simple: what goes out in one arc, comes back in through the corresponding arc with the help of the transformations A, B, etc.; the incoming wave becomes

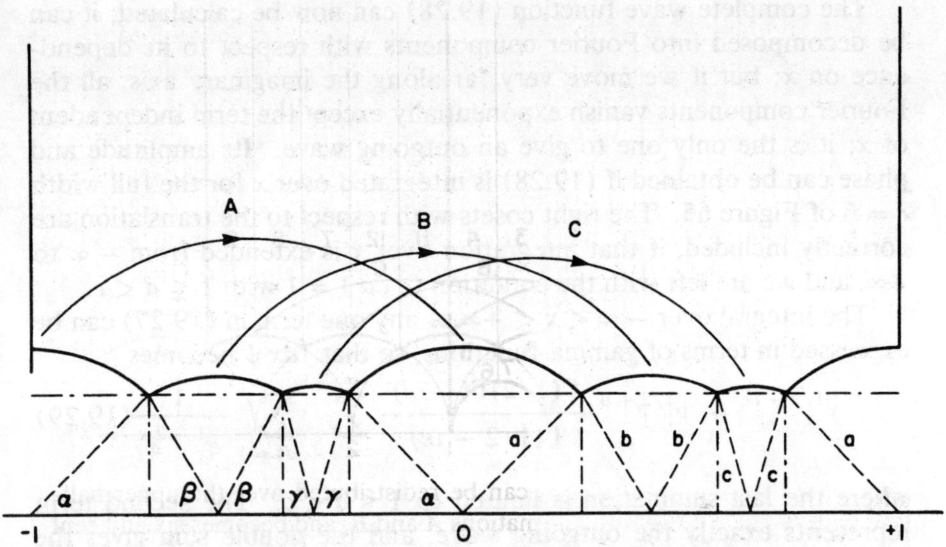

Figure 66 The singular quadrangle D_4 in the representation corresponding to Figure 65 for the singular square D_3; the sum of the angles α, β, and γ equals π so that the inner angles on the vertices add up correctly to a full 2π.

$$y^{1/2 - i\kappa} \rightarrow y'^{\,1/2 - i\kappa} = \frac{y^{1/2 - i\kappa}}{|cz + d|^{1 - 2i\kappa}}, \tag{19.27}$$

where we use (19.7). The first scattered waves get scattered again, and so on; the boundary conditions are satisfied by

$$\psi(x, y) = y^{1/2 - i\kappa} + \sum_{g \in G} \frac{y^{1/2 - i\kappa}}{|cz + d|^{1 - 2i\kappa}}, \tag{19.28}$$

where the unit-element of the group G is left out of the summation, and there are two additional modifications.

When any element $g = [\,a, b\,;\, c, d\,]$ is multiplied on the left by the translation $R = [1, r\,;\, 0, 1]$ of (19.26), one finds $Rg = [\,a + rc, b + rd\,;\, c, d\,]$ so that the scattered wave (19.27) remains the same; therefore, it is enough to have only one representative for each left coset with respect to the translations in the summation (19.28). When we multiply with a translation on the right to get the right cosets, we find $gR = [\,a, b + ra\,;\, c, d + rc\,]$; in the case of the singular square, c and d are integers, and relatively prime because $ad - bc = 1$, while r is a multiple of 6. It is not trivial to show that every pair of integers with the greatest common divisor $(c, d) = 1$, and $c \geq 1$, occurs exactly once in the summation (19.28).

The complete wave function (19.28) can now be calculated; it can be decomposed into Fourier components with respect to its dependence on x; but if we move very far along the imaginary axis, all the Fourier components vanish exponentially except the term independent of x; it is the only one to give an outgoing wave. Its amplitude and phase can be obtained if (19.28) is integrated over x for the full width $r = 6$ of Figure 65. The right cosets with respect to the translation are correctly included, if that integration over x is extended from $-\infty$ to $+\infty$, and we are left with the condition $(c, d) = 1$ with $1 \leq d < c$.

The integral over $-\infty < x < +\infty$ of any one term in (19.27) can be expressed in terms of gamma-functions, so that $\int dx\, \psi$ becomes

$$y^{1/2 - i\kappa} + y^{1/2 + i\kappa} \frac{\Gamma(1/2)\Gamma(-i\kappa)}{\Gamma(1/2 - i\kappa)} \sum_{c=1}^{\infty} \sum_{(c,d)=1} \frac{1}{c^{1-2i\kappa}} \tag{19.29}$$

where the last summation is limited to $1 \leq d < c$. The second term represents exactly the outgoing wave, and the double sum gives the amplitude and phase as a function of κ.

The second sum in (19.29) is equal to the number of integers, traditionally designated by $\phi(c)$, between 0 and c, which are prime relative to c. The reader will find the explicit formula for $\phi(c)$ in any elementary textbook on number theory; it involves the decomposition of c into primes. The first summation in (19.29) becomes, therefore, expressed as a sum over all combinations of prime numbers, and it is not difficult to see that one ends up eventually with the Euler product (17.26) for the Riemann zeta-function.

With some minor manipulations, one finds that

$$\int dx\, \psi = y^{1/2 - i\kappa} + \frac{Z(1 + 2i\kappa)}{Z(1 - 2i\kappa)} y^{1/2 + i\kappa} , \tag{19.30}$$

where $Z(z) = \pi^{-z/2}\, \Gamma(z/2)\, \zeta(z)$. This function of the complex variable z is real on the real axis; the numerator and the denominator are complex conjugates. The factor of $y^{1/2 + i\kappa}$ in (19.29) can, therefore, be written as $\exp(2i\beta(\kappa))$ where $\beta(\kappa)$ is a real number dependent on κ, called the *phase shift* of the outgoing wave with respect to the incoming wave. (The peculiar combination $Z(z) = \pi^{-z/2}\, \Gamma(z/2)\, \zeta(z)$ was shown by Riemann to be symmetric with respect to the critical line $\mathrm{Re}(z) = 1/2$, i.e., $Z(z) = Z(1 - z)$; therefore, $Z(1/2 + 2i\kappa)$ is real-valued for real values of κ; and the famous zeros of the Riemann zeta-function are the zeros of this 'ordinary' function, which can be plotted like any other.)

A more detailed, and mathematically complete derivation can be found in the monographs of Kubota (1973), and of Lax and Phillips (1976). While the formula (19.30) was obviously known to the mathematicians for some time, they did not realize its importance for the understanding of chaotic features in wave mechanics; the author (Gutzwiller 1983) was the first to study the scattering in the singular square from this viewpoint.

19.9 Chaos in Quantum-Mechanical Scattering

Before comparing this quantum-mechanical result to its classical analog, the phase-shift $\beta(\kappa)$ will be examined more closely. The function $Z(z)$ is to be evaluated on the line $\text{Re}(z) = 1$; Riemann's zeta-function $\zeta(z)$ does not vanish on this parallel to the critical line $\text{Re}(z) = 1/2$; this fact is the essential step in the proof of the Prime Number Theorem (cf. Section 16.1). On the other hand, $\beta(\kappa)$ is basically the logarithm of the $\zeta(z)$, so that the zeros of $\zeta(z)$ on the critical line become singularities in the phase shift, and make themselves felt. The random nature of the zeros (cf. Section 17.9) is found again in the phase shift $\beta(\kappa)$, although smoothened by the distance of $1/2$, which guarantees a Taylor expansion in $\kappa - \bar{\kappa}$ near any point $\bar{\kappa}$, with a convergence radius of $1/4$ at least.

A plot of the phase shift as a function of κ bears out this impression; since the factors $\pi^{-z/2}$ and $\Gamma(z/2)$ in $Z(z)$ make only additive, monotonically increasing contributions to $\beta(\kappa)$, we have plotted in Figure 67 only the phase angle of $\zeta(1 + 2i\kappa)$ for large values of κ. Although the Riemann zeta-function is an analytic function with the deceptively simple definition (17.26), it keeps bouncing around almost randomly without settling down to some regular asymptotic pattern. *The Riemann zeta-function displays the essence of chaos in quantum mechanics, analytically smooth, and yet seemingly unpredictable.*

This chaotic feature has been demonstrated more dramatically by Reich (1980) and Good (1981), in the form of the following amazing *theorem:*

> Let D be a disk of radius $r > 0$ in the complex plane, centered on one of the points $\mu + i\,m\,\nu$, where $1/2 < \mu < 1$ and $\nu > 0$ are fixed, while the integers $m > 0$; the size r of these disks is limited by the condition that they have to be inside the strip $1/2 < z < 1$; then choose an arbitrary non-vanishing, holomorphic function $f(z)$, whose Taylor expansion around 0 converges inside a circle of radius r. Now consider the set M of integers $m > 0$, for which the difference $|f(z - \mu - im\nu) - \zeta(z)| < \varepsilon$ in the whole disk around $\mu + i\,m\,\nu$ for some fixed $\varepsilon < 0$. This set is proved to have a

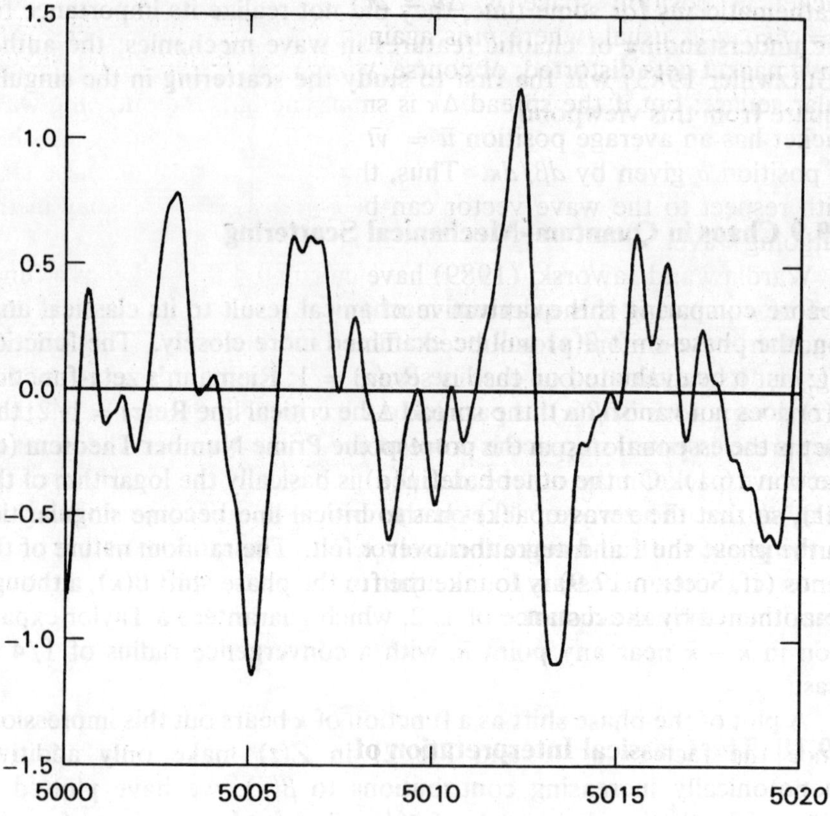

5000.00000000 0.05000000 400

Figure 67 Phase angle of $\zeta(1 + 2i\kappa)$ in the range $5000 \le \kappa \le 5020$; although this function is analytic (having a converging Taylor expansion at every point), it looks entirely unpredictable; it represents a smooth form of chaos typical of quantum mechanics.

> non-vanishing density, i.e., the number of points in M below some large $N > 0$ is a non-vanishing fraction of N with a lower bound greater than 0; this fraction depends, of course, on the function $f(z)$ and on ε.

In a more intuitive language, the Riemann zeta-function is capable of fitting any arbitrary smooth function over a finite disk with arbitrary accuracy, and it does so with comparative ease, since it repeats the performance like a good actor infinitely many times on a designated set of stages.

Instead of a wave entering the singular square with a fixed wave vector κ, one can construct a wave packet from the superposition of such waves, with a spread of $\Delta\kappa$ around $\bar{\kappa}$. The width of the wave packet in position space is $\Delta u \simeq 2\pi/\Delta\kappa$, and the average speed $\bar{v} = \hbar\bar{\kappa}/m$ as usual, where m is again the mass of the particle. The wave packet gets distorted, of course, when it wanders around the singular square; but if the spread $\Delta\kappa$ is small enough, the outgoing wave packet has an average position $\bar{u} = \bar{v}t - \sigma(\bar{\kappa})$, with a backward shift in position σ given by $d\beta/d\kappa$. Thus, the derivative of the phase shift with respect to the wave vector can be interpreted as a delay in the outgoing wave.

Wardlaw and Jaworski (1989) have calculated this delay as a function of κ ; since it is the derivative of $\beta(\kappa)$, the random nature of this function is even more pronounced. This interpretation is very appealing; but its validity in our case is severely limited. In order to see the strong variations of $\beta(\kappa)$, the spread $\Delta\kappa$ has to be quite small; but then the spread in position space is correspondingly large, and the center of the wave packet may not be defined precisely enough to detect the delay $d\beta/d\kappa$. The wave packet has to be calculated more carefully by carrying out the full integration over κ, rather than simply finding its center; it is then necessary to take the full complication of the Riemann zeta-function into account.

19.10 The Classical Interpretation of the Quantal Scattering

The scattering on the singular square has been worked out in detail because the final result can be written down explicitly, and its surprising properties can be seen very clearly. Whether they are typical of scattering in systems with hard chaos cannot be demonstrated at this time, although the author is convinced of this fact. It is, therefore, important to show that the critical formula (19.29) has a simple interpretation that goes directly back to classical mechanics. Indeed, the basic idea can be readily expressed for scattering problems in a more physical context, such as an electron scattered from a molecule, assumed to be rigid for the discussion's sake (cf. Section 20.1).

The wave function (19.28) is valid for any singular quadrangle, provided we understand correctly which terms are left out of the summation; instead of the highly symmetric domain D_3 in Figure 65, we now deal with the more general domain D_4 in Figure 66. The translation R in the definition of the cosets is given by the rhs of (19.26); if it is transformed to the standard form $[1, r\,; 0,1]$, the explanations following (19.28) are still applicable. Only one representative for a left coset with respect to R is used, and the integration over x in (19.29) can be carried out the same way, provided we take again only one representative in each right coset with respect to R. Thus, we can write for $\int dx\, \psi$ the general expression,

$$y^{1/2 - i\kappa} + y^{1/2 + i\kappa} \frac{\Gamma(1/2)\Gamma(-i\kappa)}{\Gamma(1/2 - i\kappa)} \sum_{\text{double cosets}} \frac{1}{c^{1 - 2i\kappa}} \quad (19.31)$$

where the unit element in G is left out of the summation.

The only quantity left over from the 2 by 2 matrix $[a, b; c, d]$ is the lower left-hand entry c, which is the same for all the group elements in the double coset; we now give a geometrical meaning to c. The reader may find it helpful to go back to the original picture (Figure 64) of an exponential horn which is attached to a torus; let us fix a ring around the horn at some distance y from the torus. In the upper half-plane, this ring becomes a line parallel to the x-axis at the arbitrary, but fixed distance y. Let us now take a string both of whose end-points are attached to this line $y =$ constant, and which follows the path corresponding to the group element.

To be more precise, we choose two points in the upper half-plane, $z = x + iy$ and $z' = x' + iy$; then we calculate $z'' = (az' + b)/(cz' + d)$, and the hyperbolic distance s between z and z'' according to (19.8). The coordinates x and x' are allowed to vary freely while y stays constant; intuitively, the string is allowed to wind around the exponential horn at both ends, although its path around the torus has to remain topologically the same; in this manner, we take advantage of the double coset with respect to the translation R.

The length s is now minimized with respect to x and x'. The computation is elementary, although the reader has to proceed with some caution so as not to get bogged down in an algebraic morass. The minimum length is found to be $s = 2\log(|c|y)$; we can, therefore, express the outgoing wave $y^{1/2 + i\kappa}/c^{1 - 2i\kappa}$ as the incoming wave $y^{1/2 - i\kappa}$ times the scattering amplitude $\exp(i\kappa s)/c$; the common factor $\Gamma(1/2)\Gamma(-i\kappa)/\Gamma(1/2 - i\kappa)$ of all outgoing waves is not relevant here.

Formula (19.31) is simply the sum over all possible shortest paths from a ring or monitoring station around the exponential horn at the distance $\log y$ and back. The contribution of each path has a phase equal to the wave vector times the length of the path. The amplitude $\exp(-\log c)$ is the stability factor for a path of length $\log c$, which is the length of the path inside the torus, rather than along the exponential horn. Thus, the expression (19.31) is the analog of the trace formula for positive energies; indeed, the reader will find in the second volume of Hejhal's treatise (1983) a complete discussion of the Selberg trace formula for a non-compact surface of negative curvature; the first term on the right-hand side is exactly (19.31).

The phase shift for the scattering from a singular quadrangle has not been evaluated, except for the few special cases that are *commensurable* with the modular group; such a group has a subgroup of finite index which is isomorphic to a subgroup of the modular group. Then, the rhs of (19.31) differs from (19.30) by a simple analytic function whose phase is periodic in κ, and does not change any of the statements at the end of the last section; cf. Gutzwiller (1987). The main problem in the general case is to generate the group elements in some effective way, other than writing down all the elements of a free group in the generators A and B, and using the relation (19.26) whenever possible. This issue will be discussed in the next chapter.

The classical interpretation of the quantal scattering as demonstarted in (19.31) shows that classical mechanics, if used properly, is able to yield a correct quantum-mechanical result. Such a conclusion may not be surprising on a surface of constant negative curvature, since Selberg's trace formula has already accomplished a similar feat for the spectrum on a compact surface of this kind. Actually, formula (19.31) is only the first term in a generalization of STF for non-compact surfaces (cf. Hejhal 1983); the singular square of Figure 63 has a discrete spectrum in addition to the scattering wave functions (19.29) and (19.31). These bound states decay exponentially toward the exceptional point at ∞; their difficult and treacherous computation was first attacked systematically by Winkler (1988) and Hejhal (1989).

The remarkable feature of (19.31) is that the classical scattering trajectories form a set of measure 0 in phase space; most trajectories neither come in from ∞ nor escape to ∞; there is no classical scattering in this sense. Nevertheless these same exceptional trajectories give a perfect account of the quantum mechanics!

The scarcity of trajectories that come in never to leave again, or that leave sometimes in the future after a whole lifetime in the box, is well known in celestial mechanics. It has been shown only quite recently, that two bodies bound in a Kepler ellipse can capture a third one, or the inverse process, the ejection of one of the bodies from a bound three-body system; such trajectories form a set of measure 0 in phase space (cf. Alekseev 1969 and Moser 1973).

The scattering problem of the last three sections was intentionally constructed to imitate the experimental situation where a probe is sent into a 'black box' along one degree of freedom to find out what is inside. Since the scattering is elastic, the only information to be obtained is the scattering phase shift β as a function of the wave vector k. Schrödinger's equation tells us how to compute the phase shift if the scattering potential V is known; but V is not well understood in nuclear and particle physics, so the problem has to be inverted: Can we determine the scattering potential once the phase shift as a function of the wave vector is known?

This problem was first discussed by Bargmann (1949), and became very important in the discussion of solitons during the 1960s; then Jost and Kohn (1952 and 1953) proposed several methods for constructing V from $\beta(k)$; Dyson (1976) wrote an interesting review. This work was done in the context of elastic scattering from a compact, spherically symmetric object in three dimensions; angular momentum is conserved in addition to the energy, and there is a phase shift for each value $L = \ell\hbar$. The most important case is $\ell = 0$, which can be reduced to the scattering in one dimension, say the positive x-axis, provided the wave function is required to vanish at the origin; one of the difficulties arises from the existence of bound states.

It is natural to ask whether the phase shift (19.30) could be due to some one-dimensional potential $V(x)$, similar to the ones constructed by Jost and Kohn. Although I have not examined which conditions are violated by (19.30) so as to prevent the construction of $V(x)$, I don't think that any reasonable function $V(x)$ can yield a complicated result like (19.30). Nor is it likely that a three-dimensional, spherically symmetric potential could do it. On the other hand, if the requirement of spherical symmetry is dropped, we will show at the beginning of the next chapter that chaotic behavior is to be expected classically. In view of the classical interpretation of the quantum calculation, one can expect results that are qualitatively similar to (19.30) quite generally. They open up a completely new vista on scattering theory.

CHAPTER 20

Scattering Problems, Coding, and Multifractal Invariant Measures

The two topics of this last chapter have only recently entered into the discussion of chaotic Hamiltonian systems; in both cases the author (Gutzwiller 1983; Gutzwiller and Mandelbrot 1988) was the first, with a safe headstart, to consider the problems involved, and to demonstrate the basic features by working out a typical example in some mathematical detail. Scattering in a surface of constant negative curvature was discussed at the end of the last chapter; but quite generally, the phase shift can be expected to depend in an analytically smooth, and yet essentially chaotic manner on the wave vector. Moreover, this behavior is the result of a straightforward approximation based on classical mechanics, even though the purely classical interpretation of scattering by itself may not yield reasonable, let alone good results.

Finding a useful code for all the classical trajectories seems to be the most important tool for making any progress. The Anisotropic Kepler Problem (AKP in Chapter 11 and the last three sections of Chapter 17) was presented in great detail, because it is coded by binary sequences, and appears, therefore, to be the simplest case of hard chaos. Codes for scattering seem easier to construct than for bound states; we will give some general examples.

A good code tells the full and unique story of a particular classical trajectory in a Hamiltonian system with hard chaos; but its primary information is totally different from the usual coordinates in phase space, or, more specifically, in the surface of section. The physics is

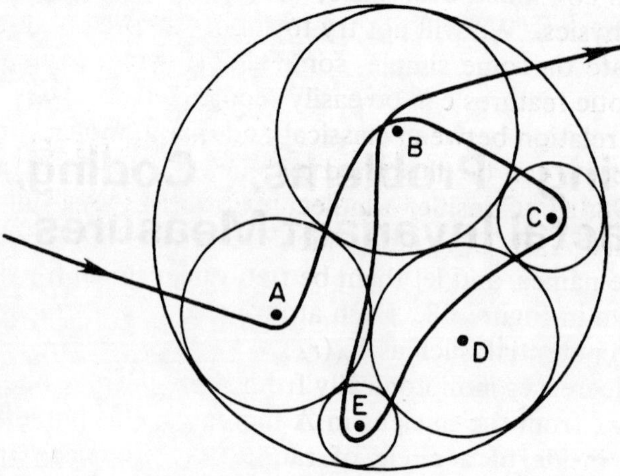

Figure 68 The two-dimensional muffin-tin molecule consists of non-overlapping circular potentials of finite range, centered on the fixed positions A, B, ... , all contained inside a large circle. A trajectory winds through the molecule as through a maze.

contained in the relation between these two equally important descriptions. Both of them define useful measures in the space of all trajectories, and both are invariant as one goes from one surface of section to the next. In order to express one in terms of the other, it is necessary to invoke the concept of multifractal sets, which was first proposed by Mandelbrot (1982).

As we approach the regime of soft chaos, however, the code either becomes more complicated, or it ceases to characterize a trajectory uniquely. We will display this phenomenon in the context of the AKP, when the mass ratio falls below 2, or when the angular momentum around the longitudinal axis is allowed to differ from zero. Whereas a *'slippery devil's staircase,'* a strictly increasing function, is found to characterize hard chaos, the usual 'devil's staircase', an increasing function with both horizontal and vertical pieces is typical of soft chaos.

20.1 Electron Scattering in a Muffin-Tin Potential

Scattering problems in quantum mechanics have a long and important history, starting with Rutherford scattering where classical mechanics gives a correct quantum result (cf. Section 12.4). Miller (1974) was perhaps the first to apply systematically the classical information, in order to extract quantum results in the theory of molecular scattering;

this effort has continued ever since, with many practical applications in chemical physics. We will not try to follow up on any of this work, but concentrate on some simple, somewhat abstract examples where the basic chaotic features can be easily recognized.

The close relation between classical and quantum mechanics in the scattering process can be illustrated by a simple physical example (cf. Gutzwiller 1985a). Consider a molecule consisting of a finite number of atoms, called A, B, C, ... etc., which are centered in fixed positions with the same names, and let them be two-dimensional for simplicity's sake, as shown in Figure 68. Each atom generates a screened, attractive Coulomb potential, such as $V_A(r_A) = -(e^2/r_A)\,\xi_A(r_A)$, where the function ξ_A decreases monotonically from some positive integer Z_A, as the distance r_A from the nucleus in A increases; this potential is confined to the inside of a circle of radius R_A. Between these non-overlapping circles, the potential is assumed to be ≤ 0 out to some large circle which surrounds the whole molecule, whereas the potential beyond is zero. This model is inspired by certain band calculations in solid-state physics, and we give it the corresponding name, the two-dimensional *muffin-tin molecule*. A muffin-tin in American folklore is a flat sheet of metal with a regular array of circular depressions; these get filled with dough, and the whole thing is heated in an oven to be baked; the dough in the depression rises, and out come the muffins, which are particularly appreciated at breakfast time.

Let us call this muffin-tin potential $V(x, y)$, and let us investigate an electron of energy $E > 0$, which comes from very far away along some direction 1, is scattered by the molecule, and leaves along some direction 2. Classically, we can just as well fix two end-points, (x_1, y_1) and (x_2, y_2), very far from the molecule, and lying along the directions 1 and 2. The classical trajectories are obtained from finding the stationary values of the Euler-Maupertuis action-integral (2.6),

$$\int_1^2 \sqrt{2m[E - V(x,y)]}\,\sqrt{dx^2 + dy^2} \,, \tag{20.1}$$

exactly as in Section 2.3. The corresponding Green's function follows immediately from the Van Vleck-type expression (12.28), in terms of the stationary values of (20.1).

The novel feature in this argument is the combination of $E > 0$ with the special shapes of $V(x, y) < 0$ inside each one of the molecular circles. The Euclidean distance $\sqrt{dx^2 + dy^2}$ is stretched by the constant factor $\sqrt{2mE}$ outside the molecule, whereas inside any of the circles the stretching is done by the larger factor $\sqrt{2m[E - V(x,y)]}$. This additional stretch is comparable to creating a hill inside each of the atoms, such that the length of the slope is in the ratio of

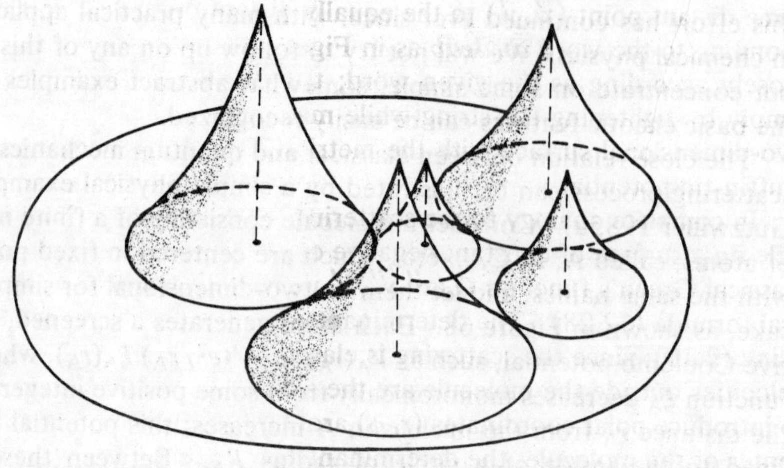

Figure 69 For positive energies, the classical trajectories in the muffin-tin potential are the geodesics on a two-dimensional Riemannian surface which can be embedded somewhat schematically in three-dimensional Euclidean space. A string is layed down loosely at first to follow a prescribed sequence of turns around the various atoms, and the corresponding trajectory is then obtained by tightening the string.

$\sqrt{2m[E - V(x, y)]}$ to $\sqrt{2mE}$ compared with its projection onto the (x, y) plane. The Gaussian curvature of this hill is negative for the screened Coulomb potentials, and there is a conical tip of opening 60° at the center of each atom.

A classical trajectory through the molecule minimizes the action (20.1); but there are many such trajectories. Each is characterized by the sequence in which the nuclei of the individual atoms are avoided by going either clockwise or counterclockwise around them. Thus, we can write down a word that consists of letters from a finite alphabet, a, $\bar{a}$, b, $\bar{b}$, ... where a indicates going clockwise and $\bar{a}$ going counterclockwise around the atom A, and so on. This sequence can be prescribed almost arbitrarily, with very few restrictions; e.g., the trajectory cannot turn more than once around a particular atom before visiting another one, and there is no use for the sequence $a\,\bar{a}$, etc. On the other hand, the trajectory can turn around B after hitting A, and go back to A; that would imply a deflection by π in B, or equivalently, hitting B with the angular momentum 0; the Coulomb singularity at the center of each atom permits the electron to make this kind of U-turn. The instability of these trajectories is obvious, as is their great proliferation, a sure sign of chaos.

The variational principle (20.1) can be given an intuitive, geometric twist. Suppose that we want to construct the classical trajectory from some distant point (x', y') to the equally distant point (x'', y'') corresponding to the word $a\bar{b}cd\bar{e}a\bar{b}$ as in Figure 68. A string is laid down loosely according as the given word; the trajectory is then obtained simply by tightening the string while making sure that it stays on the two-dimensional surface with the metric (20.1) corresponding to the muffin-tin potential.

In complete analogy to the scattering formula (19.31) for the particle on a surface of constant negative curvature, we can construct the classical Green's function $G_c(x'' y'', x' y', E)$ with the help of the general formula (12.28). The determinant $D(x'' y'', x' y', E)$ is calculated using (2.10); since the scattering is elastic, the incoming and outgoing velocities outside the molecule are the same, $(2E/m)^{1/2}$. It is natural to introduce polar coordinates (r, ϕ) around some conveniently chosen center of the molecule; the determinant D of second-order derivatives acquires an extra factor $1/r'' r'$ thereby; the derivatives at right angles to the trajectory are now obtained simply by differentiating the action integral with respect to the polar angles ϕ' and ϕ''. Thus,

$$G_c = \frac{2\pi}{(2\pi i\hbar)^{3/2}} \sum_{cl.traj} \sqrt{m/2E\, r'' r'}\, \sqrt{-\,\partial^2 S/\partial\phi''\partial\phi'}$$

$$\exp\!\left[\frac{i}{\hbar} S(r''\,\phi'', r'\,\phi', E)\right] . \qquad (20.2)$$

The second derivative in the amplitude of (20.2) has a simple interpretation: quite generally, one finds that the angular momentum $M = \partial S/\partial\phi$; therefore, $\partial S/\partial\phi' = M'$, so that $\partial^2 S/\partial\phi''\phi' = \partial M'/\partial\phi''$. This rate of change in the initial angular momentum M' with respect to the final angle ϕ'' is smaller the more complicated the trajectory through the molecule. It actually decreases exponentially with the length of the trajectory inside the molecule; the straight-line trajectories coming in and going out do not contribute.

Just as the general trace formula (17.13), the scattering formula (20.2) is now seen as special case of the more general formula (12.28) for positive energies. The exact scattering wave function (19.31) for a singular quadrangle could have been obtained from this classical approximation; the expression (20.2) turns out to be correct quantum-mechanically on a surface of constant negative curvature, although it is hard to understand why, since only the leading term in the limit of a small Planck's quantum is considered.

The simple analysis of this section has recently been worked out in great mathematical detail by various authors, including Jung and Scholz (1987), Eckhardt (1987), Blümel and Smilansky (1988), Gaspard and Rice (1989), as well as Bleher, Ott, and Grebogi (1989). The models differ from the muffin-tin molecule mostly because the scattering potential is assumed to be repulsive rather than attractive; but there is again a molecule which consists of at least three distinct mountains. The code gives the sequence in which the scattered particle bounces off one or the other; there is again an exponential proliferation and exponential divergence of neighboring trajectories.

The same analysis applies to a number of recent experiments with so-called ballistic electron transport (cf. van Houten et al. 1988, as well as Beenakker and van Houten 1989). A two-dimensional metal can be made between two layers of semiconductors, and given various shapes on a scale of less than a micrometer = 10^{-6} m $\simeq$ 10000 atomic distances. One common structure has four arms, two for the electric current to enter and exit, and two to measure the voltage transverse to the current. A magnetic field perpendicular to the plane of the device deflects the electrons, yielding a so-called Hall voltage. At very low temperatures, the only scattering comes from the boundaries; their concave shape makes the classical electron-motion chaotic. The voltage in the transverse direction is found to fluctuate (reproducibly) as a function of the energy in the same manner as the phase shift in Figure 67.

The first step in the study of scattering problems is to understand the classical trajectories and to extract some kind of scattering probability. We will not discuss this topic, except to remark again that purely classical, as opposed to quantal, scattering is fraught with conceptual difficulties. Also referring to the end of Section 10.5, but without further explanation, the metric entropy is in general larger than the topological entropy; the difference can be directly interpreted as as an escape rate for the particle to avoid getting caught in the molecular tangle.

If the scattering potential corresponds to a pinball machine, with circular obstacles of infinite height, the energy of the particle does not matter classically; only the geometry is important. In quantum mechanics, however, the de Broglie wave-length is the essential parameter, exactly as in the scattering on a surface of constant negative curvature. If the potential is smooth and of finite height less than E_0, the transition from the free motion for energies $E > E_0$, to the increasingly chaotic, classical motion for $E < E_0$ is studied exhaustively by Bleher, Grebogi, and Ott (1989). The scattering angle as a function of the incoming angular momentum, for a fixed incoming direction, has a striking fractal structure which can be explained as a sequence of bifurcations as the energy E decreases. Of course, such fractals appear also in the billiard-type scattering, just as they are present in the muffin-tin molecule. The classical Green's function (20.2) provides the most reasonable, and presumably quite effective approximation to the corresponding quantal scattering.

Although these fractals are closely related to the ones that will be discussed in the remainder of this chapter, none of the above quoted authors has explicitly investigated the relation between the code on the one hand and the physical variables on the other. So far, this last issue has been discussed only by Gutzwiller and Mandelbrot (1988); although it is a purely classical problem, it may well be at the bottom of the connection between classical and quantum mechanics. Thus, it seems to be an appropriate topic for the last sections of this book.

The classical Green's function (20.2) is part and parcel of the general trace formula (17.13), exactly as the scattering formula (19.31) for a singular quadrangle constitutes a generalization of Selberg's trace formula (19.20). Therefore, we can expect quite generally the same kind of chaotic behavior in the phase shift as was shown in Figure 67. The local variations (called *Ericson fluctuations* in nuclear physics; cf. Ericson 1960, 1963, and with Mayer-Kuckuk 1966) can be ascribed to nearby singularities ('resonances') in the complex wave-number plane; in the special example of Section 19.8, they are the zeroes of Riemann's zeta-function. As discussed in Section 17.9, they are distributed like the eigenvalues of the Gaussian Unitary Ensemble (GUE); according to Blümel and Smilansky (1988), the scattering resonances quite generally are distributed as the eigenvalues of the appropriate random matrix ensemble. There is obviously a wide field waiting to be studied.

20.2 The Coding of Geodesics on a Singular Polygon

The motion of a particle on a surface of constant negative curvature will again be taken up in this section, and the following two. It provides the simplest, and most striking example for the relation between the two complementary ways of looking at classically chaotic systems, phase space on one hand, and the coding of trajectories on the other.

In order to evaluate the phase shift on a singular quadrangle, the double cosets in (19.31) have to be enumerated in some effective manner. In looking for such a scheme, our model is the binary code which we found for the AKP in Chapter 11, and used in Section 17.11 to calculate the right-hand side of the trace formula. We would have liked to accomplish this feat for the double torus, i.e., the octogon in the hyperbolic plane; but the coding is much easier for the singular polygons; for the double torus, it will be discussed very briefly at the end of this section. Here, we will concentrate on the special case of the singular square, rather than some general polygon, because it shows all the interesting features, and the generalization is straightforward.

The idea for this code comes from a number of papers of Series (1985a, 1985b, and 1986); but as in the other instances of this chapter, the mathematical results were apparently derived with the intention of understanding hyperbolic geometry, rather than grasping the relation between classical and quantum mechanics in a system with hard chaos. It was then left to the author (Gutzwiller and Mandelbrot 1988) to recast the basic ideas of the mathematicians into a physically oriented picture.

The singular square D_3 is shown in Figure 63; its four singular corners are on the real x-axis in -1, 0, $+1$, and ∞; opposite sides are mapped into each other by the transformations $A = [e, e; e, (1+e^2)/e]$ and $B = [e, -e; -e, (1+e^2)/e]$ where $-\infty < e < +\infty$, as mentioned at the end of Section 19.7. The simple scattering formula (19.28) was based on $e = 1$; but nothing in this section ties us to this special case.

A geodesic is represented in the upper half-plane by a Euclidean circle whose center lies on the x-axis; its intersections with the x-axis are called ξ and η; the motion is always from ξ to η, as was explained already in Section 19.2, and shown in Figure 54. The four corners of the singular square divide the real axis into four intervals which will be called $\{1\} = (-\infty, -1)$, $\{2\} = (-1,0)$, $\{3\} = (0,1)$, and $\{4\} = (1,+\infty)$; the four sides of D_3 are given the same names. The necessary and sufficient condition for a geodesic to cut D_3 is for the two endpoints to belong to two different intervals; the interval to which ξ belongs will be called the entry, and the interval of η the exit.

To each exit interval belongs one of the four operations which maps the corresponding side into its opposite: $\{1\}$ calls for A which maps $\{1\}$ into $\{3\}$, $\{2\}$ calls for B^{-1}, which maps $\{2\}$ into $\{4\}$, $\{3\}$ calls for A^{-1}, and $\{4\}$ for B; a similar pairing goes with the entry intervals. The exit map into the opposite side changes η into η_1, while the entry map changes ξ into ξ_1; this procedure can be continued, yielding η_2 at the next exit, and ξ_2 at the preceding entry, and so on. The sequence $\eta, \eta_1, \eta_2, \ldots$ describes the forward motion, and depends only on the initial value η, while the sequence $\xi, \xi_1, \xi_2, \ldots$ describes the backward motion, and depends only on ξ.

In the following discussion, we will concentrate on the forward sequence $\eta, \eta_1, \ldots$; everything works out the same way for the backward sequence $\xi, \xi_1, \ldots$. This complete separation of the future and the past is remarkable; it comes about because the broken linear transformation $[a, b; c, d]$ acts according as (19.7) on ξ and η separately. The independence of past and future applies in this simple form only to the singular polygons; it is, of course, a symptom of chaos just as when a die is thrown, and nothing in the past of the gambler permits any conclusion concerning the next move.

The sequence of intervals corresponding to the consecutive values of η constitutes a good code for the forward trajectory; but there are obvious limitations. If $\eta \in \{3\}$, i.e., the geodesic leaves D_3 through the side {3}, the map A^{-1} will transform the exit point into an entry point on the side {1}. The new exit point is bound to be either on {2}, or on {3}, or on {4}; but it cannot be on {1}. The triple choice to be made can be characterized more intuitively as right-center-left, since looking from {1} into the singular square, the side {2} is on the right, {3} is straight ahead, and {4} is on the left.

This triple choice is seen directly in Figure 63; instead of moving the geodesic back into the original domain D_3 with the help of A^{-1}, we make a copy $A\,D_3$ of D_3 with the help of A. The three new exit sides are three contiguous, small Euclidean circles centered on the real axis, which fill completely the previous exit side {3}; they are in the correct order right-straight-left. If $\eta_1 \in \{2\}$, i.e., the geodesic makes a right turn upon leaving the original D_3, it will hit the first of the small circles inside {3}. If we were still in the original domain, the side {2} would have to be mapped into {4} with the help B^{-1}; instead, we can make a new copy $AB\,D_3$ of D_3 by mapping first with B and then with A. The small circle corresponding to the side {2} on $A\,D_3$ is divided into three even smaller circles which correspond to the three new possible exits, {1}, {2}, or {3}, again in the correct order.

The continued subdivision into three contiguous subintervals is quite obvious in Figure 63, and leads naturally to the *ternary code* corresponding to the triple choice right-center-left at each stage. There is no limitation on the words that can be formed in this manner from the letters R (right), S (straight), and L (left). Given η, the algortihm for finding the ternary code is straightforward: in the example above, one starts with $\eta \in \{3\}$; applying the map A^{-1} one obtains η_1; since it turns out that $\eta_1 \in \{2\}$, the first letter in the code word is R; the exit side {2} dictates the map of η_1 by B^{-1} which yields η_2; according as $\eta_2 \in$ {1}, {2}, or {3}, the second letter in the code is R, S, or L; an so on. Thus, η generates the first exit side, {3}, plus a unique ternary code word starting with R.

Conversely, if an initial exit side and a ternary code word is given, Figure 63 suggests how to find the subinterval on the real axis where η is located. It takes some rather obvious mathematics to show that a code word of finite length leads to a subinterval of non-vanishing length, and that this subinterval shrinks to a point as the code word

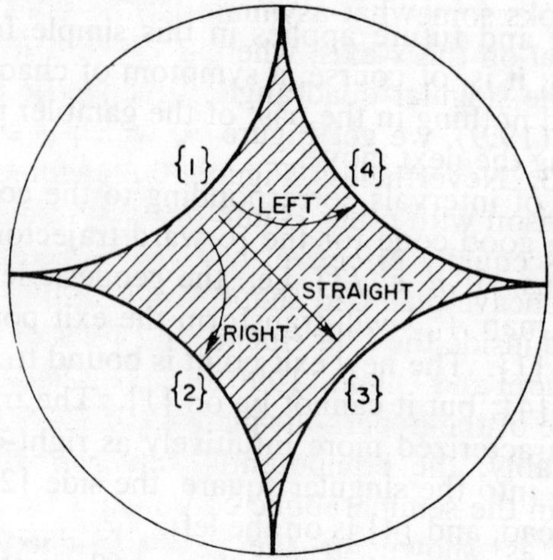

Figure 70 The singular square in the unit-disk model of the hyperbolic plane; the concave diamond shape of the domain can be found also in the diamagnetic Kepler problem, and is the cause for the chaos there as well. The ternary code simply records whether the trajectory entering, say, in {1} turns right, goes straight, or turns left.

becomes infinitely long. The reader should be warned, however, that the distribution of these subintervals is very uneven; a sharp, quantitative statement of this fact will be given in the next section.

The code word for a given endpoint η is easy to compute, if the above algorithm is used. If η is known to a certain precision, say $\Delta\eta$, the length of the code word is limited roughly to $-\log(\Delta\eta)/\log 3$; indeed, the consecutive maps have the effect of spreading η_j from one of the four intervals, {1}, {2}, {3}, or {4}, to one of the three possible subintervals for each of them; the uncertainty $\Delta\eta$ then gets multiplied by 3 approximately.

The reverse process of obtaining η from the given initial exit side and the given ternary code word, however, is quite different. First off, the sequence of mappings has to be determined: again, suppose that the initial exit interval is {3}, and the code word begins with RL...; {3} requires A^{-1} to get us to {1}; R then leads to {2} which requires B^{-1} to get us to {4}; L then leads to {3} which requires A^{-1}, and so forth; the computation so far is purely logical. Secondly, the original domain D_3 has to be mapped by the inverse sequence, i.e., $...ABA$ in this case; the resulting copy $...ABA\ D_3$ spans a short interval on the real axis which is most easily obtained by calculating the four corners, i.e., by finding the images of $-1, 0, +1$, and ∞. The value of η is found in this interval; its total length is roughly given by as many powers of $1/3$ as letters in the code word.

Figure 63 looks somewhat asymmetric; the sides {1} and {4} get an infinite interval on the x-axis, whereas {2} and {3} get an unit interval each. If the singular quadrangle is mapped into the unit circle with the help of (19.9), we get Figure 70 which is completely equivalent to Figure 63. Nevertheless, Figure 70 immediately inspires us to make the comparison with other Hamiltonian systems where the same kind of coding is equally effective, because the equipotential curves have the same concave diamond shape. Indeed, one could consider a Euclidean billiard inside the four Euclidean circles of Figure 70; in the same vein, Giannoni and Ullmo (1990) have studied the motion inside a singular triangle of the hyperbolic plane.

Most remarkably, the equipotential curves for the diamagnetic Kepler problem in the semiparabolic coordinates (μ, ν) have the same diamond-shape, as shown in the potential-energy term of the Schrödinger equation (18.13), or more explicitly in the article by Reinhardt and Farrelly (1982). That fact is the basis for the successful coding of the trajectories by Eckhardt and Wintgen (1990). Finally, the planar-metal device mentioned at the end of the last section (cf. Beenakker and van Houten 1989) owes the fluctuations in its Hall voltage to the same chaotic scattering in a diamond shape. The chaotic features can also be seen in the conductance calculations of Avishai and Band (1990).

20.3 The Geometry of the Continued Fractions

A particularly simple example for this kind of code is due to Series (1985a and b), although it goes back to Artin (1924). It will be used in the next section to demonstrate some novel features in chaotic Hamiltonian systems. The singular triangle D_2 was defined in Section 19.7; it is bounded by the vertical lines $x = 0$ and $x = 1$, as well as the Euclidean circle $(x - 1/2)^2 + y^2 = 1/4$, and constitutes one-half of the singular square D_3.

Exactly the same construction of a code can be applied; but there are only three intervals on the real axis: $\{1\} = (-\infty, 0)$, $\{2\} = (0,+1)$, and $\{3\} = (+1,+\infty)$. The mappings which are needed to bring a geodesic back into D_2, are not as natural as in the singular square; the following definitions differ from what was said in Section 19.7. The geodesic is assumed to enter through {1}, and leave through {2} or {3}, so that $\xi < 0$ and $\eta > 0$; we will discuss only the forward map. If $\eta \in \{2\}$, then {2} is mapped into {1} by $\eta \to \eta_1 = S(\eta) = \eta/(1 - \eta)$; if $\eta \in \{3\}$, then {3} is also mapped into {1}, using the translation

$\eta \to \eta_1 = T(\eta) = \eta - 1$. In Section 19.7, the group of maps for D_2 was generated differently; nevertheless, the tilings of the upper half-plane are the same, although the sides of the singular triangle are not identified in the same way.

The code is now binary, with the letter R (right) for {2}, and the letter L (left) for {3}; the exit coordinate η determines the binary sequence, and vice versa. The same procedure could be used to go from one to the other as above; but Series proposed a more direct method which has been known in a different guise for some time (cf. Adler and Flatto 1982).

Let us describe the binary sequence by a sequence of integers $(n_0, n_1, n_2, \ldots)$, where $n_0 \geq 0$ and $n_j > 0$ for $j > 0$; it indicates n_0 times L, followed by n_1 times R, followed by n_2 times L, and so on. Almost trivially, although quite unexpectedly, one has

$$\eta = n_0 + \cfrac{1}{n_1 + \cfrac{1}{n_2 + \cfrac{1}{\cdots}}}; \tag{20.3}$$

indeed, suppose that η has been represented by this continued fraction (cf. Section 9.5); if $n_0 > 0$, it requires n_0 translations T to get from η in {3} to $\eta_{n_0} = \eta - n_0$ in {2}; now the map S has to be used n_1 times, in order to find $\eta_{n_0 + n_1}$ given by an expression similar to (20.3), but n_j replaced by n_{j+2}; a second cycle can now start as before.

The map between the binary code and the continued fractions came rather naturally out of a purely geometric problem, namely the geodesics in a singular triangle. The binary code describes how the geodesic winds itself through the tiling of the hyperbolic plane, whereas the continued fraction gives the exit point η for the geodesic. These two aspects of the same geodesic are uniquely determined by two entirely different mathematical objects, the purely logical code word on one hand, and the purely arithmetical continued fraction on the other. The explicit relation between these two objects is a model for describing chaotic Hamiltonian systems, such as we found already in the AKP.

Since a substantial part of Chapter 19 was spent on studying the double torus (compact two-dimensional manifold of genus 2) of constant negative curvature, the question of coding for its geodesics will be taken up briefly. In principle, the four generators A, B, C, D and their inverses could be used, in exact analogy to the generators A, B with their inverses in the case of the singular quadrangle. The difficulty with this scheme is the relation (19.11) between the four generators which eliminates an increasing number of words as they become longer; the restrictions on the alphabet are not easy to implement, and the corresponding partitioning of phase space is very complicated.

The relation (19.11) goes directly back to the definition of the homotopy group for the double torus, and is based on Figure 57. The resulting octogon (Figure 56) tiles the hyperbolic plane; but a geodesic which bounds one particular octogonal tile, in general pierces a neighboring tile when this geodesic is produced indefinitely. This unfortunate circumstance can be avoided with the help of a construction by Nielsen (1927), which was discussed in more detail by Bowen and Series (1979) as well as Series (1986), and finally worked out completely by Adler and Flatto (1989).

The double torus is sliced open by cutting along four periodic geodesics; if we refer to Figure 58, we could choose the two geodesics which define the perimeter of each torus, plus the geodesic defining the distance between them, and the circumference of the right-hand torus. The 'skin' presents itself now as a *dodecagon* in the hyperbolic plane; only four 12-sided tiles meet in each vertex; the twelve sides of the original dodecagon can be produced, until they intersect the limiting unit circle which gets divided thereby into 24 adjoining intervals, to be named with the first 24 letters of the alphabet, a, b, ... , w , x.

The endpoints ξ and η of an arbitrary geodesic are now specified by the interval on the unit circle to which they belong. Upon leaving the dodecagon, the geodesic undergoes the appropriate bilinear transformation; there are now six of them plus their inverses, and the endpoints of the geodesic, ξ and η, get transformed accordingly. The code registers the successive intervals to which ξ and η belong; any particular interval cannot be succeeded by all the 23 others, however, but only by a subset. There is a 24 by 24 matrix indicating which two letters can follow one another. In the language of ergodic theory, we have a *subshift* (cf. Section 10.1). Although this scheme looks more complicated than the homotopy group with the relation (19.11), the subshift sets conditions only between two consecutive letters, while the relation (19.11) involves eight symbols in sequence.

20.4 A New Measure in Phase Space Based on the Coding

The vertical line $x = 0$ in the Poincaré upper half-plane would normally serve as a surface of section for the flow of geodesics. As an alternative for the coordinates in a surface of section, however, we can use the coordinates ξ and η of the endpoints of a geodesic, as shown in Figure 54. When the trajectory intersects the vertical $x = 0$, its vertical momentum v and its position y can be expressed in terms of ξ and η. Thus, the invariant volume $d\omega$ on this usual surface of section can be ex-

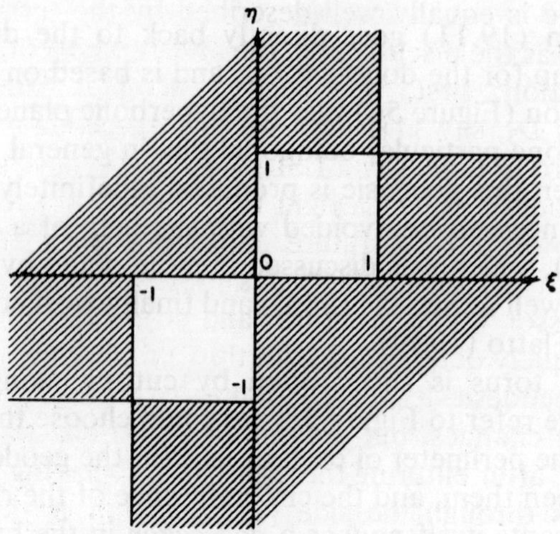

Figure 71 Surface of section of the singular square (shaded region) with the endpoint coordinates (ξ , η) for the geodesics; the lines η = constant are the stable manifold, and the lines ξ = constant the unstable ones.

pressed in terms of the endpoint coordinates (ξ, η). The calculation yields the expression

$$d\omega \;=\; dv\,dy \;=\; \frac{d\xi\, d\eta}{(\eta \;-\; \xi)^2}\;. \tag{20.4}$$

If both ξ and η are subjected to the same transformation $[\,a\,,b\,;c\,,d\,]$, it is easy to check that this volume element remains the same. We will use (20.4) as the *Liouville measure* for the motion of a particle on a surface of constant negative curvature.

In the motion on the singular polygon, the endpoint coordinates ξ and η belong to different intervals on the x-axis. Figure 71 shows the admissible points in the shaded region. The total area of this region with the measure (20.4) is infinite, reflecting the non-compact nature of this dynamical system. The lines ξ = constant are the unstable manifolds, while the lines η = constant are the stable manifolds. Indeed, all the geodesics with the same η approach one another exponentially with distance as they move toward the real axis of Poincaré's upper half-plane; similarly, the trajectories with the same ξ emerge from the same point, as the particle on them is followed backwards in time.

Everytime the geodesic tries to leave the domain D_3, it has to be mapped by A, B, or their inverses. The same mapping applies to ξ and η; the resulting points in the (ξ, η) plane represent the geodesic in the surface of section as usual.

The geodesic is equally well described by the ternary code, however; according as the discussion in the preceding section, we adopt the following notation: first, we specify the side through which the geodesic enters D_3, say {1}; then we give the code word forward as well as backward in time, say ...LSL{1}RLS...; it says that after crossing {1}, the geodesic makes a right turn, then a left turn, then moves straight, and so on into the future. Similarly, in order to hit {1}, the geodesic made a left turn, which was preceded by a straight motion, whose predecessor was a left turn, and so on out of the past. The geodesic is uniquely determined with this information.

When the geodesic leaves D_3, it has to be mapped as explained above; its code changes, but only in a very simple way. To take the above example, after entering through {1} and making a right turn, the geodesic leaves through the side {2}; the map B^{-1} makes it enter through {4}; the new code becomes ...LSLR{4}LS..., where the sequence {1}R in the center of the word has been replaced by sequence R{4}, while the remaining letters are the same as before. If we disregard the replacement of {1} by {4}, all that has happened is to shift the first letter in the word for the future into first position of the word for the past.

The code word can be given a measure in the standard manner: a ternary digit c is associated with each letter; 0 for R, 1 for S, and 2 for L. The real number $0 \leq \theta \leq 1$ has the ternary representation given by the code word for the future; similarly, the real number $0 \leq \Theta \leq 1$ has the ternary representation given by the code word for the past;

$$\Theta = \sum_{j=0}^{\infty} c_{-j} 3^{-j-1} \; , \theta = \sum_{j=1}^{\infty} c_j 3^{-j} \; . \tag{20.5}$$

If necessary, there can be four kinds of such pairs (Θ, θ) to accommodate the four possibilities for the entry-interval.

This ternary representation defines another measure on the surface of section, $d\Theta \, d\theta$. As in the binary sequences for the AKP, the Poincaré map of the surface of section into itself is equivalent to shifting all ternaries c_j to the left, $c'_j = c_{j+1}$. The scale of Θ gets contracted by a factor three, while the scale of θ gets expanded by the same factor three. The square $-1 \leq \theta, \Theta \leq 1$ undergoes a ternary bakers' transformation which preserves the area, every time the geodesic has to be mapped in order to stay inside the singular square.

The endpoint coordinates (ξ, η) can now be mapped into the number pairs (Θ, θ) representing the code word; the argument in the preceding section shows that this map is essentially one-to-one and continuous. Therefore, a new invariant measure in a surface of section, or more generally in phase space, is naturally associated with the code for the trajectories. How is this new measure related to the traditional Liouville measure (20.4)? In spite of the one-to-one and continuous relation between the two kinds of coordinates, each measure will now be shown to be fractal in terms to the other.

The demonstration will be simplified by investigating only how the exit coordinate η is related to the ternary representation θ of the code word for the future. We discussed in the preceding section the algorithms which allow us to go from one to the other, in either direction. Figure 72 gives a graph of the function $\theta(\eta)$ for different values of the parameter e in the transformations of the singular square, as explained at the end of Section 19.7.

These functions are obviously highly singular, with both very steep and very flat portions; but they are not ordinary 'devil's staircases' which are constant except at a denumerable set of points where they jump discontinuously (cf. Bak and Bruinsma 1982). The designation as *'slippery devil's staircase'* seems quite appropriate for this new kind of function (cf. Gutzwiller and Mandelbrot 1988). The almost constant portion of these functions indicate that a large change in η does not alter much the code word; the geodesic gets trapped, and, as a closer inspection shows, it winds for many turns around the exponential horn, making either many consecutive left or right turns. The nearly vertical portions, on the other hand, display the extreme sensitivity of the geodesic's qualitative behavior with respect to small changes in η; they occur mainly when there are many straight motions, so that the geodesic wanders around the torus proper rather than on the attached exponential horn.

20.5 Invariant Multifractal Measures in Phase Space

The analytical character of the functions $\theta(\eta)$ is best understood in terms of a *multifractal set;* we will give the most primitive version of this modern concept which was first proposed by Mandelbrot (1982), and has since been studied extensively (cf. Frisch and Parisi 1985; Halsey, Jensen, Kadanoff, Procaccia, and Shraiman 1986). In order to simplify the arithmetic, we shall only discuss the example of the geodesics in the singular triangle at the end of the last section.

A function $\beta(\eta)$ is defined as follows: $0 < \eta < 1$ is represented as the continued fraction (20.3) where $n_0 = 0$, while $0 < \beta < 1$ is the real number whose binary expansion consists of $n_1 - 1$ times 0, followed by n_2 times 1, followed by n_3 times 0, and so on. This function and its inverse, calculated separately, are shown in Figure 73; it has the

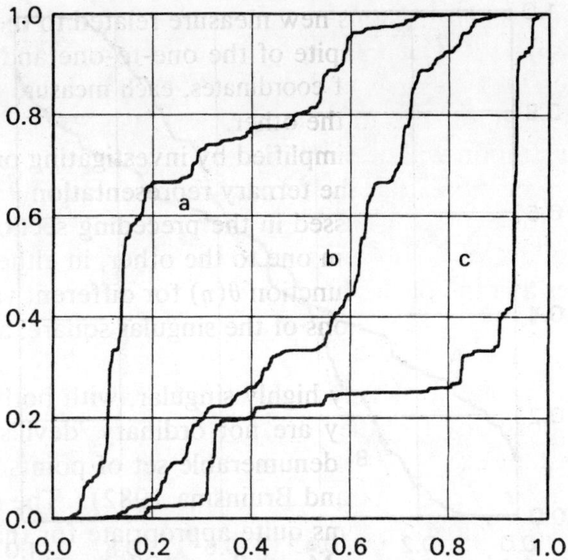

Figure 72 The functions $\theta(\eta)$ relating the coordinate η in the surface of section with the ternary representation of the corresponding code word θ; the values of the parameter e in the matrices A and B are $1/\sqrt{7}$ for (a), 1 for (b), and $\sqrt{7}$ for (c).

same character as the functions $\theta(\eta)$ in Figure 72. If anything, the latter are even more irregular than the former.

The nearly flat portions of $\beta(\eta)$ are found on an everywhere dense set, but not in whole intervals. The bottom of the curve shows this feature quite clearly: small values of η have $n_1 \simeq 1/\eta$, and therefore

$$\beta \simeq \exp(-n_1 \log 2) \simeq \exp[-(\log 2)/\eta] \,. \qquad (20.6)$$

All derivatives vanish, although the function is not constant. The same behavior is found every time η gets close to a rational number, because there is then a large integer n_j in its continued fraction. This remarkable property characterizes a *slippery devil's staircase.*

The more detailed analysis of the function $\beta(\eta)$ is based on assuming that $\Delta\beta \simeq (\Delta\eta)^\alpha$ in some small neighborhood. The *Hölder exponent* α is calculated from the finite differences

$$\alpha = \log[\beta(\eta + \Delta\eta) - \beta(\eta)]/\log(\Delta\eta) = \log\Delta\beta/\log\Delta\eta \,. \quad (20.7)$$

In practice, the whole interval for η is divided into 100,000 equal pieces $\Delta\eta$, and the corresponding $\Delta\beta$ is calculated.

The distribution of the α's is extremely singular; the points where α has a fixed value is a fractal subset of the interval $0 < \eta < 1$ whose dimension will be called $f(\alpha)$. For its calculation, the following scheme

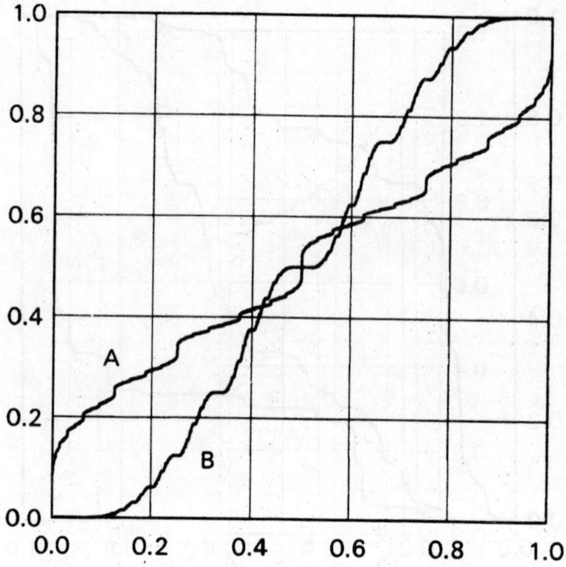

Figure 73 The functions $\beta(\eta)$ and $\eta(\beta)$ for the relation between the coordinate η and the binary code β in the singular triangle; they are typical 'slippery devil's staircases.'

is adopted: the α's obtained from (20.7) are put into 100 bins of equal size $\Delta\alpha$; if the number in the bin around α is called $N(\alpha)$, then

$$f(\alpha) = \frac{\log [N(\alpha)/\Delta\alpha]}{\log [1/\Delta\eta]} . \tag{20.8}$$

The idea behind this formula is quite elementary, and follows one of the definitions of the *Hausdorff dimension:* The points in the interval $0 \leq \eta \leq 1$ which are characterized by the exponent α are covered by the small intervals of length $\Delta\eta$; there are $N = N(\alpha)/\Delta\alpha$ of them. The fractal dimension for this subset is the exponent $f = f(\alpha)$, which guarantees the relation $N(\Delta\eta)^f = 1$, and yields (20.8).

This function is plotted in Figure 74; it stops abruptly at 6.6 for the following reason: the calculations were carried out in quadruple precision, i.e., with 112 significant bits or an accuracy of $\Delta\beta \simeq 10^{-33}$. About 10% of these differences, however, are even smaller and were put equal to 0; therefore, no data are available for $\alpha > 33/5 = 6.6$ since $\Delta\eta = 10^{-5}$. The lower end of the $f(\alpha)$ curve comes from the point with the largest difference $\Delta\beta$. It arises for $\eta = \gamma$, the golden mean, and $\beta = 2/3$, and is given by $\alpha_{\min} = \log 2/\log(1/\gamma^2) = 0.7202$.

The inverse function $\eta(\beta)$ can be investigated in a similar manner. Its $f(\alpha)$ curve is plotted in Figure 75. Instead of the long tail in Figure 74, we now find a long head. The beginning portion in Figure 75 is

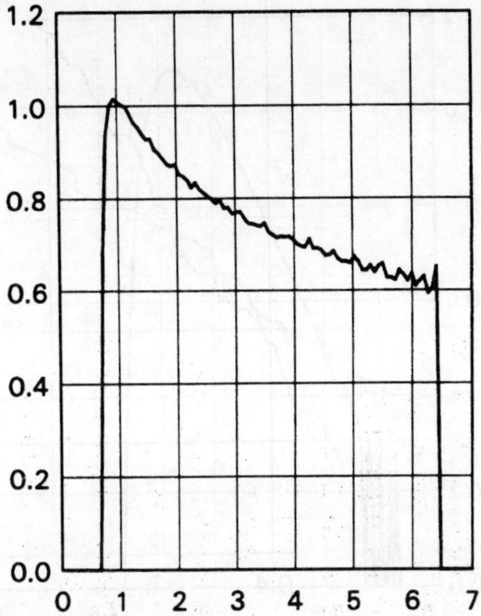

Figure 74 The fractal dimension $f(\alpha)$ of the point-set in $0 < \eta \leq 1$ where the Hölder exponent of the function $\beta(\eta)$ equals α; no data are available for $\alpha > 6.6$ because the difference $\Delta\beta$ falls below 10^{-33}.

actually linear with slope 1, as nearly as the finite statistics of the computational data permit to show this feature. The constant difference $\Delta\beta$ was chosen as 2^{-26}; the largest jump occurs at $\beta = 0$ and 1 where $\Delta\eta = 1/26$, so that according as (20.7) we find $\alpha \geq \log 26/26 \log 2 = .1786$. On the other hand, the minimum value of α in the preceding paragraph now yields $\alpha_{max} = \log(\gamma^{-2})/\log 2 = 1.3885$.

The foregoing analysis of the functions $\beta(\eta)$ and $\eta(\beta)$ is a most primitive first effort which is probably beset by poor statistics. By the standards of the experts in this multifractal trade, the above example seems pathological, just as ordinary fractals seemed at one time questionable products of some mathematician's distorted fantasy. On the contrary, the above functions appear to be typical of Hamiltonian systems with hard chaos, as the discussion of the Anisotropic Kepler Problem in the next sections will show.

The invariant measure in phase space that is based on the coding of the trajectories has in general this multifractal character with respect to the usual Liouville measure. Although this feature has not been exploited as yet, it is bound to be important for the evaluation of the trace formula. The $f(\alpha)$ curve is only a first, rather crude, but very telling indication of the hidden complications which are found even in a simple Hamiltonian system with hard chaos, such as the geodesics in a singular polygon.

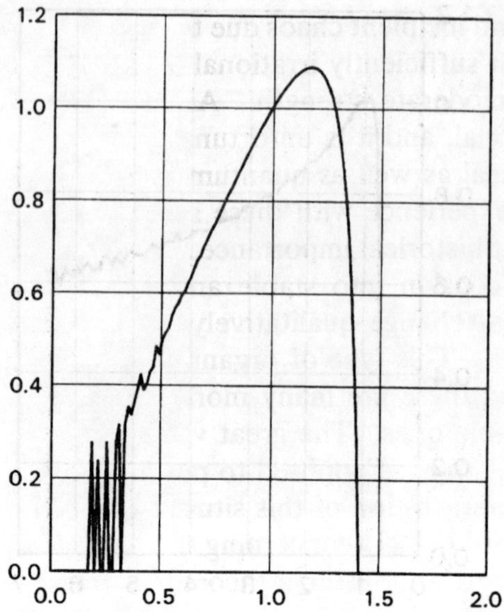

Figure 75 The fractal dimension $f(\alpha)$ of the point set in $0 < \beta < 1$ where the function $\eta(\beta)$ has the Hölder exponent α; all data are in the plot.

20.6 Multifractals in the Anisotropic Kepler Problem

The multifractal analysis of the preceding section is a first step toward the transition from hard to soft chaos. The last two sections of this book are designed to start an overdue discussion of this topic. In order to understand the urgency of such a project, we have to go back to some of the earlier sections, in particular Chapter 9 on Soft Chaos and the KAM Theorem and Chapter 10 on Entropy.

The two extremes of a Hamiltonian system are described by two particularly simple types of organization in phase space: on the one hand, in an integrable system there are as many integrals of motion as degrees of freedom, and phase space gets foliated into a set of invariant tori. On the other hand, in a system with hard chaos there is a double foliation into (stable and unstable) submanifolds, each with as many dimensions as degrees of freedom, and each trajectory is the intersection of a submanifold from one foliation with a submanifold from the other foliation. Figure 26 shows the surface of section in the latter case, for the Anisotropic Kepler Problem (AKP) with mass ratio 5.

The foliation for the integrable Hamiltonians cannot maintain itself, even under a weak perturbation, e.g., in the potential energy. The KAM theorem tells us to what extent the invariant tori resist the destructive effect and incipient chaos due to such small changes; as a rule, only the tori with sufficiently irrational frequency ratios remain for a perturbation of moderate strength. Accordingly, integrable systems are very exceptional, and it is unfortunate that our whole intuition in mechanics, classical as well as quantum, and perhaps even statistical, is based on the experience with these special cases, although nobody doubts their great historical importance.

The double foliation into stable and unstable manifolds, on the contrary, does not change qualitatively, even under perturbations of moderate strength. This type of organization in phase space is structurally stable, and there are many more Hamiltonian systems of this kind than integrable ones. The great variety of Riemannian surfaces with constant negative, as opposed to positive, curvature provides the most striking demonstration of this situation. Also, the original arguments of Hadamard (1898) concerning the chaotic nature of geodesics are all based on the Gauss-Bonnet theorem (19.10), and are, therefore, valid for any surface whose Gaussian curvature is strictly negative, although it may vary from place to place. In contrast, strictly positive curvature without rotational symmetry, like on the surface of a potato, does not guarantee the integrability of the geodesics.

Generic Hamiltonian systems belong to the category of soft chaos where some parts of phase space have foliations into invariant tori while others have the double foliation into stable and unstable manifolds. The two complementary regions penetrate into each other in a most intimate and fractal fashion. Rather than to describe the newly created layers of chaos between the remaining invariant tori, it seems more reasonable to try to understand how isolated islands of tori can arise somewhere inside the double foliation of a system with hard chaos.

The AKP presents a system where this transition can be followed in function of its two main parameters: the mass ratio, μ/ν in the normalization (11.4), and the angular momentum M around the longitudinal (heavy) axis. In this section, we will discuss some recent calculations of the author (Gutzwiller 1989) concerning the effect of changing the mass ratio, along the lines of the preceding section. Only a few preliminary calculations were carried out for $M \neq 0$; they will be mentioned shortly in the next section.

Figure 26 shows in graphic detail how the numbers (ξ, η), representing the binary coding for the trajectories in the AKP, are related to the points in the surface of section that is the heavy axis. There seems little doubt that this relation is one-to-one and continuous, if the mass ratio is 5. The leaves in each of the two foliations are labeled by equidistant values of ξ and η; nevertheless, they do not seem equally spaced in the surface of section. This situation was described in terms of a Hölder exponent α in formula (11.14), an idea that will be pursued more systematically in this section.

A multifractal analysis of the AKP has to deal with the whole two-dimensional surface of section of Figure 26; but the ingredients for such an investigation are not in place as yet, it seems. The singular polygons of constant negative curvature (cf. the preceding three sections) are exceptional, because on the one hand the stable and unstable manifolds in the surface of section are simply the horizontal and the vertical lines as shown in Figure 71, while on the other hand the (ternary) coding clearly separates past from future. Therefore, the two-dimensional map between the surface of section and the coding decays into a product of two one-dimensional maps like Figure 72.

The task for the AKP will be simplified by considering only the special trajectories that are symmetric with respect to the heavy axis; they intersect the x-axis at time $t = 0$ at a right angle. Their binary sequence is symmetric such that $a_{-j} = a_j$; the symmetry of the potential with respect to the y-axis allows us to consider only the case where $a_0 = \mathrm{sign}(x_0) > 0$. Moreover, since the momentum in the x-direction $u_0 = 0$ for these trajectories, we find that $x_0 = X_0$ according as (11.9).

The calculations become quite simple: the initial coordinate x_0 is increased in equal steps over its whole interval from 0 to 2. The equations of motion are integrated numerically out to some large number of intersections with the surface of section. With the programming skill and computing resources available, x_0 was increased in steps of .0002, and 48 intersections were obtained for each trajectory. The mass ratio was chosen to be 1.5, 2.0, 3.0, 5.0, 8.0, and 13.0, imitating a Fibonacci series.

Figure 76 gives the corresponding plots of the binary label η from (11.10) as a function of x_0. Before going to a more detailed analysis, the first striking feature is the monotonic increase: with the glaring exception of a short interval near $x_0 = .01$ for the mass ratio 1.5, the value of η was always found to be strictly increasing, although it sometimes did so only with excruciating slowness as x_0 increased in steps of equal size. On the other hand, the value of η remained constant in the exceptional interval for mass ratio 1.5; but it registered a strictly monotonic increase elsewhere, in particular in the long, seemingly flat portion around $x_0 = 1$.

The multifractal analysis of these curves follows the same recipes (20.7) and (20.8) as were used to obtain Figure 74 from Figure 73. The reader has to make a slight shift in nomenclature; the coordinate η on the surface of section in Figures 73 and 74 becomes x_0 in Figures

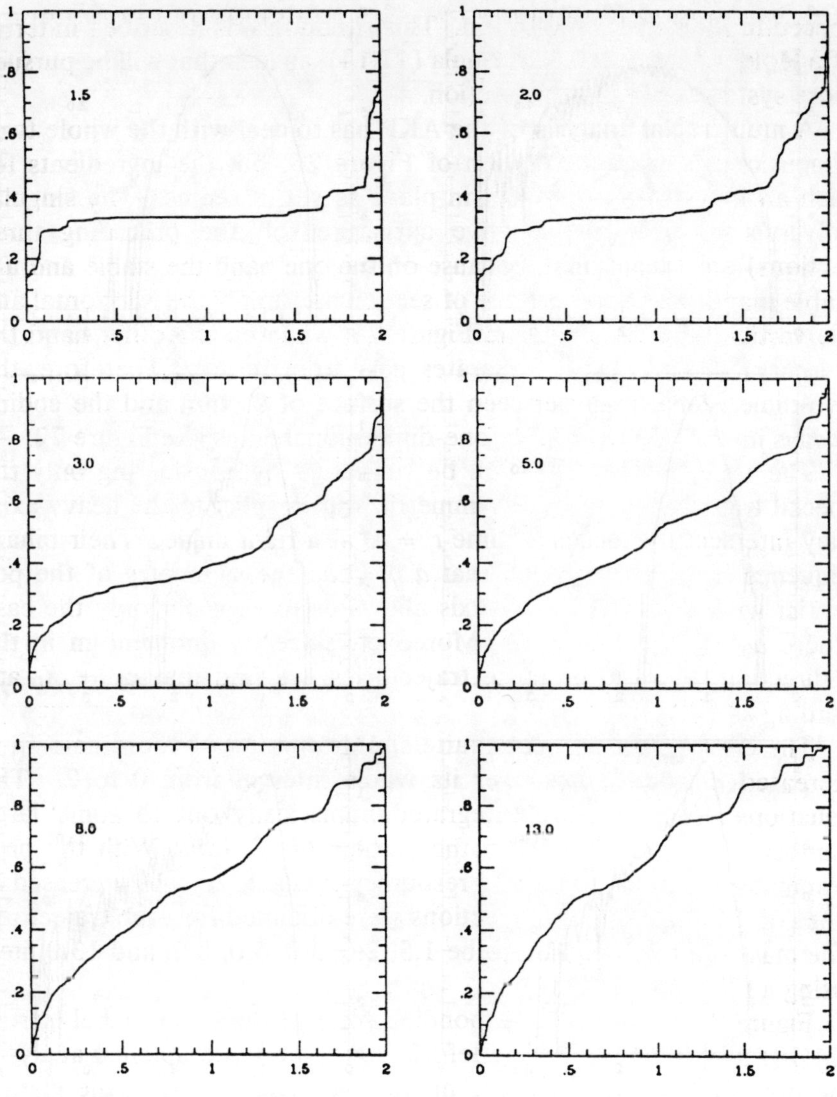

Figure 76 Plot of the binary characteristic η versus the initial value x_0, with $u_0 = 0$ (time-symmetric trajectories) for the mass ratios 1.5, 2.0, 3.0, 5.0, 8.0, and 13.0.

76 and 77, while β (the binary characteristic) is now written as η. The statistics are now rather poor, since we have only 10,000 sample points compared to 100,000 in Figure 74.

As in the earlier example, the distribution of the values for α over the interval $0 < x_0 < 2$ is extremely irregular, and must reflect some very subtle numerological features hidden in the dynamical system. A first stab was made in the earlier example by showing with formula

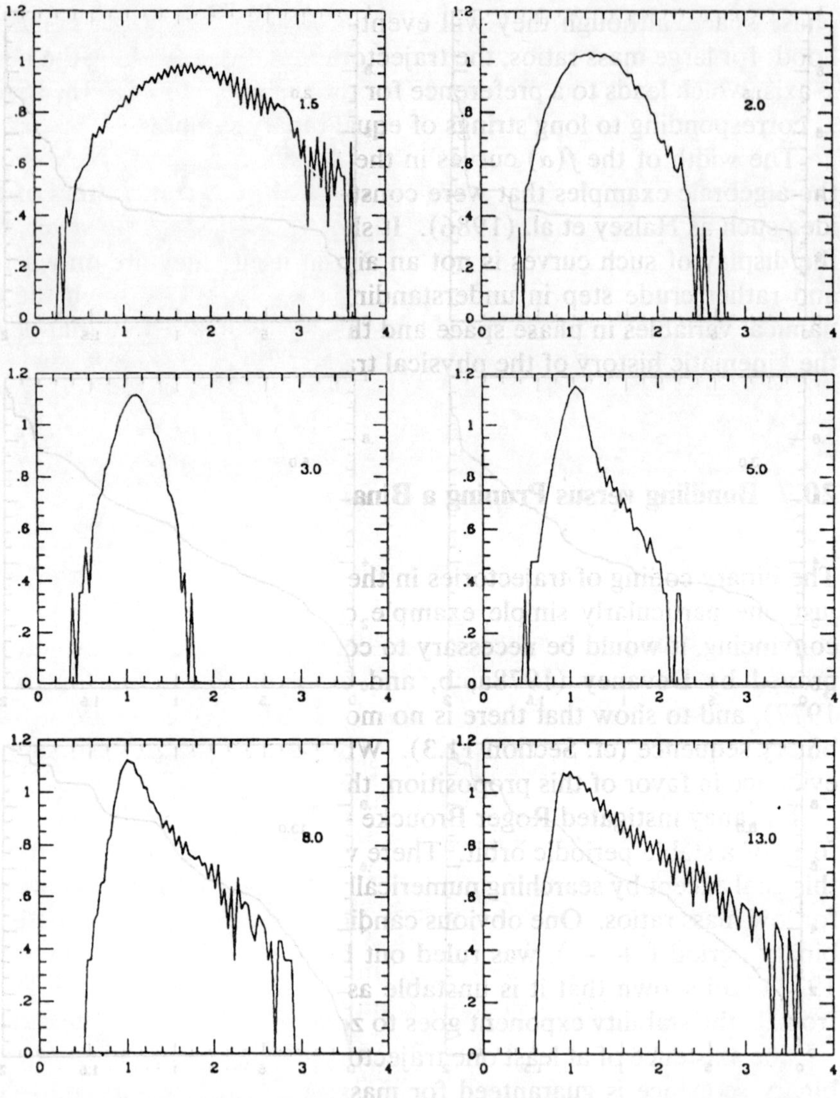

Figure 77 Fractal dimension $f(\alpha)$ of the subset in $0 < x_0 < 2$ where the Hölder exponent of the function $\eta(x_0)$ is α, computed for the mass ratios 1.5, 2.0, 3.0, 5.0, 8.0, and 13.0.

(20.6) that the function $\beta(\eta)$ has all derivatives vanishing wherever η is rational. The physical reason for the similar phenomenon in the singular squares was the trapping of the trajectories near the exit-entry at ∞.

The horizontal portions in the AKP are again due to trapping: for small mass ratios, the 'Kepler' orbit, with the binary period $(+\,-)$ corresponding to $\eta = 1/3$, attracts trajectories from a large part of phase space, although they will eventually again leave that neighborhood; for large mass ratios, the trajectories get trapped along the heavy x-axis, which leads to a preference for the values $\eta = 3/4, 7/8, 15/16,$ ... corresponding to long strings of equal binary symbols.

The width of the $f(\alpha)$ curves in the AKP again seems to exceed all the algebraic examples that were constructed by the promoters of this idea such as Halsey et al. (1986). It should be stressed, however, that the display of such curves is not an aim in itself; they are only a first and rather crude step in understanding the relation between the dynamical variables in phase space and the symbol sequences that define the kinematic history of the physical trajectories.

20.7 Bundling versus Pruning a Binary Tree

The binary coding of trajectories in the Anisotropic Kepler Problem is just one particularly simple example of hard chaos. To be entirely convincing, it would be necessary to complete the theorem which was proved by Devaney (1978a, b, and c) and the author (Gutzwiller 1977), and to show that there is no more than one trajectory for each binary sequence (cf. Section 11.3). While there is plenty of numerical evidence in favor of this proposition, the proof is missing.

Devaney instigated Roger Broucke to find a counterexample in the form of a stable periodic orbit. There was no other way to accomplish this goal except by searching numerically through the surface of section for low mass ratios. One obvious candidate, the 'Kepler' orbit with the binary period (+ −), was ruled out because the author (Gutzwiller 1971) had shown that it is unstable as soon as the mass ratio differs from 1; the stability exponent goes to zero linearly with the mass ratio.

The existence of at least one trajectory corresponding to a particular binary sequence is guaranteed for mass ratios greater than 9/8. The challenge, to find a stable periodic orbit for $\mu/\nu > 9/8$, was met by Broucke (1985) who published the parameters for one particular case of mass ratio 1.7218. Its binary sequence is the indefinite repetition of (+ − − + − −); it intersects the x-axis perpendicularly at a place ever closer to the origin as the mass ratio decreases; in a manner of speaking, it gets driven toward the origin by the expanding 'basin' of the 'Kepler' orbit.

The periodic orbit (+ − − + − −) becomes unstable when the mass ratio goes up to 2; but it is present without interruption in an interval starting at 1. It is obviously interesting to know what happens in its neighborhood; although the relevant calculations have been per-

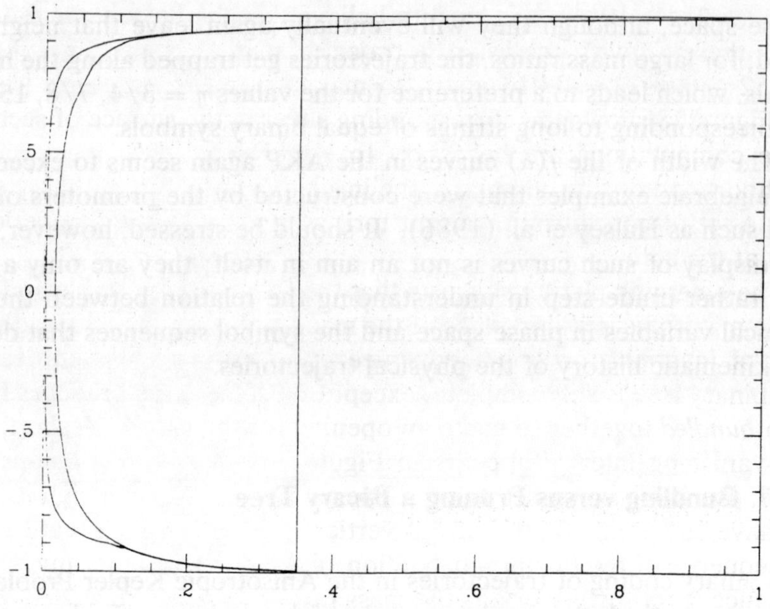

Figure 78 Broucke's Island in the surface of section (X, U) for the mass ratio 1.5, as obtained from the collision trajectories on either side; the coordinate U is divided by $\sqrt{\mu}\ \pi/2$.

formed only at a few mass ratios, in particular for 1.5, the results are assumed to be valid throughout this interval.

First, a search was undertaken along the X-axis of the surface of section, using the coordinates (U, X) in the rectangle defined by the formulas (11.9). All trajectories starting with $0.0022 \leq X \leq 0.0152$ and $U = 0$ have the same binary sequence, namely $(+ - - + - -)$ repeating indefinitely, although they are not periodic. A similar search, varying X while keeping U constant, always yields an interval of this kind, as long as $U < 0.9$. These intervals can be fitted together to form an elongated, narrow island, to be called *Broucke's island* for its discoverer.

Now it becomes crucial to find out how the AKP manages to produce all the trajectories whose binary sequences differ only in very late binaries from the indefinite repetition of $(+ - - + - -)$. For this purpose we revert to the construction of the stable and unstable manifold in Section 11.4. With sufficient patience we find the two collision trajectories that start with either $| - - + - - +)$ or with $| - + - - +)$ to be followed by seven repetitions of $(- - + - - +)$. The 48-th intersection of the first lies to the left, whereas the 47-th intersection of the second lies to the right of Broucke's island in the surface of section.

Since $2^{-48} \simeq 10^{-15}$, it is hard to believe that the unstable manifolds for these two collision trajectories succeed in being $0.0152 - 0.0022 = 0.013$ apart, to make space for Broucke's island; but Figure 78 shows the corresponding sliver in the surface of section. More computations are necessary to establish what happens in the neighborhood of this boundary; the inside of this island probably contains a set of concentric invariant tori, with Broucke's stable periodic orbit at the core.

The emerging picture suggests that the foliation into stable and unstable manifolds is still in place at the mass ratio 1.5; but an island with a set of invariant tori has been squeezed in between adjacent leaves. The binary tree is still complete, except that some of its branches have been *bundled* together to make an opening for the island. A short, but non-vanishing interval appears in Figure 76 where η has a constant value; the staircase ceases to be slippery at this point, although it does not have any point where it drops vertically. The bundling feature is a consequence of the theorem in Section 11.3, which prevents any binary sequence from getting lost as the mass ratio goes down to 9/8.

If the angular momentum M around the heavy axis is allowed to differ from zero, however, the y-coordinate has to be interpreted as the radial distance from the heavy x-axis, as indicated in the Hamitonian (11.6). The centrifugal repulsion overwhelms the Coulomb attraction; $y = \rho$ can no longer vanish. The surface of section has to be redefined, along with a criterion for defining a binary sequence to go with every trajectory. Long sequences of identical binaries no longer occur; they get *pruned out,* and the devil's staircase shows (vertical) jumps. Again, soft chaos is seen to yield the usual kind of devil's staircase, whereas hard chaos leads to the slippery kind.

References

Abraham R and Marsden JE (1978) *Foundations of Mechanics,* 2d edn. Reading, Mass: Benjamin/Cummings

Abramson E, Field RW, Imre D, Innes KK, and Kinsey JL (1985) J Chem Phys 83: 453

Adachi S, Toda M, and Ikeda K (1988) Phys Rev Lett 61: 655 and 659

-----, -----, and ----- (1989) J Phys A 22: 3291

----- (1989) Ann Phys (New York) 195: 45

Adams JC (1878) Mon Not RAS 38: 181

Adler RL, Konheim AG, and McAndrew MH (1965) Trans Am Math Soc 114: 309

----- (1987) IBM J Research and Development 31: 224

----- and Flatto L (1982) In: A Katok (ed) *Ergodic Theory and Dynamical Systems II.* Boston: Birkhäuser p 103

----- and ----- (1988) IBM Research Report RC 13575

Alekseev VM (1969) Uspekhi Mat Nauk 24:185

Albeverio S and Höegh-Kron R (1977) Inventiones Mathematicae 40: 59

-----, Blanchard P, and Hoegh-Krohn R (1982) Comm Math Phys 83: 49

Alhassid Y and Levine RD (1986) Phys Rev Lett 57: 2879

Allen CW (1962) *Astrophysical Quantities.* London: Athlone Press

Anosov DV (1969) *Geodesic Flows on Closed Riemann Manifolds with Negative Curvature.* In: Proc Steklov Inst Math 90. Providence: Am Math Soc

Arnold VI (1963) Usp Mat Nauk SSSR 18: 13, Russian Math Surveys 18: 9

----- (1964) Dokl Akad Nauk SSSR 156: 9

----- and Avez A (1967) *Méthodes Ergodiques de la Mécanique Classique.* Paris: Gauthier-Villars. Translated as *Ergodic Problems of Classical Mechanics.* New York: WA Benjamin 1968

----- (1978) *Mathematical Methods of Classical Mechanics.* New York: Springer

Artin E (1924) Abh Math Sem Hamburg 3: 499

Atela P (1988) Contemporary Mathematics 81: 43

Auerbach A and Kivelson S (1984) Phys Rev Lett 53: 41

-----, -----, and Nicole D (1985) Nucl Phys B 257:799

Aurich R, Sieber M, and Steiner F (1988) Phys Rev Lett 61: 483

----- and Steiner F (1988) Physica D 32: 451

----- and ----- (1989) Physica D 39: 169

----- and ----- (1990) Physica D

Avishai Y and Band YB (1990) Phys Rev B 41: 3253

Bai YY, Hose G, Stefanski K, and Taylor HS (1985) Phys Rev A 31: 2821

Bak P and Bruinsma R (1982) Phys Rev Lett 49: 249

Baker BB and Copson ET (1950) *The Mathematical Theory of Huygens' Principle,* 2d edn. Oxford: Clarendon Press

Balazs N and Voros A (1986) *Chaos on the Pseudosphere.* Physics Reports 143: 109

----- , Schmit C, and Voros A (1987) J Stat Phys 46: 1067

----- and Voros A (1989) Ann Phys (New York) 190: 1

Balian RB and Bloch C (1970) Ann Phys (New York) 60: 401

----- and ----- (1971) Ann Phys (New York) 63: 592 and 64: 271; Errata in Ann Phys (1974) 84: 559

----- and ----- (1972) Ann Phys (New York) 69: 76

----- and ----- (1974) Ann Phys (New York) 85: 514

Baranger M and Davies KTR (1987) Ann Phys (New York) 177: 330

Bargmann (1949) Phys Rev 75: 301 and Rev Mod Phys 21: 488

Barton D (1966) The Astronomical Journal 71: 438

----- (1967) The Astronomical Journal 72: 1281

Battin RH (1964) *Astronautical Guidance.* New York: McGraw-Hill p 70

Bayfield J and Koch P (1974) Phys Rev Lett 33: 258

----- (1987) Studies of the sinusoidally driven weakly bound atomic electron in the threshold region for classically stochastic behavior. In: Pike ER and Sarben Sarkar (eds) *Quantum Measurement and Chaos.* New York: Plenmum p 1

Beardon AF (1983) *The Geometry of Discrete Groups.* New York: Springer-Verlag

Beenakker CWJ and van Houten H (1989) Phys Rev Lett 63: 1857

Benettin GC, Galgani L, Giorgilli A (1985). In: Livi R and Politi A (eds) *Advances in Nonlinear Dynamics and Stochastic Processes.* Singapore: World Scientific.

-----, -----, -----, and Strelcyn JM (1984) Nuovo Cimento B 79: 201

Bennet CH (1987) Demons, Engines, and the Second Law. Scientific American, November 108

Berry MV and Mount KE (1972) Semiclassical Wave Mechanics. Rep Prog Phys 35: 315

----- and Tabor M (1976) Proc Roy Soc London A 349: 101

----- and ----- (1977a) J Phys A 10: 371

----- and ----- (1977b) Proc Roy Soc London A 356: 375

Berry MV (1977a) Phil Trans Roy Soc 287: 237

----- (1977b) J Phys A 10: 2083

Berry MV (1981) Ann Phys (New York) 131: 163

----- and Robnik M (1984) J Phys A 17: 2413

----- and ----- (1986) J Phys A 19: 649 and 669

Berry MV (1985) Proc Roy Soc London 400: 229

----- (1986) In: Seligmann TH and Nishioka H (eds) *Quantum Chaos and Statistical Nuclear Physics.* Lecture Notes in Physics 263. Berlin: Springer-Verlag p 1

----- (1988) Nonlinearity 1: 399

----- (1989) Proc Roy Soc London A 423: 219

Berry RS (1986) In: de Boer J, Dal E, and Ulfbeck O (eds) *The Lesson of Quantum Theory.* Amsterdam: Elsevier Sc Publ p 241

Bethe HA and Salpeter EA (1957) *Quantum Mechanics of One- and Two-Electron Atoms.* Berlin: Springer-Verlag

Birkhoff GD (1913) Trans Am Math Soc 14: 14

----- (1927) *Dynamical Systems.* New York: American Mathematical Society Colloquium Publications Vol IX

----- (1935) Mem Pont Acad Sci Novi Lyncaei 1: 85

Bleher S, Ott E, and Grebogi C (1989) Phys Rev Lett 63: 919

-----, Grebogi C, and Ott E (1989) University of Maryland Preprint

Blümel R and Smilansky U (1988) Phys Rev Lett 60: 477

Bogomolnyi EG (1984a) Sov Phys JETP 59: 917

----- (1984b) Physica D 13: 281

----- (1988) Physica D 31: 169

----- and Steiner F (1990) Physica D

Bohigas O, Haq RU, and Pandey A (1983). In: Böckhoff KM (ed) *Nuclear Data in Science and Technology.* Dordrecht: Reidel p 809

----- and Giannoni MJ (1984) Chaotic Motion and Random Matrix Theory. In: Dehesa JS, Gomez JMG, and Polls A (eds) *Mathematical and Computational Methods in Nuclear Physics.* New York: Springer-Verlag Lect Not Phys 209: 1

-----, -----, and Schmit C (1984a) Phys Rev Lett 52: 1

-----, -----, and ---- (1984b) J Physique Lett 45: L1015

-----, Haq RU, and Pandey A (1985) Phys Rev Lett 54: 1645

-----, Pandey A, and Giannoni MJ (1989) J Phys A 22: 4083

Born M (1925) Engl translation (1927) *The Mechanics of the Atom.* London: Bell & Sons, republished (1960) New York: Frederick Ungar

----- and Wolf E (1959) *Principles of Optics.* New York: Pergamon Press

Bowen R and Series C (1979) Inst Hautes Etudes Sc, Publ Math 50: 153

Brody TA, Flores J, French JB, Mello PA, Pandey A, and Wong SSM (1981) Rev Mod Phys 53: 385

Broucke R (1985) In: Szebehely V and Balazs B (eds) *Dynamical Astronomy.* Austin: University of Texas Press pp 9-20

Brouwer D and Clemence GM (1961) *Methods of Celestial Mechanics.* New York: Academic Press

Brown EW (1896) *An Introductroy Treatise on the Lunar Theory.* Cambridge: University of Cambridge Press; reprinted in New York: Dover Publications 1960

----- (1897 to 1908) Mem Roy Astr Soc 53 (1897) 39-116, 53 (1899) 163-202, 54 (1900) 1-63, 57 (1905) 51-145, 59 (1908) 1-103.

Brown R, Ott E, and Grebogi C (1987) J Stat Phys 49: 511

Brumer P and Shapiro M (1980) Chem Phys Lett 72: 528

----- (1981) Adv Chem Phys 47: 201

Bunker D L (1962) J Chem Phys 37: 393

Bunimovich LA (1974) Funct Anal Appl 8: 254

----- (1979) Commun Math Phys 65: 295

----- and Sinai YG (1980) Comm Math Phys 78: 247

Burns G (1985) *Solid State Physics.* Orlando, Florida: Academic Press

Caldeira A O and Leggett A J (1983) Ann Phys (NY) 149: 374

Camarda HS and Georgopoulos PD (1983) Phys Rev Lett 50: 492
Carathéodory C (1935) *Variationsrechnung und Partielle Differentialgleichungen Erster Ordnung.* Leipzig: BG Teubner
Cary JR and Skodje RT (1989) Physica D 36: 287
Casati G and Ford J (ed) (1979) *Stochastic Behavior in Classical and Quantum Hamiltonian Systems, Proc Como Conf 1977.* Lect Notes Phys 93, Berlin: Springer-Verlag
Casati G, Chirikov B, Izraelev F and Ford J (1979) in preceding reference p.334
-----, Ford J, Vivaldi F, and Visscher WM (1984) Phys Rev Lett 52: 1861
-----, Guarneri I, and Valz-Gris F (1980) Lettere Nuovo Cimento 28: 279
-----, -----, and ----- (1984) Phys Bev A 30: 1586
-----, Chirikov BV, and Guarneri I (1985) Phys Rev Lett 54: 1350
Castro JC, Zimmermann ML, Hulet RG, Kleppner D, and Freeman RR (1980) Phys Rev Lett 45: 1780
Casayas J and Llibre J (1984) Memoirs of Am Math Soc 312
Chang SJ and Friedberg R (1988) J Math Phys 29: 1537
Chang YT, Tabor M, and Weiss J (1982) J Math Phys 23: 531
Chapman SC, Garrett BC, and Miller WH (1976) J Chem Phys 74: 502
Chazarain J (1974) Formule de Poisson pour les Variétés Riemaniennes. Inventiones Mathematicae 24: 65
----- (1980) Comm Part Diff Eq 5(6): 595
Chen Y, Jonas DM, Hamilton CE, Green PG, Kinsey JL, and Field RW (1988) Ber Bunsenges Phys Chem 92: 329
Chirikov BV (1979) A Universal Instability of Many-Dimensional Oscillator Systems. Physics Reports 52: 263
Choodnovsky GV (1979) C R Acad Sc Paris 288: A 607 and A 965
Choquard P (1955) Helv Phys Acta 28: 89
Clark CW and Taylor KT (1980) J Phys B 13: L737
----- and ----- (1982) J Phys B 15: 1175
Coffey SL, Deprit A, Miller B, and Williams CA (1987) Ann New York Ac Sc 497: 22
Cohen H (1972) Math Ann 196: 8
Colin de Verdière Y (1973) Spectre du Laplacien et Longueurs des Géodésiques Périodiques I and II. Compositio Mathematica 27:83 and 159
Contopoulos G (1960) Z Astrophys 49: 275
----- (ed) (1966) *The Theory of Orbits in the Solar System and in Stellar Systems.* Symposium No. 25 of the International Astronomical Union, London and New York: Academic Press
----- (1970) Astronomical Journal 75: 96, 108
----- (1979) In: Casati G and Ford J (eds) *Stochastic Behavior in Classical and Quantum Hamiltonian Systems.* Berlin: Springer-Verlag pp 1-17
Courant R and Hilbert D (1924) *Methoden der Matematischen Physik,* vol 1. Berlin: Springer. 2d ed (1931)
Courant R and Hilbert D (1953) *Methods of Mathematical Physics,* vol 1. New York: Interscience Publishers

Coxeter HSM (1961) *Introduction to Geometry.* New York: Wiley
Creagh SC, Robbins JM, and Littlejohn RG (1990) Phys Rev A
Cvitanovic P (1984) *Universality in Chaos.* Bristol: Adam Hilger
----- (1988) Phys Rev Lett 61: 2729
----- and Eckhardt B (1989) Phys Rev Lett 63: 823
Dana I and Reinhardt WP (1987) Physica D 28: 115
Darboux G (1882) Bull Sci Math (2) 6: 14, 49
Dashen RF, Hasslacher B, and Neveu A (1974) Phys Rev D 12, 4114
Daubechies I and Klauder JR (1982) J Math Phys 23:1806
----- and ----- (1985) J Math Phys 26: 2239
Davis MJ and Heller EJ (1979) J Chem Phys 71: 3383
-----, Stechel EB, and Heller EJ (1980) Chem Phys Lett 76: 21
----- and Heller EJ (1981) J Chem Phys 75: 3916
de Aguiar MAM, Malta CP, Baranger M, Davies KTR (1987) Ann Phys (New York 180: 167
Dehn M (1911) Math Ann 71: 116
Delande D and Gay JC (1981) Phys Lett 82A: 399
----- and ----- (1984) J Phys B 17: L335
----- and ----- (1986) J Phys B 19: L173
----- and ----- (1986a) Phys Rev Lett 57: 2006
Delaunay C (1860) *Théorie du Mouvement de la Lune, Premier Volume.* Mém Acad Sci Paris XXVIII
----- (1867) *Théorie du Mouvement de la Lune, Second Volume.* Mém Acad Sci Paris XXIX
Delos JB and Swimm RT (1977) Chem Phys Lett 47: 76
----- , Knudson SK, and Noid DW (1983) Phys Rev A 28: 7
----- , Knudson SK, Sikora SD, Waterland RL, Whitworth S (1988) Phys Rev A 37: 4582
Deprit A (1969) Cel Mech 1: 12
-----, Henrard J, Price JF, and Rom A (1969) Cel Mech 1: 222
-----, -----, and Rom A (1971) Astronomical Journal 76: 269
----- and Ferrer S (1990) submitted to J Phys B
Devaney R (1978a) J Diff Equ 29: 253
----- (1978b) Inventiones Math 45: 221
----- (1978c) *Lecture Notes in Mathematics 668.* New York: Springer-Verlag
----- (1980) Inventiones Mathematicae 60:249
----- (1981) Comm Math Phys 80: 465
----- (1982) Celestial Mechanics 28:25
----- (1985) *An Introduction to Chaotic Dynamical Systems.* Reading Mass: Addison-Wesley
Dicke RH and Wittke JP (1960) *Introduction to Quantum Mechanics.* Reading Mass: Addison-Wesley
Dirac PAM (1933) Phys Zeits Sowjetunion 3:64
----- (1935) *The Principles of Quantum Mechanics,* 2d edn. Oxford: The Clarendon Press
Drobot S (1964) *Real Numbers.* Englewood Cliffs: Prentice-Hall
Du ML and Delos JB (1987) Phys Rev Lett 58: 1731

----- and ----- (1988) Phys Rev A 38: 1896 and 1912
Dumont RS and Brumer P (1988) J Chem Phys 88: 1481
Duru IH and Kleinert H (1979) Phys Lett 84B: 185
Dyson FJ and Mehta ML (1963) J Math Phys 4: 701
----- (1976) Lieb EH, Simon B, and Wightman AS (eds) *Studies in Mathematical Physics.* Princeton University Press p 151
Eckert WJ, Jones R, and Clark HK (1954) Construction of the Lunar Ephemeris. In: *Improved Lunar Ephemeris 1952-1959.* Washington: US Government Printing Office: p 242-363
-----, Walker MJ, and Eckert D (1966) Astronomical Journal 71: 314
-----, and Eckert D (1967) Astronomical Journal 72: 1299
Eckhardt B (1986) J Phys A 19: 2961
----- (1987) J Phys A 20: 5971
----- (1988) Phys Rep 163: 205-297
-----, Hose G, and Pollak E (1989) Phys Rev A 39: 3776
-----, and Wintgen D (1990) J Phys B 23: 355
Edmonds AR (1970) J de Physique Colloque C4 Tome 31 p 71
Edwards HM (1974) *Riemann's Zeta Function.* New York: Academic Press
Einstein A (1917) Verh Dtsch Phys Ges 19: 82
Ericson T (1960) Phys Rev Lett 5: 430
----- (1963) Ann Phys (New York) 23: 390
----- and Mayer-Kuckuk T (1966) Ann Rev Nucl Sc 16: 183
Escande DF and Doveil F (1981) Phys Lett 83A: 307, J Stat Phys 26: 257
----- (1985) *Stochasticity in Classical Hamiltonian Systems: Universal Aspects.* Physics Reports 121: 165
Euler L (1744) *Methodus Inveniendi Lineas Curvas Maximi Minimive Proprietate Gaudentes: Additamentum II.* Lausanne & Geneva: Bousquet. Also Carathéodory C (ed) (1952) *Leonhardi Euleri Opera Omnia: Series I, vol 24.* Zurich: Orell F&uessli
Fatou P (1906) C R Acad Sc Paris 143: 546
Faulkner RA (1969) Phys Rev 184: 713
Feit MD and Fleck JA Jr. (1983) J Chem Phys 80: 2578
Feingold M and Peres A (1985) Phys Rev A 31: 2472
Feynman RP (1948) Space-Time Approach to Non-Relativistic Quantum Mechanics. Rev Mod Phys 20:367-387
Feynman RP and Hibbs AR (1965) *Quantum Mechanics and Path Integrals.* New York: McGraw-Hill
Fishman S, Grempel DR, and Prange RE (1982) Phys Rev Lett 49: 509
-----, -----, ----- (1984) Phys Rev A 29: 1639
Flaschka H (1974) Phys Rev B 9: 1924, Prog Theor Phys 51: 703
Floquet (1883) Ann Ec Norm Sup (2) 12: 47
Fock V (1935) Z Phys 98: 145
Fonck RJ, Roesler FL, Tracy DH, and Tomkins FS (1980) Phys Rev A 21: 861
Ford J, Stoddard SD, and Turner JS (1973) Prog Theor Phys 50:1547
Founargiotakis M, Farantos SC, Contopoulos G, and Polymilis C (1989) J Chem Phys 91: 1389

Freed K (1972) J Chem Phys 56: 692
Friedrich H and Wintgen D (1989) Physics Reports 183: 37-79
Frisch U and Parisi G (1985) In: Ghill M, Benzi R, and Parisi G (eds) *Turbulence and Predictability in Geophysical Fluid Dynamics and Climate Dynamics.* Amsterdam: North-Holland p 84
Froeschlé C (1968) C R Acad Sc Paris 266: 747
----- (1970) Astron Astrophys 4: 115 and 5: 177
----- and Scholl H (1982) Astron Astrophys 111: 346
Garret BC and Truhlar DG (1983) J Chem Phys 79: 4931
-----, Abusalbi N, Kouri DJ, and Truhlar DG (1985) J Chem Phys 83: 2252
Garrod C (1966) Rev Mod Phys 38: 483
----- (1968) Phys Rev 167: 1143
Garton WRS and Tomkins FS (1969) Astrophys J 158: 839
Gaspard P and Rice SA (1989) J Chem Phys 90: 2225, 2242, 2255
Gay JC, Delande D, and Biraben F (1980) J Phys B 13: L729
----- (1984) In: Beyer HJ and Kleinpoppen H (eds) *Progress in Atomic Spectroscopy, Part C.* Plenum p 177-246
----- (1985) In: McGlynn SP et al (eds) *Photophysics and Photochemistry in the Vacuum Ultraviolet.* Amsterdam: Reidel p 631-705
Geisel T, Radons G, and Rubner J (1986) Phys Rev Lett 57: 2883
Giannoni MJ and Ullmo D (1990) Physica D
Goldstein H (1950) *Classical Mechanics.* Reading, Mass: Addison-Wesley
Gomez Llorente JM, Zakrzewski J, Taylor HS, and Kulander KC (1989) J Chem Phys 90: 1505
Good A (1981) Acta Arithmetica 28: 347
Grabert H and Weiss U (1984) Phys Rev Lett 53: 1787
----- and ----- (1984) Z Phys B 56: 171
Graffi S, Paul T, and Silverstone HJ (1987) Phys Rev Lett 59: 255
Grebogi C, Ott E, and Yorke J (1987) Science 238: 632
Greendlinger M (1961) Comm Pure Appl Math 12:414 and 13: 641
Greene JM (1979) J Math Phys 20: 1183
Greene J, Tabor M, and Carnevale G (1983) J Math Phys 24: 522
Grosche C and Steiner F (1988) Ann Phys (New York) 182: 120
Grünbaum B and Shepherd GC (1987) *Tilings and Patterns.* San Francisco: WH Freeman
Guckenheimer J and Holmes P (1983) *Non-Linear Oscillations, Dynamical Systems, and Bifurcations of Vector Fields.* New York: Springer-Verlag
Gustavson FG (1966) Astronomical Journal 71: 670
Gutzwiller MC (1967) J Math Phys 8: 1979
----- (1969) J Math Phys 10: 1004
----- (1970) J Math Phys 11: 1791
----- (1971) J Math Phys 12: 343
----- (1973) J Math Phys 14: 139
----- (1977) J Math Phys 18: 806
----- (1979) Astronomical Journal 84: 889
----- (1980) Phys Rev Lett 45: 150

----- (1981a) In: Devaney RL and Nitecki ZH (eds) *Classical Mechanics and Dynamical Systems.* New York: Marcel Dekker p 69
----- (1981b) Ann Phys (New York) 133: 304; an earlier version (1980) Ann Phys (New York) 124: 347 is less transparent.
----- (1982) Physica D 5: 183
----- (1983) Physica D 7: 341
----- (1985a) In: Casati G (ed) *Chaotic Behavior in Quantum Systems.* New York: Plenum p 149
----- (1985b) Physica Scripta T9: 184
----- and Schmidt D (1986) *The Motion of the Moon as Computed by the Method of Hill, Brown, and Eckert.* Astronomical Papers prepared for the use of the American Ephemeris and Nautical Almanac vol XXIII Part I. Washington: US Naval Observatory pp 1-272
----- (1986a) In: Gutzwiller MC, Inomata A, Klauder JR, and Streit L (eds) *Path Integrals from meV to MeV.* Singapore: World Scientific p 119
----- (1986) Contemp Math 53: 215
----- (1987) In: Chudnovsky DV, Chudnovsky GV, Cohn H, and Nathanson MB (eds) *Number Theory.* Lecture Notes in Math 1240. Berlin: Springer-Verlag p 230
----- (1988a) J Phys Chem 92: 3154
----- and Mandelbrot BB (1988) Phys Rev Lett 60: 673
----- (1988c) In: Lundqvist S, Ranfagni A, Sa-yakanit V, and Schulman LS (eds) *Path Summation: Achievements and Goals.* Singapore: World Scientific p 47
----- (1989) Physica D 38: 160
Hadamard J (1898) J Math Pure Appl 4: 27; Soc Sci Bordeaux Proc Verb 1898: 147
Haller E, Köppel H, and Cederbaum LS (1983) Chem Phys Lett 101: 215.
-----, -----, and ----- (1984) Phys Rev Lett 52: 1665
Halsey TC, Jensen MH, Kadanoff LP, Procaccia I, and Shraiman PE (1986) Phys Rev A 33: 1141
Hamilton WR (1834, 1835) *On a General Method in Dynamics.* Phil Trans Roy Soc 1834: 307, 1835: 95
Hannay JH and Ozorio de Almeida AM (1984) J Phys A 17: 3429
----- (1985) In: Casati G (ed) *Chaotic Behavior in Quantum Systems Theory and Applications.* New York: Plenum p 141
Haq RU, Pandey A, and Bohigas O (1982) Phys Rev Lett 48: 1086
Harada A and Hasegawa H (1983) J Phys A 16: L259
Hasegawa H (1969) *Effects of High Magnetic Fields on Electronic States in Semiconductors - The Rydberg Series and the Landau Levels.* In: Physics of Solids in Intense Magnetic Fields. New York: Plenum Press p 246
----- , Adachi S, and Harada A (1983) J Phys A 16: L503
----- , Harada A, and Okazaki Y (1984) J Phys A 17: L883
----- , Robnik M, and Wunner G (1989) Prog Theor Phys Suppl no 98 pp 198-286
Heiss WD and Sannino AL (1989) submitted to J Phys A

Hejhal DA (1976a) *The Selberg Trace Formula for PSL(2,R).* Volume 1, Lecture Notes in Mathematics 548. Berlin: Springer-Verlag
----- (1976b) Duke Math J 43: 441
----- (1983) *The Selberg Trace Formula for PSL(2,R).* Volume 2, Lecture Notes in Mathematics 1001. Berlin: Springer-Verlag
----- (1989) preprint University of Minnesota Supercomputer Institute
Helleman RHG (1978) Am Inst Phys Proc 46: 264
Heller EJ (1975) J Chem Phys 62: 1544
----- (1978) J Chem Phys 68: 2066
----- (1986) In: Seligmann TH and Nishioka H (eds) *Quantum Chaos and Statistical Nuclear Physics.* Lecture Notes in Physics 263. Berlin: Springer-Verlag p 162
----- (1987) Phys Rev A 35: 1360
-----, O'Connor PW, and Gehlen J (1989) Physica Scripta 40: 354
Hénon M and Heiles C (1964) The Applicability of the Third Integral of Motion: Some Numerical Experiments. Astronomical Journal 69: 73
Hénon M (1966) Annales d'Astrophysique 28: 499, 29: 992
----- (1967) Bull Astr Paris 1 fasc. 1: 57, 1 fasc. 2: 49
----- (1969-70) Astron Astrophys 1: 223, 9: 25
----- (1969) Quarterly Appl Math XXVII: 291
----- (1974) Phys Rev B 9: 1921
----- (1976) Comm Math Phys 50: 69
----- (1988) Physica D 33: 132
Herrick DR (1982) Phys Rev A 26: 323
----- and Sinanoglu O (1975) Phys Rev A 11: 97
Herring C (1962) Rev Mod Phys 34: 631
Hilbert D (1901) Trans Am Math Soc 2: 87
Hill GW (1877) reprinted in Acta Mathemica 8 (1886) 1-36
----- (1878) Am J Math 1: 5-26, 129-147, 245-260
----- (1905) *The Collected Mathematical Works.* Introduction by H Poincaré, Washington: Carnegie Institution
Ho R and Inomata A (1982) Phys Rev Lett 48: 231
Hocking JG and Young GS (1961) *Topology.* Reading, Mass: Addison-Wesley
Hönig A and Wintgen D (1989) Phys Rev A 39: 5642
Holle A, Wiebusch G, Main J, Hager B, Rottke H, and Welge KH (1986) Phys Rev Lett 56: 2594
-----, -----, -----, Welge KH, Zeller G, Wunner G, Ertl T, and Ruder H (1987) Z Phys D 5: 279
-----, Main J, Wiebusch G, Rottke H, and Welge KH (1988) Phys Rev 61: 161
Hori G (1966) Publ Astron Soc Japan 18: 287
Huang ZH, Feuchtwang TE, Cutler PH, and Kazes E (1990) Phys Rev A 41: 32
Huber D, Heller EJ, and Littlejohn R (1988) J Chem Phys 89: 2003
Huber H (1959) Math Annalen 139: 1

Husimi K (1940) Proc Phys Math Soc Japan 22: 264
Hutchinson JS and Wyatt RE (1980) Chem Phys Lett 72: 378
Ishikawa T and Yukawa T (1985a) Phys Rev Lett 54: 1617
----- and ----- (1985b) KEK-TH 109 (July 1985) unpublished
Itzykson C, Moussa P, and Luck JM (1986) J Phys A 19: L 111
Jacobi CGJ (1842) *Vorlesungen über Dynamik, gehalten an der Universität Königsberg im Wintersemester 1842-1843.* A Clebsch (ed). Berlin: Reimer 1866
Jaffé C and Reinhardt WP (1979) J Chem Phys 71: 1862
----- and Watanabe M (1988) J Chem Phys 89: 6329
José JV (1988) In: Hao B-L (ed) *Directions in Chaos,* vol II. Singapore: World Scientific
Jost R and Kohn W (1952) Phys Rev 87: 977 and 88: 382
----- and ----- (1953) Kgl Danske Vidensk Selsk Mat-fys Medd 29 no 9
Jost R and Lombardi M (1986) Lecture Notes in Physics 263. Berlin: Springer-Verlag p 72
Julia G (1918) J Math Pure Appl 4: 47
Jung C and Scholz H-J (1987) J Phys A 20: 3607
Kac M (1959) *Probability and Related Topics in the Physical Sciences.* London: Interscience Publishers
----- (1966a) Bull Am Math Soc 72: 52
----- (1966b) Am Math Monthly 73: 1
----- and Moerbeke P van (1974) Proc Nat Acad Sci USA 71: 2350, 72: 2627 and 2879
Katok A and Strelcyn JM (1986) *Invariant Manifolds, Entropy and Billiards, Smooth Maps with Singularities.* Springer Lecture Notes in Mathematics 1222
Keating JP and Berry MV (1987) J Phys A 20: L1139
Keen L (1966a) Ann Math (Princeton) 84: 404
----- (1966b) Acta Mathematica 115: 1
Keller JB (1958) Ann Phys (New York) 4: 180
----- and Rubinow SI (1960) Ann Phys (New York) 9: 24
Khintchine AY (1963) *Continued Fractions.* Groningen: P Noordhoff
Kittel C (1967) *Introduction to Solid State Physics.* 3rd edn New York: Wiley
Klauder JR (1988) Ann Phys (NY) 188: 120
----- (1989) In: Saya-kanit V and Sritrakool W (eds) *Path-Integrals from meV to MeV.* Singapore: World Scientific p 48
Klein MJ (1970) *Paul Ehrenfest vol I The Making of a Theoretical Physicist.* Amsterdam: North-Holland
Kohn W and Luettinger JM (1954) Phys Rev 96: 1488
Kolmogoroff AN (1954) Dokl Akad Nauk SSSR 98: 527, cf. also Proc Int Congr Math 1954: 315 which is reprinted in Abraham and Marsden (1978)
Kook H and Meiss JD (1989) Physica D 35: 65 and 36: 317
Kovalevsky J (1982) In: Szebehely V (ed) *Applications of Modern Dynamics to Celestial Mechanics and Astrodynamics.* Dordrecht-Holland: Reidel p 59

Koval'chik IM (1963) The Wiener Integral. Russian Math Surveys 18: 97
Kramer P and Saraceno M (1981) Springer Lecture Notes in Physics 140
Kubota T (1973) *Elementary Theory of Eisenstein Series.* New York: Wiley
Lagrange JL de (1760-1) Miscellanea Taurinensia II: pp 173 and 196
----- (1764) *Oeuvres,* tome 6: 5
----- (1780) *Théorie de la Libration de la Lune.* Mém Ac Sc Berlin; and *Oeuvres,* tome 5: 5
Lakshmanan M and Hasegawa H (1984) J Phys A 17: L889
Lanczos C (1949) *The Variational Principles of Mechanics.* Toronto: University of Toronto Press p 253
Landau LD and Lifschitz EM (1973) *Mechanics.* Moscow: NAUKA
Landauer R (1951) J Appl Phys 22: 87; Phys Rev 82: 80
----- (1952) Am J Phys 20: 363
Langer RE (1937) Phys Rev 51: 669
Lax P (1968) Comm Pure Appl Math 21: 467
----- and Phillips RS (1976) *Scattering Theory for Automorphic Functions.* Princeton University Press
Lefschetz S (1949) *Introduction to Topology.* Princeton, NJ: Princeton University Press
Leggett A J (1978) J de Phys C6: 1264
----- (1980) Prog Theor Phys Suppl 69: 80
Leimanis E (1965) *The General Problem of the Motion of Coupled Rigid Bodies about a Fixed Point.* New York: Springer
Levit S and Smilansky U (1977) Ann Phys (New York) 108: 165
Lichtenberg AJ and Liebermann MA (1981) *Regular and Stochastic Motion.* New York: Springer-Verlag
Lin WA and Reichl LE (1987) Phys Rev A 36: 5099
----- and ----- (1988) Phys Rev A 37: 3972
Littlejohn RG (1986) Phys Rep 138: 193-291
----- and Robbins JM (1987) Phys Rev A 36: 2953
----- (1990) J Math Phys
Llave R and Rana D (1990) In: Meyer K and Schmidt D (eds) *Computer Aided Proofs in Analysis.* IMA Proceedings, Springer-Verlag
Ludwig D (1975) SIAM Review 17: 1
MacKay RS and Meiss JD (1987) *Hamiltonian Dynamical Systems.* Bristol and Philadelphia: Adam Hilger
McDonald SW and Kaufmann AN (1979) Phys Rev Lett 42: 1189
----- (1983) (Ph.D. Thesis, U of Cal Berkeley) *Wave Dynamics of Regular and Chaotic Rays.* Lawrence Berkeley Laboratory, University of California
----- and Kaufmann AN (1988) Phys Rev A 37: 3067
McKean HP (1972) Comm Pure Appl Math 25: 225
Magnus W, Karass A, and Solitar D (1976) *Combinatorial Group Theory.* New York: Dover
Magyari E, Thomas H, Weber R, Kaufman C, and Müller G (1987) Z Phys B 65: 363
Main J, Wiebusch G, Holle A, and Welge KH (1986) Phys Rev Lett 57: 2789

----- , Holle A, Wiebusch G, and Welge KH (1987) Z Phys D 6: 295
Mandelbrot BB (1980) Ann New York Acad Sc 357: 249
----- (1982) *The Fractal Geometry of Nature.* New York: Freeman
Markus L and Meyer KR (1974) Mem Am Math Soc 144
Martinyan SG, Prokhorenko EB, and Savvidy GK (1988) Nucl Phys B 298: 414
Maslov VP (1972) *Théorie des Perturbations et Méthodes Asymptotiques.* Paris: Dunod
----- and Feodoriuk MV (1981) *Semi-Classical Approximations in Quantum Mechanics.* Boston: Reidel
Maupertuis PLN de (1744) *Accord de différentes lois de la nature qui avaient jusqu'ici paru incompatibles.* Mém As Sc Paris, p 417
----- (1746) *Les lois du mouvement et du repos, déduites d'un principe de métaphysique.* Mém Ac Berlin, p 267
Mehta ML (1967) *Random Matrices and the Statistical Theory of Energy Levels.* New York: Academic Press
Meredith DC, Koonin SE, and Zirnbauer MR (1988) Phys Rev A 37: 3499
Miller SC and Good RH (1953) Phys Rev 91: 174
Miller WH (1972) J Chem Phys 56: 38
----- (1974) Adv Chem Phys 25: 69
----- (1975) J Chem Phys 63: 996
----- (1979) J Phys Chem 83: 960
Milnor J (1983) Am Math Monthly 90: 353
Möhring K, Levit S, and Smilansky U (1980) Ann Phys (New York) 127: 198
Morette C (1951) Phys Rev 81: 848
Morse M (1934) *The Calculus of Variations in the Large.* New York: Am Math Soc Colloquium Publ 18
Moser J (1962) Nachr Akad Wiss Göttingen 1
----- (1973) *Stable and Random Motions in Dynamical Systems.* Princeton, NJ: Princeton University Press
----- (1975a) Springer Lecture Notes in Physics 38: 467
----- (1975b) Adv Math 16: 197
----- (1980) In: W Y Hsiang et al (eds) *The Chern Symposium.* New York, Springer-Verlag p 147
Nakamura K and Lakshmanan M (1986) Phys Rev Lett 57: 1661
Navarro H, Haller EE, and Keilmann F (1988) Phys Rev B 37:10822
Newell AC, Tabor M, and Zeng YB (1987) Physica D 29:1
Nielsen J (1927) Acta Math 50: 189
Nieto MM (1972) *The Titius-Bode Law of Planetary Distances: Its History and Theory.* Oxford: Pergamon
Noid DW and Marcus RA (1977) J Chem Phys 67: 559
-----, Koszykowski ML, and Marcus RA (1977) J Chem Phys 67: 404
-----, -----, Tabor M, and Marcus RA (1980) J Chem Phys 72: 6169
-----, -----, and Marcus RA (1981) Ann Phys Rev Chem 32: 267
Norcliffe A and Percival IC (1968) J Phys B 1: 774 and 784
-----, -----, and Roberts MJ (1969) J Phys B 2: 578 and 590

Nordholm KSJ and Rice SA (1974) J Chem Phys 61: 203 and 768
O'Connor PW, Gehlen J, and Heller E (1987) Phys Rev Lett 58: 1296
Odlyzko AM (1987) Math of Comp 48:273
Ollongren A (1965) *Theory of Stellar Orbits in the Galaxy.* In: Annual Review of Astronomy and Astrophysics. Palo Alto (Cal)
Ozorio de Almeida AM (1982) Physica A 110: 501
----- and Hannay JH (1982) Ann Phys (New York) 138: 115
----- (1983) Ann Phys 145: 100
----- (1986) Lecture Notes in Physics 263. Berlin: Springer-Verlag p 197
----- and ----- (1987) J Phys A 20: 5873
----- (1988) *Hamiltonian Systems: Chaos and Quantization.* Cambridge: Cambridge University Press.
----- (1989) Nonlinearity 2: 519
Pandey A, Bohigas O, and Giannoni MJ (1989) J Phys A 22: 4083
Papadopoulos G (1975) Phys Rev D 11: 2870
Pauli W Jr (1922) Ann Phys (Leipzig) IV 68: 177
----- (1926a) Z Physik 36: 336
----- (1926b) *Quantentheorie.* Handbuch der Physik, vol 23. Berlin: Springer-Verlag pp 1-278
----- (1929) *Allgemeine Grundlagen der Quantentheorie des Atombaues.* Müller-Pouillet's Lehrbuch der Physik, vol 2 part 2. Braunschweig: Vieweg pp 1709-1842
Pauli W (1933) *Die Allgemeinen Prinzipien der Wellenmechanik.* Handbuch der Physik, vol 24 part 1. Berlin: Springer-Verlag pp 83-272.
----- (1958) Handbuch der Physik, vol 5 part 1. Berlin: Springer-Verlag pp 1-168
Pechukas P (1972) J Chem Phys 57: 5577
----- (1983) Phys Rev Lett 51: 943
Peitgen H-O and Richter PH (1986) *The Beauty of Fractals.* Berlin-Heidelberg: Springer-Verlag
Percival IC (1973) J Phys B 6: L229
----- (1977) Adv Chem Phys 36: 1
(1979) J phys A 12: L 57
----- and Vivaldi F (1987) Physica D 25: 105
Pesin Ya B (1977) Russian Math Surveys 32(4): 55-114
Pique JP, Chen Y, Field RW, and Kinsey JL (1987) Phys Rev Lett 58: 475
Piro O and Feingold M (1988) Phys Rev Lett 61: 1799
Poincaré H (1892) *Les Méthodes Nouvelles de la Mécanique Céleste Tome I.* Paris: Gauthier-Villars
----- (1899) *Les Méthodes Nouvelles de la Mécanique Céleste Tome III.* Paris: Gauthier-Villars
----- (1907) *Leçons de Mécanique Céleste Tome II.* Paris: Gauthier-Villars
----- (1908) *Sur les petits diviseurs dans la théorie de la Lune.* Bull Astr 25: 321-360
----- (1912) Rendiconti del Circolo Matematico di Palermo 33: 375
Pomphrey N (1974) J Phys B 7: 1909
Pullen RA and Edmonds AR (1981) J Phys A 14: L477

Radons G and Prange R (1988) Phys Rev Lett 61: 1691
----- , Geisel T, and Rubner J (1988) Adv Chem Phys 73: 891
Randol B (1984) In: Isaac Chavel (ed) *Eigenvalues in Riemannian Geometry.* New York: Academic Press
Reich A (1980) Arch Math 34: 440
Reichl LE and Lin WA (1986) Phys Rev A 33: 3598
Reinhardt WP and Farrelly D (1982) J de Physique Colloque C2 Tome 43 p 29
----- (1983) J Phys B 16: L635
----- (1985) In: Casati G (ed) *Chaotic Behavior in Quantum Systems.* New York: Plenum Press p 235
----- and Gillilan RE (1986) *Semi-Classical Quantization on Adiabatically Generated Tori, or Einstein on the Brink.* In: Gutzwiller MC, Inomata A, Klauder JR, and Streit L (eds) *Path Integrals from meV to MeV.* Singapore: World Scientific p 154
----- and Dana I (1987) Proc Roy Soc A 413: 157
Rice S O (1944) Bell Syst Tech J 23: 282
----- (1945) Bell Syst Tech J 24: 46
Richens PJ (1982) J Phys A 15: 2101
Robnik M (1981) J Phys A 14: 3195
----- (1982) J de Physique Colloque C2 Tome 43 p 45
----- (1984) J Phys A 17: 1049
Robbins JM and Littlejohn RG (1987) Phys Rev Lett 58: 1388
----- (1989) Phys Rev A 40: 2128
Rosenzweig N (1963) *Statistical Mechanics of Equally Likely Quantum Systems.* In: Brandeis Summer Institute 1962, vol 3: Statistical Physics. New York: WA Benjamin p 91
Rowe EGP (1987) J Phys A 20: 1419
Ruelle D (1986) J Stat Phys 44: 281
Savvidy GK (1982) Nucl Phys B 246: 302
----- (1983) Phys Lett 130B: 203
Schmidt D (1979) Cel Mech 19: 279
Schulman LS (1981) *Techniques and Applications of Path Integrals.* New York: Wiley
----- (1988) In: Lundqvist S, Ranfagni A, Sa-yakanit V, Schulman LS (eds) *Path Summation: Achievements and Goals.* Singapore: World Scientific p 3
Schuster HG (1988) *Deterministic Chaos.* Weinheim FRG: VCH Verlagsgesellschaft
Seifert H and Threlfall W (1934) *Lehrbuch der Topologie.* Reprinted New York: Chelsea Pub Co, translated New York: Academic Press 1980
Seligman TH and Verbaarschot JJM (1987) J Phys A 20: 1433
-----, -----, and Zirnbauer MR (1984) Phys Rev Lett 53: 215
-----, -----, ----- (1985) J Phys A 18: 2751
Series C (1985a) J London Math Soc (2) 31: 69
----- (1985b) Math Intelligencer 7:20
----- (1986) Erg Th Dyn Sys 6: 601

Shapiro M, Ronkin J, and Brumer P (1988) Chem Phys Lett 148: 177
Sieber M and Steiner F (1990) Phys Lett A 144: 159
Siegel CL and Moser JK (1971) *Lectures on Celestial Mechanics.* Grundlehren, vol 187. New York: Springer-Verlag
Sinai YG (1968) Funct Anal Appl 2: 61 and 245
----- (1970) Russ Math Surveys 25: 137
Sklyanin E K (1985) In: *Non-Linear Equations in Classical and Quantum Field Theory.* Lecture Notes in Physics. Heidelberg: Springer-Verlag
Skodje RT, Borondo F, and Reinhardt WP (1985) J Chem Phys 82: 4611
Smale S (1965) D iffeomorphisms with Many Periodic Points. In: Cairns SS (ed) *Differential and Combinatorial Topology.* Princeton NJ, Princeton University Press p 65
----- (1967) Bull Am Math Soc 73: 747-817
Solovev EA (1981) Sov Phys JETP Lett 34: 265
----- (1982) Sov Phys JETP 55: 1017
Sommerfeld A (1919) *Atombau und Spektrallinien.* Braunschweig: Vieweg. Complete revisions were made in the 3d (1922) and in the 5th editions (1931)
----- (1942) *Vorlesungen über Theoretische Physik Band I Mechanik.* Wiesbaden: Dieterich'sche Verlagsbuchhandlung
Souriau J-M (1970) *Structure des Systèmes Dynamiques.* Paris: Dunod
Srivastava N, Kaufman C, Müller, Weber R, and Thomas H (1988) Z Phys B 70: 251
Stark HM (1970) *An Introduction to Number Theory.* Chicago: Markham
Steiner F (1987a) Fortschritte der Physik 35: 87
----- (1987b) Phys Lett B 188: 447
----- and Trillenberg P (1990) J Math Phys
Strand MP and Reinhardt WP (1979) J Chem Phys 70: 3812
Stratt RM, Handy NC, and Miller WH (1979) J Chem Phys 71: 3311
Swiatecki WJ (1988) Nucl Phys A 488: 375
Swimm RT and Delos JB (1979) J Chem Phys 71: 1706
Szebehely V (1967) *Theory of Orbits (The Restricted Problem of Three Bodies).* New York: Academic Press
Tabor M (1983) Physica D 6: 195
----- (1989) *Chaos and Integrability in Nonlinear Dynamics: An Introduction.* New York: Wiley
Terras A (1985) *Harmonic Analysis on Symmetric Spaces and Applications I.* New York: Springer-Verlag
Thiele E and Wilson DJ (1961) J Chem Phys 35: 1256
Thomas LH (1942) J Chem Phys 10: 532 and 538
Titchmarsh EC (1951) *The Theory of the Riemann Zeta-Function.* Oxford: Carendon Press
Toda M (1967) J Phys Soc Japan 22: 431
----- (1970) Prog Theor Phys 45: 174
Tomaschitz R (1989) Physica D 34: 42
Uhlenbeck K (1976) Am J Math 98: 1059

Van Houten H, Van Wees BJ, Mooij JE, Beenakker CWJ, Williamson JG, and Foxon CT (1988) Europhysics Letters 5 (8): 721
Van Vleck JH (1926) *Quantum Principles and Line Spectra.* Bull Natl Res Council, vol 10 no 54 pp 1-316
----- (1928) Proc Natl Acad Sci USA 14: 178
Verbaarschot JJM (1987) J Phys A 20: 5589
Vivaldi F (1987) Proc Roy Soc London A 413: 97
Voros A (1975) In: Colloques Internationaux CNRS no 237 pp 277-286
----- (1977) Ann Inst Henri Poincaré 26: 343
----- (1980) Nucl Phys B 165: 209
----- (1983) Ann Inst H Poincaré 39A 211
----- (1986) In: Gutzwiller MC, Inomata A, Klauder JR, and Streit L (eds) *Path Integrals from meV to MeV.* Singapore: World Scientific p 173
----- (1987) Comm Math Phys 110: 439
----- (1988) J Math Phys A 21: 685
Waff CB (1975) *Alexis Clairaut and His Proposed Modification of Newton's Inverse-Square Law of Gravitation.* In: *Avant, Avec, Après Copernic.* Paris: Blanchard p 281
----- (1976) Vistas in Astronomy 20: 99
----- (1977) Centaurus 21: 64
Walker G and Ford J (1969) Phys Rev 188: 416
Wang MC and Uhlenbeck GE (1945) Rev Mod Phys 17:323
Wardlaw DM and Jaworski W (1989) J Phys A 22: 3561
Waterland RL, Yuan JM, Martens CC, Gillilan RE, and Reinhardt WP (1988) Phys Rev Lett 61: 2733
Weiss U and Haeffner W (1983) Phys Rev D 27: 2916
Weissman Y and Jortner J (1981) Chem Phys Lett 78: 224
Wheeler JA and Zureck WH (eds) (1983) *Quantum Theory and Measurement.* Princeton, NJ: Princeton University Press
Whittaker ET (1904) *A Treatise on the Analytical Dynamics of Particles and Rigid Bodies with an Introduction to the Problem of Three Bodies.* Cambridge: Cambridge University Press; New York: Dover Publications 1944
Wigner EP (1932) Phys Rev 40: 749
----- (1957) Proc Fourth Canadian Math Congress p 174
Wilkinson M (1986) Physica D 21: 341
----- and Hannay JH (1987) Physica D 27: 201
----- (1987) J Phys A 20: 2415
----- (1988) J Phys A 21: 1173
Winkler A (1988a) J Reine Angew Math 386: 187
----- (1988b) Comm Pure Appl Math 41: 305
Wintgen D, Holle A, Wiebusch G, Main J, Friedrich H, and Welge KH (1986) J Phys B 19: L557
----- and Friedrich H (1986) Phys Rev Lett 57: 571
----- (1987) Phys Rev Lett 58: 1589
----- and Friedrich H (1987a) Phys Rev A 35: 1464
----- and ----- (1987b) Phys Rev A 36: 131

-----, Marxer H, and Briggs JS (1987) J Phys A 20: L 965
----- (1988) Phys Rev Lett 61: 1803
----- and ----- (1988) Phys Rev Lett 60: 971
Yafet Y, Keyes RW, and Adams EN (1956) J Phys Chem Solids 1: 137
Yang CN and Mills RL (1954) Phys Rev 96: 191
Yoshida H (1983) Cel Mech 31: 363 and 381
----- (1987a) Phys Lett A 120: 388
----- (1987b) Physica D 29: 128
----- (1987c) Cel Mech 40: 51
-----, Ramani A, and Grammaticos B (1988) Physica D 30: 151
Yukawa T (1985) Phys Rev Lett 54: 1883
Zaslavskii GM and Chirikov BV (1972) Sov Phys Uspekhi 14: 549
----- (1977) Sov Phys JETP 46: 1094
Zimmermann ML, Kash MM, and Kleppner D (1980) Phys Rev Lett 45: 1092
Zimmermann T, Cederbaum LS, Meyer H-D, and Köppel H (1987) J Phys Chem 91: 4446
-----, Köppel H, Cederbaum LS, Persch G, and Demtröder W (1988) Phys Rev Lett 61: 3

Index